Ministers of Reform

MINISTERS

OF

REFORM

The Progressives' Achievement

in American Civilization

1889–1920

———————————

ROBERT M. CRUNDEN

———

Basic Books, Inc., Publishers   *New York*

The quotation by Arnold Schonberg on page 133 is taken from *Charles E. Ives Memos* (New York, 1972) and is used by permission of Belmont Music Publishers, Los Angeles, California 90049.

Library of Congress Cataloging in Publication Data

Crunden, Robert Morse.
  Ministers of reform.

  Includes bibliographical references and index.
  1. Progressivism (United States politics) 2. United States—Politics and government—1865–1933. I. Title
E661.C945 1982   973.91   82–70848
ISBN 0–465–04631–2

For Frank Freidel

Contents

Preface   ix

Prologue: Growing Up Progressive   3

Chapter 1:   The Conversion to Social Reform   16

Chapter 2:   Two Visions of Democracy   39

Chapter 3:   A Hull-House of the Mind   64

Chapter 4:   Innovative Nostalgia in Literature and
Painting   90

Chapter 5:   Innovative Nostalgia in Music and
Architecture   116

Chapter 6:   It Is Sin to Be Sick: The Muckrakers and
the Pure Food and Drug Act   163

Chapter 7:   Compromise at Armageddon   200

Chapter 8:   A Presbyterian Foreign Policy   225

A Methodological Note   274
Acknowledgments   279
Notes   280
Index   301

Preface

THIS BOOK is an extended answer to the question, What was progressivism? The central argument is that progressivism was a climate of creativity within which writers, artists, politicians, and thinkers functioned. Progressives shared no platform, nor were they members of a single movement. In general they shared moral values and agreed that America needed a spiritual reformation to fulfill God's plan for democracy in the New World. As the methodological note at the end of the book indicates, I have examined one hundred progressives and stressed the contributions of twenty-one. Some of this smaller group are there because of their intrinsic significance; others had representative experiences; a few were simply present during key events. My choices are mostly progressives with national roles, but I believe that state and local progressives were roughly similar. In an earlier volume, *A Hero In Spite of Himself: Brand Whitlock in Art, Politics and War,* I examined local activity in detail, and I did not wish to repeat myself.

Born between 1854 and 1874, the first generation of creative progressives absorbed the severe, Protestant moral values of their parents and instinctively identified those values with Abraham Lincoln, the Union, and the Republican Party. But they grew up in a world where the ministry no longer seemed intellectually respectable and alternatives were few. Educated men and women demanded useful careers that satisfied demanding consciences. They groped toward new professions such as social work, journalism, academia, the law, and politics. In each of these careers, they could become preachers urging moral reform on institutions as well as on individuals. In time they tried to impose their moral assumptions about American democracy on Latin America and Europe, only to find to their bewilderment that people living in other climates of creativity did not wish to adopt American values. The structure of the book thus shifts back and forth from individual need to institutional realization, and I stress the psychological origin of progressive achievements in every area of cultural ferment.

As the progressive ethos took shape and spread its influence, it appealed to young people of disparate backgrounds. Roman Catholics, Jews, and people of no religious affiliation found progressive goals attractive and came to regard themselves as progressives. Because of this broad appeal, the progressive ethos often seems amorphous, inchoate, and difficult to define. The subject is not all that complicated. Protestantism provided the chief thrust and defined the perimeters of discourse, but the civil religion of American mission soon transcended its origins and became a complex of secular democratic values.

The Progressive era began almost a century ago and the intervening time has produced an enormous body of scholarly analysis. I have thus felt free to ignore many conventional topics and to concentrate on figures, ideas, and events that usually escape notice. I have included sections on politics and diplomacy merely to indicate how the thesis works in those well-known areas. My own greatest interest has been in the arts, and the number of pages devoted to the arts does not strike me as out of proportion to their long-term importance to world history. The names of Charles Ives and Frank Lloyd Wright grow in importance around the world with every passing year, while the names of politicians sink unmourned into oblivion. Close behind the artists, the leading social scientists seem likely to survive for a long time. John Dewey has long been famous, but George Herbert Mead is emerging from obscurity, Jane Addams will long outlast the feminist movement, and men such as W. I. Thomas and Robert Park, once little known outside professional circles, are reappearing in print and receiving serious study. History is not publicity, and I believe that time will vindicate the proportions I have chosen here.

In trying to sum up the position of the artists, I have chosen the term "innovative nostalgia." It is a concept that tries to convey some of the complexities of the years between 1889 and 1920. Progressive artists were neither derivative nor original in any pure sense of those terms. They looked backward for emotional support and secure ideas even as they bravely experimented with federal regulation, instrumentalism, and polyrhythms. Too many analysts for too long have assumed that creative people look forward, and that everyone else is, in some sense, reactionary. Yet a clear sense of the past can often lead to the most original of ideas that will influence the future. The sections on Ives and Wright are where I work out these ideas in greatest detail, but they apply everywhere in the book. No one should finish this volume without sensing the ironies involved in the word "progressive."

Progressivism left a complex legacy to modern America. Many early historical accounts stressed the laws that led to the New Deal and the welfare-state aspects of life in the last half of the twentieth century. In the whiggish sense that the past generally does evolve into the present, this analysis had its merits. As cultural history, however, it was misleading.

Progressives were not collectivists of the New Deal variety. If they lived into the 1930s, they either changed significantly in order to support the measures of the New Deal, or they flatly opposed those measures as being contrary to the reform spirit as they understood it. In exceptional environments where non-Protestant influence was strong, such as Chicago and New York, progressivism could provide valuable lessons to later reformers. Most of the time, though, progressivism left later generations with examples of ineffective legislation and superficial analysis.

The real legacies of progressivism were in the arts, philosophy, diplomacy, and cultural empathy; and the record is a mixed one. Charles Ives could teach later generations not only his technical innovations, but also his means of coping with the problem of growing up a genius in a culture that had no place for it; he also inadvertently taught later composers that the cost of such adjustments could be a complete physical breakdown and decades of creative sterility. The achievements of progressive architecture include not only structures such as "Fallingwater," but also the automobile-dependent sprawl of the typical American suburb. Frank Lloyd Wright probably deserves more credit, *and more blame,* for what modern America looks like than any other single figure in American history. The writings of John Dewey and George Herbert Mead have helped liberate both philosophy and education from provincial backwardness, and both men continue to have many intelligent disciples. Yet the long-range impact on ethics and education has been decidedly mixed; progressivism in education has become an especially inviting target for the attacks of later educators.

The record in diplomacy has been the worst. American assumptions of moral and political superiority hardly mattered in the world arena before 1898. By 1917 they represented the balance of power. America's unwillingness to participate in world affairs did at least some damage between the world wars. The shift toward participation not only won World War II, it led to American primacy for years thereafter. This primacy had many advantages for citizens in a number of countries, but it is hard to see that progressivism played much role in obtaining or sustaining these advantages. Often tolerant at home, progressives left a legacy of intolerance abroad. An extraordinary collection of presidents and secretaries of state from both parties followed the basic course of action outlined by Wilson and Bryan; they assumed the virtues of American ideals and political procedures and insisted that nations accept American leadership; they treated all people as possible immigrants rather than as responsible adults holding different views; they intervened, wherever feasible, whenever local peoples strayed too far from American wishes. Merely to name figures such as John Foster Dulles and Dean Rusk is enough to emphasize the precedents of Presbyterianism in foreign policy; to name figures such as Dean Acheson and Henry Kissinger demonstrates the way these assumptions have become so endemic in the American civil religion that it no longer matters whether a diplomat is Presbyterian by background—or even Christian.

Finally, the larger cultural framework of modern America sometimes bears an odd resemblance to the years around the turn of the century. Both the student left of the late 1960s and the resurgent right of the early 1980s wanted above all to remoralize society, to make a political entity resemble an essentially religious ideal. In watching them both, a detached observer could be pardoned for having a sense of *déjà vu* and for reflecting on how American traditions seemed to have co-opted everyone. Attitudes once strictly religious had changed into assumptions chiefly cultural. In this context the most compelling problem of both the 1880s and the 1980s was how a young person could find a meaningful place for himself in a world that seemed hostile to every creative impulse and deep-seated need. The progressives solved the problem by inventing new professions to cope with the new problems in a personally meaningful way. Whether the young of the 1980s will do as well remains to be seen. Problems of this sort have an eternal relevance.

Ministers of Reform

Prologue

Growing Up Progressive

THE PROGRESSIVES left a number of autobiographies that evoke the atmosphere in which they grew up. They routinely noted the pervasiveness of Calvinist influences in their homes and made few objections to their import. They frequently rebelled as soon as they were old enough to succeed, yet they remained devoted to their parents and tolerant of discipline as an inevitable part of their emotional worlds.

They could be proud of their psychological ancestry. Theodore Roosevelt looked back at the doctrinal mixtures that composed his extraordinary personality and wrote as "one proud of his Holland, Huguenot, and Covenanting ancestors, and proud that the blood of that stark Puritan divine Jonathan Edwards flows in the veins of his children." Roosevelt's father remained for him "the best man I ever knew"; he was "the only man of whom I was ever really afraid," and yet "we children adored him." From childhood Roosevelt insensibly connected a life of social reform and charitable service with Protestant doctrine. His father was deeply committed to both. He heard regular morning prayers in the home and conducted a Sunday-school class. "Under the spur of his example I taught a mission class myself for three years before going to college and for all four years that I was in college." Half a continent away, William Jennings Bryan was growing up in a home where father was a Baptist and mother, a Methodist. He therefore "went to both Sunday schools" until he attended a revival conducted in a Presbyterian church and "was converted." One of his

3

fondest memories was of his family gathered around the piano in the parlor on Sundays, "singing Sunday-school songs and church hymns." His father called it "the Bryan Choir." The Bryan home even had its own altar. Bryan was as devoted to his parents as Roosevelt and recorded almost superfluously: "My father's attitude on spiritual matters made a very deep impression on me." Neither Bryan saw any "necessary conflict . . . between the principles of our government and the principles of the Christian faith."[1]

The political heritage that informed this religious heritage was usually abolitionist. Not everyone could speak of "Aunt Harriet Stowe in Hartford" and stress her pride in being the great-granddaughter of Lyman Beecher the way Charlotte Perkins Gilman could. "By heredity I owe him much," especially "the Beecher urge to social service," she wrote. But many progressives experienced the psychological equivalent of such a background. Brand Whitlock began his autobiography with an extended and affectionate portrait of his grandfather and namesake, Joseph Carter Brand. Of Scottish descent Brand had left Kentucky and a patrimony of slaves to move to free territory. Under the influence of Salmon P. Chase and his own puritan conscience, he became an abolitionist, an abetter of fugitive slaves, and a delegate to the first convention of the Republican Party in 1856. William Allen White lovingly recalled his mother: from a Roman Catholic family, she nevertheless grew up as a Congregationalist. Married to an antislavery Democrat, she herself was "a black abolitionist Republican" who "fell madly, platonically, but eternally, in love with Abraham Lincoln." Having given "her heart and soul to the abolition movement," she continued after the war to help Negroes; when she taught school, she insisted, even against violent threats, to permit Negro children to attend. She finally lost her teaching position, but to her son "she was a heroine."[2]

Though a Jacksonian Democrat, White's father came to share his wife's affection for Lincoln. Whitlock's grandfather, Whitlock proudly remembered, once had an interview with Lincoln that became part of family folklore. Whitlock's law teacher, General John M. Palmer, had known Lincoln "intimately." To progressives, Lincoln was not only a legendary hero who haunted their dreams and talked to their ancestors, but he was a living ideal against which they constantly measured themselves, their friends, and their leaders. They, too, needed a cause like abolitionism and a charismatic leader like Lincoln. Whitlock described his political mentor Samuel M. "Golden Rule" Jones as a man who "summed up in himself, as no other figure of our time since Lincoln, all that the democratic spirit is and hopes to be." William Borah echoed the Lincoln theme in his speeches. He could think of no higher praise for his president and party leader than to compare Theodore Roosevelt to the greatest western president: "No man, since Lincoln left his Western prairie home in Illinois, has understood so well or has been so anxious to give to the West all which is hers as the present leader of the Republican Party." Roosevelt understood such frames of reference. When he sought to praise his young supporter Albert J. Beveridge, he referred to him as one "than whom I have never heard anyone expound the principles for which Abraham Lincoln lived and died more ably."

Jane Addams provides the most extreme example. Lincoln rated an entire chapter in her autobiography. "My father always spoke of the martyred President as Mr. Lincoln," she assured her readers, "and I never heard the great name without a thrill." Throughout her life she always tended "to associate Lincoln with the tenderest thoughts of" her father. Figures such as Lyman Trumbull provided living links between Lincoln and the progressives, and the admiration such men expressed for the former President "was quite unlike even the best of the devotion and reverent understanding which has developed since." She made Carl Schurz's "Appreciation of Abraham Lincoln" a gift to twenty-five boys during the first Christmas at Hull-House and insisted always that Lincoln was the man who "cleared the title to our democracy." He was the leader who "made plain, once for all, that democratic government, associated as it is with all the mistakes and shortcomings of the common people, still remains the most valuable contribution America has made to the moral life of the world."[3]

Lincoln's death began "the present" for progressive youth. "Although I was but four and a half years old when Lincoln died, I distinctly remember the day when I found on our two white gateposts American flags companioned with black," Jane Addams wrote. "I found my father in tears, something that I had never seen before." Ida Tarbell could never forget the time when her usually brisk father came drooping up the front steps of their house. Her mother ran out in a panic to ask what was the matter. "I shall always see my mother turning at his words, burying her face in her apron, running into her room, sobbing as if her heart would break. And then the house was shut up, and crape was put on all the doors, and I was told that Lincoln was dead." The interest so implanted always remained with Tarbell. In her journalistic work, she found that her boss, S. S. McClure, revered Lincoln and thought him "the most vital factor in our life since the Civil War." Tarbell wrote one of the few significant lives of Lincoln in the period, and even in old age reported that Lincoln "had come to mean more to me as a human being than anybody I had studied."

Charles Evans Hughes' father was so upset at Lincoln's death that he told his wife: "Mary, I could not feel worse if you had died."[4]

This allegiance to Lincoln led the loyal to the Republican Party. Albert Beveridge's earliest memory was of Union veterans marching home after the war. He taught himself to give political speeches by imitating the "bloody shirt" heroics of John A. Logan condemning the Democrats for disunion. Whitlock's grandfather had never abandoned his early Republicanism. President Grant had rewarded him with a German consulship, and the family retained the allegiance. The atmosphere of Ohio made it "natural to be a Republican; it was more than that, it was inevitable that one should be a Republican; it was not a matter of intellectual choice, it was a process of biological selection." The Republicans were not merely an organized faction; they constituted "an institution like those Emerson speaks of in his essay on Politics, rooted like oak-trees in the center around which men group themselves as best they can." The party "was a fundamental and self-evident thing, like life, liberty, and the pursuit of happiness, or the flag, or the federal judiciary." The party was "elemental, like gravity, the sun, the stars, the ocean. It was merely a synonym for patriotism, another name for the nation." A person in his town became a Republican "just as the Eskimo dons fur clothes. It was inconceivable that any self-respecting person should be a Democrat."

In the long run, if progressives became disillusioned by the control that so many wealthy conservatives exercised over the party of Lincoln, they broke ranks reluctantly, often bringing what they thought was the spirit of Lincoln with them. Theodore Roosevelt remained Republican except for the brief period when he led the Progressive Party. William White managed to be a Republican, a Progressive, and a follower of Franklin D. Roosevelt except at election times. Whitlock became a Gold Democrat, an Independent, and a Wilsonian Democrat. Lincoln and abolitionism often seemed to be living in groups not labeled "republican," but the spirit and the sense of heritage remained the same.[5]

II

A few of the more literate progressives, like Brand Whitlock, chose never to go to college and instead tried journalism or studied in a law office. Others, like John R. Commons, tried farming, teaching school, and even tramping before taking a conventional route—attendance at a small, church-oriented college. Named after puritan martyr John Rogers by his Presbyterian mother and Quaker father, Commons grew up in the Western Reserve of Ohio thrilling to family tales of the underground railway and defiance of the Fugitive Slave Law. An ardent feminist and prohibitionist, his mother came to Oberlin College with Commons. Together they founded an antisaloon publication in 1887, and for twenty years thereafter, he could not bring himself to smoke or drink. He referred later to his college experience as his "salvation," and there can be little doubt that Oberlin and the other evangelical colleges marked the young people who attended them. They also may have contributed to a little-noticed pattern of nervous breakdowns that reappears in the lives of a number of progressives. Unable to meet the psychological demands made by strict parental supervision, they collapsed when life became too much for them. Charles Evans Hughes and Woodrow Wilson are two notable examples.[6]

Oberlin remains the most thoroughly studied college in this context. The faculty came largely from rural and small-town New England stock that had carried its puritan values as it migrated west before the Civil War. The college had a heritage of fierce loyalty to the romantic reforms that came out of that era, from millennialism and abolitionism to women's suffrage and coeducation. Most of the students came from farming or clerical families, and more Oberlin students chose the ministry as a possible career than did students elsewhere. Some classes were so Republican politically that not a single student would admit to voting for the Democrats. In the overwhelmingly religious atmosphere, strict rules regulated sexual contacts and no one could dance or gamble. Even the library was open to men and women only at alternate times. Evangelism was as much a part of life as were regular meals. With little effort a student could spend the greater part of his time out of class in some sort of religious service. As one visitor reported: "Why, if anyone walking along the sidewalks of Oberlin catches his foot and stumbles, nine chances out of ten, he stumbles into a prayer meeting." Such an environment easily led to smug bigotry. The students suspected the poor and the immigrant of excessive alcoholic consumption and inadequate moral fiber, and they frequently confused patriotism with their own variety of Protestantism.

Commons left for graduate study and then returned briefly to join the faculty. He helped introduce American history and sociology to the students of the early 1890s. He also pioneered in bringing his message to the local community, speaking to groups such as the Congregational Club of Cleveland on topics such as "The Christian Minister and Sociology." The published version of the talk urged Christian ministers to involve themselves deeply in the social problems of their time and not merely to dwell on the life hereafter. Commons soon left Oberlin, but even without him, Oberlin had become something of a center for conferences on Christian Sociology by the middle 1890s. One of these, the Institute of Christian Sociology, met at the campus in November 1894. The hosts included faculty member William I. Thomas and Washington Gladden, the social-gospel pastor of a Columbus church. Guests included Josiah Strong, Graham Taylor, and J. H. W. Stuckenberg, three of the best-known names in Christian reform activities. But revivals soon became less frequent, fewer students chose the ministry for a career, and expressions of social ethics replaced any remaining tendency toward religious quietism. Professor of Philosophy Simon MacLennan summed up the thinking at the school in 1911 when he said that "the religion of Christ was really democracy," and that "all religion as well as all government should be by the people and for the people." By this time visitors to the college included such seemingly secular reformers as Lincoln Steffens, Charles A. Beard, and Robert M. La Follette, not to mention outright socialists such as Jack London and John Spargo. Theodore Roosevelt summed it all up in his 1912 campaign when he remarked: "This is the community of the applied square deal. . . . What I preach you put into practice."[7]

III

Just as not every talented young progressive went to an Oberlin, not every graduate went on to graduate school. The lines separating educational institutions from political, missionary, journalistic, and settlement work were never all that clear, and students often wandered from one institution to another, finding out who they were. But once the graduate schools established themselves, they proved irresistibly attractive to many progressives. Success in scholarship, politics, or publishing seemed obtainable once a person had gone to a place like Johns Hopkins, mastered the new German research methodologies, and produced an original piece of research.

The most famous student and lecturer at the Hopkins was Woodrow Wilson, a man whose life with but a single exception provided a model of progressive personality development. Wilson grew up in a family stuffed with Presbyterian divines. His "incomparable father" was an eminent church official and professor. The Scots-Irish family was descended from antislavery Whigs in Ohio, but Wilson's father went South in the 1850s to Virginia and Georgia, became friendlier to slavery, and actually had slaves serving him. Obsessed as were many Scots Presbyterians by notions of a covenant between God and man and between God and society, he defended slavery as a part of that covenant and became a high official in the Confederate Presbyterian Church. He held a number of ministerial, professorial, and administrative jobs after the Civil War, moving about the South and often finding himself embroiled in theological controversy. Despite his overbearing nature, he retained the devotion of his son. Six years after his death in 1903, Wilson could remark that a boy never got "over his boyhood" and could never "change those subtle influences which have become a part of him, that were bred in him when he was a child."[8]

Only in its tolerance of slavery did Wilson's upbringing fail to resemble the classic progressive pattern. Few Southerners who were genuinely committed to the Confederacy produced progressive children. In Wilson's case he professed to rejoice in the failure of the Confederacy even while he remained loyal to the South. Perhaps more relevant to progressivism, he remained politically a Southern conservative Democrat until long after he moved north, and did not move toward recognizably progressive positions until he had for some years been a professor at Princeton in safely Northern New Jersey. In his case as with so many others, the key influences were church and family. His was a home of daily devotions, prayers, Bible reading, hymn singing, and holy Sabbath days; intellectually, the world was Manichaean. Everything a person did had other-worldly implications. Wilson underwent the expected conversion experience and found himself subject to pressure that he enter the family profession of the ministry. He never seemed to experience a genuine "call" and never showed much interest in theology. He went off to Davidson College in North Carolina and a period of religious turmoil from which he emerged with a political rather than a religious vocation. The great liberal moralist William E. Gladstone edged John Calvin aside in Wilson's consciousness, and he envisaged a career giving sermons in the political arena to a nation in need of moral regeneration. The passion for covenants, so much a part of Scots Presbyterianism, lived on into his early adulthood in the most varied ways. He made "a solemn covenant" with a college friend "that we would school all our powers and passions for the work of establishing the principles we held in common," that "we would acquire knowledge that we might have power," that "we would drill ourselves in all the arts of persuasion, but especially in oratory," and that "we might have facility in leading others into our ways of thinking and enlisting them in our purposes." He even courted his best girl as if she were literally his salvation: "What I longed for was simply a loving covenant that we would do all that we could towards a marriage in June."⁹

Wilson first tried the law as a way of entering politics. He attended the University of Virginia Law School, but found it demanding and depressing: "This excellent thing, the Law, gets as monstrous as that other immortal article of food, *Hash,* when served with such endless frequency." He remained fascinated with Gladstone and the British constitutional system and made over the years numerous constitutions for student clubs. Convinced that "every civilized nation has taken the foundation of its laws" from the Old and New Testaments, he nevertheless found it increasingly difficult to devote himself to their application, no matter how venerable their pedigree might be. He might think he could embark on a political career by pursuing legal studies, but as time passed, he found himself attracted to literary and oratorical studies that had little relevance to the law. "I've fallen fairly in love with speech-making," he wrote a friend on New Year's Day of 1881. "My *end* is a commanding influence in the councils (and counsels) of my country—and *means* to be employed are writing and speaking." When he tried to pursue a legal career in Georgia, he did not so much fail as bore himself out of a job. He decided that a literary academic career was more suitable. For an extended period, both literature and political science seemed to have equal appeal, meeting at times in the middle ground of historical writing. His "appetite is for general literature" and his "ambition is for writing," as he told a close friend.¹⁰

Wilson went to Johns Hopkins only to find that the atmosphere of archival, institutional research encouraged in the seminars of Herbert Baxter Adams and Richard Ely repelled him. For a brief period, he confessed that he was "downcast at finding that there was no line of study pursued there that could quite legitimately admit under it such studies as have been my chief amusement and delight during leisure hours for the past five or six years, namely, studies in comparative politics." He was most interested in comparing the British and American political systems, and perhaps in pursuing his analysis into France and Germany. But he discovered that the typical Hopkins professors "wanted to set everybody under their authority to working on what they called 'institutional history,' to digging, that is, into the dusty records of old settlements and colonial cities, to rehabilitating in authentic form the stories, now almost mythical, of the struggles, the ups and downs, of the first colonists." He found this work "dry" and "tiresome," and resolved to demand some freedom, from Adams in particular. Adams proved more amenable than Wilson had anticipated, and so the young rebel was soon free to pursue his " 'constitutional' studies." Despite his criticisms, he thought the Hopkins "the best place in America to study, because of its freedom and its almost unrivalled facilities, and because one can from here, better than from anywhere else in the country, command an appointment to a professorship." The only other alternative was study in Germany, and Wilson's German was not up to that.[11]

Despite several periods of blinding headaches and related illnesses, many of them probably psychosomatic, Wilson managed to take his Ph.D. Strictly speaking, he never completed his doctoral studies any more than he completed his legal studies, but his work and publications were so impressive that Professor Adams finally put him through a pro-forma exam, gave him the degree, and accepted *Congressional Government,* already in print, as his dissertation. As a number of subsequent analysts have pointed out, both the pattern of Wilson's behavior and the substance of his publications presaged much of his future career. He frequently would begin brilliantly at an assigned task, overwhelming people by his mental and oratorical abilities, only to lapse into psychosomatic illness, lassitude, and sometimes failure before the task was done—whether at Princeton, in the New Jersey statehouse, or in the presidency. Likewise his book has proved disappointing. At least some praise for it still remains customary in secondary treatments, but the book illustrates all too well the abstract, antiempirical nature of Wilson's mind and the superficiality of his research. Although he lived close to Washington, he never visited the capital city to see congressional government in action while writing the book. Despite problems that might well have included dyslexic vision, he preferred to use his books and develop abstract principles to foreordained conclusions.

While Johns Hopkins enabled Wilson to enter upon a brilliant academic career, his longtime friend Walter Hines Page demonstrated how Hopkins could solve career problems in nonacademic ways. Page was also Southern, but from a Whig Unionist family. His uncles and aunts tended either toward academic or ministerial pursuits, and throughout his adolescence, Page seems to have planned on a career in the Methodist Church. His Evangelical enthusiasm waned, however, and he began to seek cultural outlets to replace his religious enthusiasms. When the Hopkins opened, Page was eager for the great adventure and enthralled to hear President Daniel Coit Gilman inform the new arrivals that the university would be "a temple of learning and upon its altar we shall light the sacred flame." Each fellow was "to light his own torch at the altar flame and to maintain it burning as brightly as possible so long as he shall live." Page gradually lost his formal religion and even became anticlerical, although never anti-Christian. Soon the life of a professor lost its appeal as well. Like several other young professors, he decided on journalism and publishing as respectable substitutes. He went on to a distinguished career editing journals and publishing books that instructed Americans on how to live. He ended his life as American ambassador to Great Britain, having become one of the more prominent men in the Wilson administration.[12]

IV

The Hopkins, however, remained primarily a graduate school for the training of scholars, and the typical job for a scholar to choose was teaching. In terms of progressive influence, Richard Ely was the preeminent figure both at the Hopkins and in his later career at the University of Wisconsin. Ely's ancestors had been puritan refugees who fled the Restoration; their descendants included a dour, prohibitionist crew of Presbyterian and Congregationalist clergy. Despite the constant air of theological bickering in his family, he failed to have the expected conversion experience, toyed with the thought of becoming a Universalist preacher, and settled for a less strenuous psychological life in the understanding arms of the Episcopal Church. But as with so many progressives, the mental habits and worldview of childhood persisted into the new professions of adulthood and Ely remained an evangelist and missionary to those unconverted to his economic, sociological, and political doctrines. He never forgot his father's strong belief in egalitarianism and reform or his childhood hero-worship of Horace Greeley. Fervently believing that education was an instrument for social reform, he sought his own at Dartmouth and Columbia and in a three-year quest for a suitable mentor in Europe. He found his man in Karl Knies, one of the major figures in the German Historical School of Economics. Through Knies, Ely absorbed many ideas outside the British laissez-faire tradition that predominated in America. He also absorbed rather idealized notions of German scholarship and bureaucratic efficiency that persisted into the twentieth century in his ideas on city management and the chances for socialism in America.

Johns Hopkins President Gilman had problems recruiting a suitable professor of political economy, and Ely, despite his youth, seemed a plausible choice. He was inexperienced and uncertain as a professor, impressing the sarcastic J. Franklin Jameson as being homely and insignificant in class, but he soon improved his presentations and began attracting students such as John R. Commons, Albion Small, Newton D. Baker, Albert Shaw, Edward A. Ross, Frederick Jackson Turner, Frederic C. Howe, and Woodrow

Wilson. Ely combined an insistence on practical research problems with an emphasis on religious idealism, and a number of his students actually chose the ministry for their careers. As Howe wrote him in 1894: "I think we all have you to thank that you disclosed to us the whole forest rather than a few trees which constituted the science of political economy in the past, and that man is something more than a mere covetous machine and that the science which deals with him in society has larger aims than the study of rent, interest, wages and value." Ely was himself always clear about this interplay between religion and sociology. He believed throughout his life that "the function of social science" was "to teach us how to fulfill the Second Commandment." For him Christianity "is primarily concerned with this world, and it is the mission of Christianity to bring to pass here a kingdom of righteousness."

In 1885 his interests took institutional form with the founding of the American Economic Association. Classical economic theory reminded him all too much of orthodox Presbyterianism, and he wanted to pioneer a new kind of scientific, Christian economics. The platform of the organization declared that the state was "an educational and ethical agency whose positive aid is an indispensable condition of human progress." "The doctrine of laissez faire is unsafe in politics and unsound in morals," and so the founders looked forward "to an impartial study of actual conditions of economic life" that would lead to a genuinely scientific economics. "Church, state, and science" should unite "in a progressive development of economic conditions" that would lead to "corresponding changes of Policy." Ely explained his efforts to President Gilman as a movement "which will help in the diffusion of a sound, Christian political economy." Not content with merely professional activity, Ely also carried the battle for a Christian social science into the countryside. He spoke regularly to summer sessions of the Chautauqua movement, where one of his devoted students was Ida Tarbell. He carried his message to various Alliance and populist groups, where one of his most enthusiastic supporters was Populist Senator William A. Peffer. He even became popular in socialist circles at home and abroad, and no less a figure than Sidney Webb helped distribute his writings to members of the Fabian Society. When examined from the perspective of someone like Ely, the common distinctions often made between progressivism, populism, and socialism no longer seem meaningful. Most of the leaders in these movements shared similar values, read similar books, and even seemed to be the same people at different times or in different circumstances.[13]

Ely made a considerable impact on life outside the university. But he never received the chair at Hopkins that he felt his achievements warranted, and in 1892 he went to the University of Wisconsin. There, he built up a distinguished group of present and former students in what became one of the most progressive and influential groupings of social scientists in America. He had particular affection for John R. Commons and Edward A. Ross and managed to bring both of them to Wisconsin to help him formulate his ideas and spread them throughout the larger community. Insofar as Wisconsin political progressivism had a brains trust, Ely was its chief and Commons a key lieutenant. Governor Robert M. La Follette was, as usual, deadly serious when he said to Ely, "You have been my teacher!" The La Follette administrations had many teachers, but most of them were from their own state university, and even those who had not been called there specifically by Ely did not escape his influence. Social scientist and politician made a potent team in probably the key state laboratory for testing progressive ideas on issues like tax reform and labor protection.[14]

Several progressives wrote up the Wisconsin experiment as an example for other progressives. Frederic Howe's *Wisconsin: An Experiment in Democracy* (1912) was in many ways the best of these. Conceived as what looks suspiciously like a campaign document for La Follette's 1912 run for the presidency, the book was an extended exercise in wishful thinking—more an analysis of what progressives wanted to believe than a report of what they really had done. Howe stressed the corrupt nature of the state before La Follette came along to underscore La Follette's successes as an influential progressive governor. Emphasizing the "scientific thoroughness" of the legislative process, Howe insisted that "the university is largely responsible for the progressive legislation that has made Wisconsin so widely known as a pioneer." Indeed, so great was the achievement that Howe's language failed him: "The politician has almost disappeared from the state-house. He does not thrive in this atmosphere." One forms an image of laws produced by immaculate conception, unsullied by lobbyists, shoddy tactics, log-rolling, and indeed, by any sort of politics. The heroic La Follette had brought democracy and cleansed the corruption; due to him Wisconsin had a direct primary, railway and public utility commissions, corporate reform, more equitable taxes, a civil service, and a more democratic system of education. But the real architects were the experts: University of Wisconsin President Charles R. Van Hise and Professors Ely, Commons, Ross, and Paul S. Reinsch in political science. "The close intimacy of the university with public affairs explains the democracy, the thoroughness, and the scientific accuracy of the state in its legislation. It, as much as any other influence, kept Wisconsin true to the progressive movement during these years."[15]

Thus, the paradigm for a progressive personality developed over the years. From its birth in the Middle West, a generation of intelligent youth, though remaining devoted to their parents, resisted efforts to compel them into the ministry or missionary work. Despite tension, uncertainty, and an occasional breakdown, these young progressives channeled their urges to help people into professions that had not existed for their mothers and fathers. Men and women found that settlement work, higher education, law, and journalism all offered possibilities for preaching without pulpits. Over the long term, their goal was an educated democracy that would create laws that would, in turn, produce a moral democracy. The place for Christianity was in this world.

Chapter 1

The Conversion to Social Reform

Progressivism began with the conversion of the young to goals of social reform. Having grown up in strict Protestant homes, survived rigorous schooling, resisted ministerial or missionary careers, many young persons found no recognized outlet for the conscience that such an upbringing had instilled. The result was often a period of depression, self-doubt, indecisive wandering, and frequent failure. Neurotic behavior was not uncommon and high intelligence permitted one to speak interminably to one's self and to others about this plight. For two-and-a-half centuries, Protestant youth had expected a conversion experience to end doubt and establish forever the role of the individual within the cosmos. Progressives often doubted the role of God or the goal of seeking entry to heaven, but they still had the mechanism of conversion even as they lost their theology. One woman—Jane Addams—and one man—George Herbert Mead— left extensive documentation concerning their painful maturation from Protestant child to progressive adult.

Jane Addams was the more important figure at the time, and her public articulation of the experience became another *Pilgrim's Progress* for progressives of both sexes. The problems facing women, however, were far worse than those facing men. Many of the best medical authorities thought they needed protection from brain fatigue fully as much as from male sexual initiatives. Women expected either to marry and have children or to become maiden aunts, cultivating beautiful thoughts, neurasthenic diseases, and the children of close relatives. No genuine careers were open to them —being a schoolteacher or governess hardly warranted serious attention, and few could stand such underpaid and undervalued positions for long. Women had to fight even for admission to many colleges and universities. Most professional schools were closed to them, and even with appropriate degrees they found few positions as doctors, professors, or architects. They not only had to break trails for those who followed, but they had to travel paths to fruitful and remunerative professional work and have that work satisfy their consciences, which were as strict as those that prostrated able men.

Jane Addams pioneered the new career of social settlement work for women and became in a larger sense the preeminent role model for the next generation. She was articulate about her own needs and efficient in mobilizing people both to assist her and to follow her example. She was the first progressive to institutionalize her own psychological needs and thus the founding of Hull-House in 1889 marks an appropriate beginning for progressivism as an important cultural phenomenon.

Jane Addams grew up as the youngest child in a motherless family. Her father John Addams was the overwhelming force in her life. On the first page of chapter one of her autobiography, Addams wrote that all of her childhood memories were "directly connected with my father"; he "was so distinctly the dominant influence" and the "cord which not only held fast my supreme affection, but also first drew me into the moral concerns of life." With no mother to intrude, Addams found it possible to monopolize her father with far less competition within the family than would otherwise have existed. One anecdote especially illustrates the intensity of her devotion and her own feelings of inadequacy in trying to be worthy of such a model. John Addams, for example, had a "miller's thumb," which had been flattened by handling grain for many years; his hands also bore marks made by the hard flints that flew off the millstones when he dressed them. Jane Addams made a great effort to spot her own hands and to flatten her own thumb. Perhaps more serious than this direct imitation of an admired adult was her marked inferiority complex. No one described her as an unattractive child, yet Addams believed she was an "ugly, pigeon-toed little girl, whose crooked back obliged her to walk with her head held very much to one side." She apparently did have a slight curvature of the spine and may have experienced pain because of it. She also held her head slightly to one side. Yet only to herself, and compared to her own ideal of her father, was she unattractive. Her shame was so great that she dreaded public walks with her father: "I simply could not endure the thought that 'strange people' should know that my handsome father owned this homely little girl."[1]

This extraordinary attachment between father and daughter becomes significant for the history of progressivism because of her associations of moral and religious ideas with him. Addams united her father's ideas about life with her growing moral concerns, and internalized her perception of his attitude. John Addams taught a large Bible class at a local church; at the same time he would never accept any denominational membership. He called himself a Hicksite Quaker, thus vaguely identifying himself with a democratic offshoot of the older Quaker church, a group with no meeting-house near their home. In practice John Addams sporadically attended several Protestant churches in the area, and had no recognizable theological beliefs. Instead, he developed his own highly personal creed from common Christian morals. He had great strength of character, courage, and kindness, and he identified these qualities with Christ. He also quietly supported his daughter when she was under great pressure to have a standard, evangelical conversion, and refused to have one. Addams later remembered his axiom "Mental integrity above everything else" as a summation of his outlook. She thus fused her image of her father with this sense of moral righteousness. Like him, she, too, would be earnest, ethical, independent, and undogmatic.

When Addams was about eight years old, her exclusive possession of her father ended abruptly with his second marriage. According to the accounts that survive, Anna Haldeman Addams was a remarkable woman who altered both the home and the relationship between Jane and her father. Her granddaughter Marcet described Anna as strong-minded, domineering, competitive, and socially ambitious. Her character tended to provoke strong reactions, mainly animus or devotion, from those who knew her. To her granddaughter, she was "a fairy godmother sort of person," who was "so alive in every fiber that I can scarcely believe she is no longer here. She was adorable, and I worshipped her." Marcet also noted that Anna could "not always control her temper." John Addams, who had admired Anna for some years while each was married to another, apparently loved her deeply. Anna, in turn, "admired—and respected" him. She was a man's woman, "the type who instinctively want men to love and protect them." She was, in short, a threatening presence that could not be defeated for the dominance of John Addams' attentions.

James Linn, Jane Addams' nephew, corroborates Marcet's impressions, although he is as sympathetic to Jane as Marcet is to Anna. To James, Anna was formidable, even overbearing. Even at ninety-three, she was "still handsome, in a terrifying sort of way." She was a constant reader of novels; she liked to gather children around her to perform parts of Shakespearean plays. Anna also played the guitar "and sang endless songs from Tom Moore, whose lyrics she knew by heart, as well as many others. She was what in those days was called 'accomplished.' " She loved to entertain and travel and displayed her disappointment when her new husband refused the political future that might have been his, perhaps because he disapproved of his wife's affinity for such social life. She was thus one example of the kind of woman approved in her day: cultivated, accomplished, appealing, and entertaining. She made no attempt at any career that might compete with men. Her life displayed no Protestant values.

Jane detested Anna. The autobiography refers to her father's remarriage only briefly and never mentions Anna by name. All surviving records indicate a prolonged and understandable hostility. Addams' relations to her beloved father were changed forever. She now had to share him with a strong-minded, domineering woman, who regarded Jane as unduly shy and retiring, and all but forced her into athletic, gregarious, and often frivolous forms of social behavior. Anna added a piano, wardrobe, good linen, and china to the Addams home, and she trained Jane in the social graces that people so admired in her later. Most important of all, though, she provided Jane with a negative model that she could use to develop her own ideas of what a woman should not be, of the femininity that had to be resisted.

Even Marcet is quite specific on this point; Anna was useless from a moral point of view. "I never saw her do anything more useful with her hands than adjust the objects in a room, care for her flowers and strum a guitar when she sang the ballads of Moore and Burns." To Jane this behavior was a constant affront to her father's values. Other girls in comfortable circumstances might cooperate with such a woman, and prepare themselves for lives of social futility. But to Jane, Anna's ideas of womanhood were proof of what happened when one ignored Tolstoy's stern warning against the "snare of preparation," the conspiracy of society to entangle the young in curious inactivity "at the very period of life when they are longing to construct the world anew and to conform it to their own ideals." When Jane finally came to feel that such a life as her stepmother led was profoundly immoral, she began the process of becoming a progressive. She had used her education in Christian ethics, reinforced by her devotion to her father, to develop a vantage point from which to criticize and then to reform society. The foundations of Hull-House were laid in one woman's moral revulsion against privileged uselessness.

MINISTERS OF REFORM

In 1877 Addams left home for the Rockford [Illinois] Seminary, a school founded by the descendants of New England Puritans to help Christianize the West. As at John Commons' Oberlin College, the faculty encouraged students to experience evangelical conversions and to enter missionary service after graduation. In this environment Addams discovered that her father's values and her own were not compatible with strict evangelical demands. On the surface nothing seemed more plausible than public conversion at one of the annual January harvests of new Christians, given her background and the social pressure of her teachers and peers. Instead, Addams explored on her own the books and lives of a number of Victorians who, like her, had found their parents' rigid faith no longer relevant, and who needed social involvement to act out the demands of their Protestant consciences.

The pressure at Rockford was strong, but once she had successfully resisted it, Addams no longer seemed upset by orthodox pressures. With paternal approval, she followed her own conscience. After several years, she felt the threat so little and had so changed in her appreciation of Christianity that she joined the Presbyterian Church to settle the problem. On 14 October 1888, Addams was baptized into the Cedarville Presbyterian Church; that church certified her as a member in good standing to Presbyterian churches in Chicago when, in 1889, she completed the move to Hull-House. "At this time there was certainly no outside pressure pushing me to such a decision." While she underwent no great conversion experience upon joining the church, she wrote: "I took upon myself the outward expressions of the religious life with all humility and sincerity." Her minister was not theologically insistent and so the primitive Christian and democratic faith that Addams developed during the 1880s proved adequate. She had "an almost passionate devotion to the ideals of democracy, and when in all history had these ideals been so thrillingly expressed as when the faith of the fisherman and the slave had been boldly opposed to the accepted moral belief that the well-being of a privileged few might justly be built upon the ignorance and sacrifice of the many?" Who was she not to "identify myself with the institutional statement of this belief, as it stood in the little village in which I was born?"

Shortly after Addams' graduation from Rockford Seminary in 1881, John Addams died from complications of appendicitis. She was grief-stricken. A letter about her father's death indicates that she equated her wounded psychological state with her own imagined physical problems. About two weeks after the event, she wrote to her closest friend: "I will not write of myself or how purposeless and without ambition I am. . . . The greatest sorrow that can ever come to me is past and I hope it is only a question of time until I get my moral purposes straightened." Her father's death left her emotions as twisted as her back: both needed to be "straightened." John Addams' presence and approval had enabled his daughter to stand tall, and she felt deformed without him.

Throughout most of the rest of the 1880s, Addams searched for a vocational solution to her problems. She briefly attended the Women's Medical College in Philadelphia, but suffered severe pain and physical collapse. Dr. S. Weir Mitchell and her stepbrother Harry Haldeman operated on her spine with some success in 1882, but whenever she tried to make plans for travel or further study, the problems returned, and she was immobilized and incapable of finding any solutions to her problems. On the one hand, many of these symptoms appear to have psychosomatic origins; her physical disabilities at this time paralleled the psychological loss of her father. On the other hand, her letters for the next dozen years, extending well into the Hull-House years of the 1890s, are full of genuine medical problems. Addams rarely possessed good health. Rheumatism, sciatica, and persistent bowel discomfort all troubled her at frequent intervals. Characteristically, she triumphed enough over her own problems to act as nurse to various afflicted relatives. When her brother Weber went insane, in the spring of 1883, she helped care for his wife and baby. Her sister Mary suffered from neuralgia, and so Addams often watched over her children. Even Anna became ill and hard to manage; Jane bore most of the burden of traveling with and caring for a stepmother she disliked. She almost seemed to be seeking out others sicker than herself, as if to avoid allowing her own body the opportunity of betraying her.

With John Addams dead and Anna often ill, Jane could hardly resist the life-style that was so dear to her stepmother. She spent much of the 1880s preparing for some vague and socially acceptable adulthood. She went to museums, balls, and parties, and during 1883–85 and 1887–88 she toured Europe—a typical remedy for the ailing members of the idle rich. John Addams had left his family financially secure, and his widow did not share his preference for rural Illinois. Jane Addams, however, remained true to her father even while she tried to please her stepmother. The tensions within her show clearly in her occasional letters to her stepbrother George Haldeman; they were almost the same age and frequent companions. She "wanted" desperately to adapt herself to this new role and to convince herself that she liked being cultivated. From Dresden, for example, she wrote to George that she was "more convinced all the time of the value of social life, of its necessity for the development of some of our best traits. There are certain feelings and conclusions which can never be reached except in an atmosphere of affection and congeniality." The friends we want to make "will not come to us without an effort of our own towards understanding them, and expressing ourselves." The tension beneath the platitude is almost palpable: she *must* find value in what she is doing; she must find friends; she must cultivate herself. John Addams' moral imperatives meant that a woman had to do what she was supposed to do, so Jane was even conscientious about being frivolous. Two weeks later, the sad, dutiful epicureanism reappeared: "I enjoy the travelling more every day. . . . I have enjoyed the architecture and pictures more than I had any idea I should." But there is no future in flogging oneself to be happy.

Between European trips the family lived in Baltimore. In the worst sort of do-gooder style, they at times even visited the "less fortunate" in their "shelters." Little points in any way to the Jane Addams of Hull-House or to the progressive thinker who suddenly loomed so large during the 1890s. Indeed, in the midst of it all, Anna made a sustained attempt to preempt any such career. Well-bred and thoroughly prepared young ladies were supposed to marry and settle down, and Anna decided that her son George would be the perfect mate for Jane, thus paralleling an earlier match between Jane's sister Alice and their stepbrother Harry Haldeman. Jane had already turned down Rollin Salisbury, a promising Beloit student, and she was not about to settle for the neurasthenic George. It was just as well. George slowly sank into insanity, never recovered, and could never have been an appropriate mate for Jane.

Yet, external appearances were deceptive. Beginning in the late 1870s, and continuing throughout the 1880s, Addams' closest confidante was Ellen Gates Starr. Their letters summarize Addams' slow shift from a purely religious outlook to the Hull-House conscience of the 1890s. The first surviving letter is from Starr to Addams, and it set a tone of religious uncertainty that provoked unceasing speculation until it achieved some form of progressivism. Starr had been reading Horace Bushnell's "Character of Jesus," and she wanted to discuss it. She found it pleasing and sensible, but admitted to doubt and confusion about the divinity of Jesus and her belief in it: "There is something in myself incomprehensible to me, in my views & feelings concerning religious subjects & especially this subject of the divinity of Christ." She would be glad to believe it, and much about Jesus seemed divine, but "I know that I don't." She thought of him instead as someone like a character out of Dickens, many of whose characters seemed quite real to her. God, she went on, "is more a reality to me every day of my life. I think more of a reality for my *thinking* about Christ as divine." She had come to the conclusion that "heaven and hell are in the condition of the soul and certainly the condition of a soul who has experienced a single good impulse, is better than that of one who has experienced none." Especially striking is the premonition of the Hull-House idea in the phrase, "a single good impulse."

Addams' reply was similarly uncertain and confused. She thought that "a people or a nation are *saved* just as soon as they comprehend their god," "comprehending your deity & being in harmony with his plans is to be saved." It did not make any difference if you realized Christ or not; it only mattered that "you realized God through Christ." You were a Christian "if God has become nearer to you, more of a reality" through Christ. She suggested that Starr read Tennyson's *In Memoriam* even though it was clear in her own letter that she had herself not found peace in reading it or anything else. The historical Jesus did not help her at all; he seemed to have been a man about whom there was a mysterious and incomprehensible beauty, but she could never get any closer to him than that. So Addams would settle for a minimal amount of theology: "My creed is now *be sincere & don't fuss.*"

The letters continued for years in this earnest, searching tone. Addams spent four years at Rockford, and Starr spent one, yet neither seemed to have retained anything resembling a lucid theology or a peaceful conscience. Literary figures such as Carlyle, Ruskin, Browning, and Dickens were far more important than religious figures, and more significant sources of ideas than were religious works. Starr's side of the correspondence was clearly more aesthetic and emotional than was Addams' side; Starr was already on the road that eventually took her to the Church of England and to Roman Catholicism. Addams' concerns were more practical and worldly. Her image of her father, her religion, and her career would have to complement each other, or she could have no peace. Early in the letters, she told Starr: "I, for my part, am convinced that the success of that work in a large degree depends upon our religion and that I can never go abroad and use my best powers until I do settle it." The key to her choice of career lay in ethical and religious needs; she could find no outlet worthy of her beloved father in the frivolities of life under Anna.

During the next revival at Rockford, a few months later, she searched her soul again. The services impressed her and she was not really rebellious, but she refused to convert. God still seemed distant. "I see more the need of knowing God but he seems further off than he did" two years earlier. "I have been trying an awful experiment. I didn't pray, at least formally, for about three months, and was shocked to find that I feel no worse for it. I can think about a great many other things that are noble & beautiful. I feel happy and unconcerned and not in the least morbid." Yet, as a later letter makes clear, she may have been talking to reinforce her courage. Her beliefs were in painful flux. "I believe, my friend, it would

MINISTERS OF REFORM

be better for us not to talk about religion any more," because "my ideas are changing."

The letters continued, full of details concerning the death of John Addams, further revivals, her medical treatments, the severe melancholy that accompanied them, and Starr's joining the Anglican Church in the summer of 1883. Addams expressed her own sense of yearning by admitting to failure in settling her religious doubts even while she congratulated Starr. When Addams toured Europe, their correspondence slackened, but even while outwardly enjoying herself with Anna (and eluding George), Addams' letters retained a tone of seriousness that indicated she had not given in. "I quite feel as if I were not following the call of my genius when I propose to devote a year's time to travel in search of a good time and this general idea of culture which some way never commanded my full respect," she wrote in 1883. "People complain of losing spiritual life when abroad. I imagine it will be quite as hard to hold to full earnestness of purpose."

The bottom of her confusion and depression came in 1885 and early 1886, but even then Addams never lost her insight. She understood that her problems revolved around her religious confusions, the demands of her father's values, and her need to stop preparing for life and start doing something that would fill her days with socially valuable activities. "I am always *blundering,* when I deal with religious nomenclature or sensations simply because my religious life has been so small," she wrote Starr in one letter. She continued: "For many years it was my ambition to reach my father's moral requirements, & now when I am needing something more, I find myself approaching a crisis, & look rather wistfully to my friends for help." Two months later, during the period when Anna was pressing her to attend parties and to think of George as a lifelong companion, she contrasted her own emptiness with Starr's rewarding choice of teaching schoolgirls in Chicago. "You do as much work as I do, and more, in addition to all the time and vitality you give to your girls and that I am filled with shame that with all my apparent leisure I do nothing at all." Doing nothing in decorous fashion could never satisfy the daughter of John Addams.

Yet, less than four years later, Addams and Starr had discovered in their plans for Hull-House the solution to their religious and psychological needs. During her last trip to Europe, Addams finally experienced enough of life, literature, and religion to choose a career. In England she discovered Canon Samuel A. Barnett and his monument to religious conscience, Toynbee Hall. Toynbee Hall enabled the men connected with it to salve their consciences while making a difference in the lives of people in need.

"It is a community for University men who live there, have their recreation and clubs and society all among the poor people, yet in the same style they would like in their own circle," she wrote Alice. "It is so free from 'professional doing good,' so unaffectedly sincere and so productive of good results in its classes and libraries so that it seems perfectly ideal." In addition she examined the People's Palace, a settlement house on a rather different plan, with facilities where working men could socialize. At the same time, she read a number of related books. In addition to John Ruskin and Leo Tolstoy, socially conscious and religious writers who were already influencing many American reformers, she also read two novels by Walter Besant, *The Children of Gibeon* and *All Sorts and Conditions of Men*, both of which dealt with wealthy women who devoted their time and money to helping the poor. The cumulative effect of these visits, this reading, and the frustrations she experienced in finding a career consonant with her religious beliefs was the decision to open Hull-House. That decision was probably made sometime between April and June of 1888.

Jane Addams' early life provides an excellent insight into several key issues of emerging progressive thought. The religious atmosphere of her father's house created ethical imperatives that could not be met by conventional roles available to women at that time. Her religious faith apparently waned, dying out altogether in the years after 1889, yet the socially productive reform activity continued, both in her life and in the institution so closely identified with her. Her conscience, in a sense, became institutionalized, and in time able even to exist without her. Other progressives seemed to share her needs and to appreciate her example. Many of them came to live for a time at her settlement house; others came to visit or to lecture and carried away the new doctrines emerging from their experiences of the slums.

II

The most available male parallel to Jane Addams was George Herbert Mead, perhaps the most original philosopher produced in the Progressive Era. Like Addams, he had an intimate, long-time correspondent, in the person of Henry Northrup Castle, to whom he could spill out his hopes and fears. Unlike Addams, he did not establish a new institution or become a role model for the young. Mead's course was more typical: he went through a crisis of conversion, emerged as an advocate of urban reform, found a place in the new profession of college professor of philosophy, and spent his life trying to fuse ideas and actions in such a way as to facilitate the making of a better world.

Mead and his best friend Henry Northrup Castle were two of the most talented students to attend Oberlin College during this period. Born in 1863 to a family densely populated by farmers and clergy, Mead moved to the campus from western Massachusetts when he was seven years old. His father was the first incumbent of a new chair of homiletics at the Oberlin theological seminary; his mother was also articulate and educated, with a career that included teaching at Oberlin and the presidency of Mount Holyoke College. Family life was suffused with piety, and Mead never escaped its influence; his sister Alice only continued the family heritage when she married a minister. Castle came from comparable if more exotic stock. His father Samuel N. Castle had gone to Hawaii two generations earlier as an advisor to the Hawaiian Protestant Mission. Over the forty years they lived in Hawaii, the family had become wealthy and well educated, deeply involved in the business and political life of the islands. Castle himself had traveled widely and was socially more sophisticated than Mead. His sister Helen, also an Oberlin student, became so much a part of the relationship that she and Mead ultimately married.

The external parts of the Mead–Castle relationship are relatively simple and straightforward: Mead entered Oberlin at the age of 16 in 1879 and soon became close to the Castles. He graduated in 1883, briefly taught primary school, and then served on the surveying crew of the Wisconsin Central Railroad Company as it worked on the track connecting Minnesota to Moose Jaw, Saskatchewan. For most of this period, Helen and George were in Hawaii or traveling. Mead and Castle then attended Harvard together briefly, Mead being in residence only for the academic year 1887–88. He then went to Germany for graduate study, doing most of his work in Berlin. The Castles were there for most of this period, and Mead married Helen in Berlin in 1891. Before Mead could complete his degree, however, he received an offer to teach from the University of Michigan. There, he formed one of the most fruitful friendships of the Progressive Era with John Dewey. When Dewey left Michigan for the new University of Chicago, Mead went with him to become part of the Chicago school of American pragmatism. Dewey himself, as well as countless colleagues and students, attested to the originality of Mead's ideas, and many of Dewey's publications may well be the fruits of their long and intimate discussions. Historians will never know because Mead had crippling psychological blocks that prevented him from being either an especially good teacher or writer. A master at evading students and their questions, he also evaded the normal academic demands of university publication and never published a book throughout his long and creative life. Most of what now bears his name was assembled from fragments or reconstructed posthumously from student class notes in the 1930s. Such is his obscurity that he never had a biographer; such is his importance that students of philosophy have made something of a cult out of studying his works. Scholars of progressivism scarcely know of his existence because by normal standards of overt behavior he was insignificant.

Mead and Castle began their intellectual relationship as scoffers at certain Christian doctrines and baiters of Oberlin President James H. Fairchild. Not only did they delight in asking Fairchild questions he seemed unable to answer, but they used him as a standard reference point—"the Prexy"—in many of their later letters. Like the great majority of progressives, they seemed to need father figures both inside and outside the home against whom to measure themselves: as models for possible careers in education or the ministry, and as representatives of orthodox Christianity against whom one could measure one's own often heterodox and fluctuating views. In Mead's case some kind of literary life always appealed to him. For many years his letters contained frequent references to classic literary works, most often British. As he wrote Castle in 1883 shortly after graduation, "I should enjoy being a minister. I should prefer that to anything except a literary life." He was still too confused and unfocused to be thinking seriously, and his widowed mother's financial condition was never far from his mind. A life resembling President Fairchild's was unattractive, and like many graduates in the liberal arts, Mead found that he had a fund of useless knowledge that both put his mind in turmoil and prevented him from making anything like a decent living outside some educational institution. He made the classic choice of many women and not a few of the men of his day: school teaching.[2]

Mead's father had died two years before, cutting off the family income. Mead waited on tables to help pay his college bills while his mother taught, but money haunted his mind for the next decade and it influenced his professional decisions: "I have 10 years work before me at least before I begin studying for any profession or fitting for any calling in life at all, for I must see that my mother does not want a cent." His eyes were already on Germany, and he envisioned a period of making money to be followed by the study of literature in Germany. But he was "miserable for the whole vacation" because he had no prospects and only when a possible teaching job appeared did his spirits pick up a bit. When it became definite, he had a feeling that the next six months could determine his future. He feared that he would not be up to the task and that he would not persevere if he found his circumstances disagreeable. He felt that his friends and family were watching him and that the pressures might either make him a man or a "pitiable wreck." By mid-October he was in harness, and the beginnings of his evolution from a reluctant Protestant to a functioning progressive were already appearing in his letters. He was "undertaking the missionary work of working up the community to some spiritual life." No orthodox Christian himself, he did not want to be misunderstood, but he wrote that the "necessity of a spiritual life" remains as much a need when one is outside Christianity as when one is within it. Despite his skepticism, he felt that Christianity "seems to be the only religion that can reach the common people" and so he thought it inevitable that he should use Christian means and rhetoric to achieve universally valid goals. He did not, however, move very far psychologically. "My mother asserts that naturally I am fitted to be a minister, and I think that she is right."[3]

Mead knew he did not satisfy his mother's religious standards, but was happy to report that he seemed promising as a creative, literary mind to her, and she was "one of the best literary judges if not the best in the Oberlin faculty." The constant reference to his mother and her opinions indicated something of his continuing emotional dependence on her in the fall of 1883. Signs soon appeared, however, marking the beginning of his conception of himself as a philosopher of some kind. He dwelt especially on the freedom of the will and on how one could advance morally, as well as on the forces of environment on people. He was impressed by the "accidental and unimportant" and thought any final standard "ridiculous for us to think of." Capable of spinning out ideas at great length, Mead then insisted irritatedly that he was "disgusted with ultimate knowledge" because it seemed to lead "so surely to agnosticism." He preferred to "play in the twilight with Kant." He suggested—the tone seems both whimsical and serious—that he and Castle "make believe we have a philosophy and that we will make everything reasonable and construct a whole fabric albeit without foundation."[4] To any student of philosophy, Mead's remarks have immediate relevance for the dawning of pragmatism and the related trends that emerged during the twentieth century. To any student of progressivism, they imply the beginnings of an attempt to construct a Kingdom of God on earth, to use the term later popularized by theologian Walter Rauschenbusch. The residue of Christian idealism and the psychological needs of the Protestant young demanded some kind of divine sanction for secular life, even if God did not exist and the society had to be imagined. Most of John Dewey's correspondence from this period has not survived, but he, too, was apparently playing "in the twilight with Kant" in his evolution into the philosophical sage of progressivism.

Despite this philosophic interest, no career in philosophy offered itself: the profession of philosopher did not yet exist in America, although it was evolving at Harvard and Johns Hopkins. Mead could still imagine, with "certain other conditions being fulfilled," a life as a "philanthropist," or a "home missionary," or a "city or foreign missionary." He also felt that he had too little "brain power" and "too much moral sentiment" to be a philosopher. He voiced again and again the conflict between his reason and his feelings. He admitted to "temptations of the most insidious and perplexing kind" that prompted him toward "becoming a Christian," and said that while his reason objected, his "sensibilities" seemed "to be clamoring for Christianity." He thought that he could possibly make "as good a Christian as De Quincey or Coleridge," but that was hardly a comfortable level of faith in small-town Ohio in the 1880s. He approvingly quoted the axiom of Tertullian, "Certum est, quia impossible est," and lamented in words that applied to most progressives: "It was not for naught that I was brought up in a Christian family, fed with the conceptions of Christianity [and] steeped in the reverence for Christianity." He positively yearned for some unreasonable faith that he would have to accept to put himself out of his misery. His life was "such a contradiction" that only a God afar off could make sense of it all. Teaching school was going badly, and soon he took refuge in "praying and reading the Bible." He admitted that he did "not know that there is a God," but he thought that if he could "pursue religions as I do other matters," he could somehow "find surety enough" and perhaps time would give him certainty in some conviction. He revolted at the very thought of becoming a theologian, yet felt that his "great danger" was that he would "swallow Christianity whole because it will give me peace." He felt that nothing satisfied the needs of mankind better than Christianity, and "why not have a little deception if need be?" He thought about revival work, but admitted being "torn this way and that." With all his heart, he longed to throw his doubts aside "and leap with my eyes shut and heart open in Christian work." But he knew he could not, and even stating the possibility seemed to be a way of avoiding the temptation. More pressing concerns were at hand: "The only good thing to work for is humanity." He might be a philosopher or a literary man by temperament, but he needed to serve humanity to give meaning and purpose to his life. Ethics and acts had to go together and he found it difficult to see how they went together teaching the rebellious young: "If I am to serve humanity by speculating and criticism it is time I climbed out of here."[5]

MINISTERS OF REFORM

III

By early March his student body had dwindled to eight, and the governing board decided to close the school. Troubled as much by the loss of his faith as by the loss of his job, Mead experimented with prayer—unfortunately to a God whose existence he doubted. He insisted to Castle that "in some way I get help," but the anxiety permeating his letters made the statement less than convincing. He continued to connect his own feelings of uselessness with the need to do good in the world, and in conceptualizing the doing of good in the world with the worship of whatever God did exist: "We both feel that the only nobility is in working for others or perhaps for a God if there is one, and to refuse to do so is suicidal." He returned to Oberlin and the presumed security of home to find himself "wallowing in the depth of agnosticism" only days after his experiments with prayer. He thought at length about his undergraduate days and felt "filled with disgust. I learned almost nothing." He lapsed constantly into biblical language and yearned for the security of his lost faith: "If I only could believe it I could find more comfort and infinite relief in the bible or the peace that passeth all understanding, than I could begin to find anywhere else under the sun." The continuing indecision about his future profession produced a related anguish. He feared that if he gave himself up to philosophy or literature, he would not be good enough, that his work might consist of writing reviews of "recent German productions in philosophy," or becoming an "art critic for some daily. Now such a life of practical inaction I must not lead." His conscience required that he also choose an ethically satisfying career, and not only did he have trouble finding one, but he doubted his talents; his religious doubts were only compounded by his secular, vocational problems. "I can not do the work that I want if I am skeptical, and I cannot become orthodox right away any way. Still I feel the immediate necessity of getting myself to work in the world, and I must do it."[6]

On 16 March Mead wrote one of the longest and most self-searching letters of his life. He laid out his alternatives as being essentially two, "if I am ever to do any good in the world." His first line of action was to give up all prospects of a career in literature or philosophy "and get out and work for men's souls." His creed was "dark and agnostic enough," but he thought Christianity the only possible means of "reaching humanity." Like William James at roughly the same time, he was groping toward the pragmatic belief that God and religion were real because they had real effects. A dogmatic belief either on his part or on humanity's part was unnecessary, because "among a good many people such work under such circumstances will bring in as much as perhaps more good than if I went and preached a necessity of Christ's salvation." He went on for pages about his various theological doubts and the conflict in himself between his skepticism and his need for emotional support. But he remained convinced that anyone who wanted to help humanity, and who did not involve himself in and use Christianity to further his goals, "must be very unreasonable unless he has found a system that will answer the needs of humanity better." He disliked having to present himself as something he was not, but "that deception is to me a small point if only by that means I may do good." He was coming to the conclusion that "the most reasonable system of the universe can be formed to myself without a God," but at the same time, "the spirit of a minister is strong in me and I come fairly by it." He wanted "to fight a valiant personal hand to hand fight with evil in the world. The inconsistencies that go with it I shall have in a measure to put up with."

His alternative course was "the line of metaphysics." He could devote himself to discovering the truth. He could teach, do research, write articles, and, if at all possible, go to Germany. In his stronger moments, he preferred the first alternative; in his weaker, the second. Like so many progressives, he was torn between the demands of his Protestant conscience, his growing inability to accept outdated dogma, and the lack of what he viewed as meaningful work in America. He effectively summed up the conflict in a scientific image: "I am made up upon two utterly different principles like the two forms of crystallization we often meet in the same stone." Whichever course he might take, "the prospect of my benefitting the world" seemed "small, but the prospect in the former seems the best." Regardless, he wrote: "I cannot count on entering the ministry," and the prospect of business, law, or medicine did not give him a feeling that he could "do much personal work." Without this emotional commitment, he feared that he might "degenerate into a mere money getting animal." He took the classic way out of any vocational dilemma: he fled.[7]

A month later he was out surveying for the railroad. His initial reaction was that the work was healthy and enjoyable; he welcomed the out-of-doors atmosphere with no need to think or feel. He still read English literature and continued old Oberlin arguments with "the Prexy"; Mead worried about Kant's transcendental idealism even as his candle guttered out and his legibility disappeared.

His world soon depressed and finally infuriated him. The men drank, the women swore, and the mosquitoes bit. Vice and debauchery seemed to permeate the atmosphere, and, of course, this led any good son of the puritans to moralize about his circumstances: "I have no courage to meet this problem of vice without a strong belief." He wished he could curse Christianity as Shelley did, but he could not. He felt weak and afraid. "I long for a fiery hell with which to startle these creatures, to shake their gross animality with thoughts beyond the reaches of their souls. Whatever the eschatology of hell in [the] future it is a blessed reality now if it could only be used as such." He wanted desperately to get out of the United States and travel the world. "I am utterly at a loss for this winter. It is disgusting. I am ashamed of myself that I should be at a loss in the matter but so it is."[8]

By the end of the year, Mead was back in Minneapolis, feeling poor and in need of work. He began tutoring and soon conceived a plan for a small, private secondary school that would emphasize subjects like Latin. He tried to enlist Castle's help, but Castle remained far away, and the few possibilities for a visit never materialized. In their letters the discussions of philosophic and religious questions continued in the same veins of dogmatic doubt and vocational sterility. Mead's mood during this period could be summed up in a few sentences: "I seem so far off from anything worth living for, and I do not see that I gain strength at all by which I can reach anything better. My life is spasmodical, uneven, without purpose. I am pushed and pulled on now by this motive and now by that." His life seemed to be running out "like an Australian river." He fretted the days away on remedial algebra and Latin tenses. He became a hypochondriac, living on milk, crackers, and eggs and pondering the possibilities of getting kidney trouble and diabetes. He worried constantly about Christianity. He remained full of doubt yet thought he should declare orthodox religious beliefs if only to please his mother. He wrote: "You understand, do you not, Henry? My soul goes out to you in a great cry that asks for sympathy in grief almost too great to bear. Where are we going? What are we leaving? What might we be?" He felt "alone in the world except for you, Henry."[9]

Over the next two years, Mead's letters grew shorter and less frequent. He tried to interest Castle in a possible appointment at the University of Minnesota but nothing came of it. He continued tutoring; he suffered various diseases, including a serious bout with the mumps, and he bewailed his poverty. He analyzed his character with something like aversion, finding himself weak, uncreative, and "rather of a feminine cast of mind." He toyed for months with the possibilities of going to Johns Hopkins or Harvard, but his mother's needs and his poverty, combined with his weak character, served to all but paralyze him. He was "disgusted with life." Everything that had ever seemed desirable "has vanished out of it. I see no prospect of success or pleasure in any field. I am discontented, disappointed, disgusted with myself." No matter how hard he tried, he could not "find a motive sufficient to inspire activity. The whole subject of the truth of Christianity "has become wearisome to the last degree," but he feared that it would inevitably "lead to practical estrangement between my mother and myself." His dark night of the soul and body began to lift only in early 1887 as he began to focus on a solution that, unknown to him, so many other young progressives in similar circumstances were also finding. He had toyed with ideas about action in the real world off and on for years, but now, as his health improved it seemed as if some kind of ethical social activity was as vital to his body as to his mind. "Especially would life seem desirable if I could find myself opening out upon some fine object upon which I could become really enthusiastic. Such a purpose is the one essential," he wrote early in May 1887. "There is plenty of evil in the world to which we can turn our energies—plenty of possible progress to which we can devote our thought. In a word there is room for action."[10]

During the summer of 1887, his letters focused on the chance of going to Harvard. No fellowship came from Johns Hopkins, and by the late 1880s, it was becoming evident that the Harvard of Josiah Royce and William James was the preeminent place to study philosophy in America anyway, and that for all its eminence in other fields, Johns Hopkins had lost its claim to first place in that discipline. Furthermore, although Mead might toy intellectually with the thought of combating evil hand to hand in the world, he was essentially correct in his judgments about his retiring personality and his predisposition for abstract thought. "What a joy it would be" to attend Harvard, he exulted. Life in Minneapolis bored him witless: "the longing to get near a city where I can have decent advantages has amounted almost to a madness. I fairly foam at the mouth with the thought of the desertlike aspect of Minneapolis musically and intellectually." He looked forward especially to study with Royce and related work in Greek and history. He saw Greek "as the means of earning my bread if necessary and as an almost essential accompaniment to thorough work in literature in any department. . . . Ye Gods and little fishes what a dry sterile time I have had of it during this and last year. You will find me quite skin and bones mentally." The two intended to room together and did so.[11]

IV

Mead's financial problems plagued him throughout his year at Harvard, but since he and Castle were together for most of the year, no written record survives of many of his academic experiences. George Herbert

Palmer was complimentary about his performance on his final exams, but Mead's greatest coup was to impress William James so much with his abilities that James came to his quarters the next day and asked him to join him and his family for the summer as tutor to his son. Mead seems to have been more interested in a summer with few expenses than he was at the chance of being in the intimate company of one of the great figures in his discipline. He described the opportunity to Castle as "a pleasant place to pass the summer without expense." James told him that several others had already applied for the position, but that Mead did so well on his exams that "he wished me especially as he would then have some one to whom he could talk metaphysics." Within two weeks Mead was writing from the James summer home in the New Hampshire mountains, part of an extended family that included four children as well as James' sister-in-law and mother-in-law; two servant girls, a hired man, three horses, and a pug dog completed the menage. Between tutoring sessions Mead buried himself in Kant, Hegel, Green, Sedgwick, and Schopenhauer.[12]

James and Mead passed long hours in conversation, but little of what they said found its way into Mead's few letters to Castle during that summer. James told him of his high regard for Charles Carroll Everett, a member of the Metaphysical Club who taught in the Divinity School, whom he considered "after Royce . . . the strongest man in the university, very subtle and original." James had filled Mead in about Royce's "foolish eating and drinking" habits and how his neglect of his health had led him to a nervous and physical breakdown. James also read him a letter from the young George Santayana, then so unknown that Mead misspelled his name. Santayana was in Europe, filled with disgust about what he was finding abroad and interested in returning to Harvard if he could retain his fellowship. Mead wrote: "James thinks he lacks the virile dogged qualities which make the successful worker in this or any direction." Such stimulating conversation heartened Mead, who soon resolved on firm professional plans to get a Ph.D. in Berlin. He wrote that James was "very kind to me" and even offered a small loan, but "what made my heart burn was that he said he might possibly be able to get the *Nation* [to] take some review of mine" and perhaps pay a decent sum for it. "He said he made a good deal of money in this way as a young man." He also offered his good offices in rounding up more students for Mead to tutor if he continued at Harvard. Much encouraged, Mead nevertheless made the break for Europe, almost a decade after he first conceived his dream to study there.[13]

Castle also went to Europe for at least part of the time Mead was there, and his letters help fill in the picture of Mead's maturation. The first semester, during the fall of 1888, does not seem to have been especially successful, but Mead did run across Professor G. Stanley Hall of Johns Hopkins, whom he described as "just the one man on our planet whom for the moment [Mead] wanted to see." In Castle's opinion Hall was "the most eminent physiological psychologist in America" and the only man with adequate European experience to advise Mead. "Poor George was utterly at a loss how to begin," because every person he consulted seemed to offer conflicting advice. He apparently decided at this time to make physiological psychology his speciality because of a great fear that the evangelical sects in America would make his life miserable if he became a philosopher who publicly expressed heterodox religious ideas. In this new field, "he has a harmless territory in which he can work quietly without drawing down upon himself the anathema and excommunication of all-potent Evangelicalism." Castle wrote his parents: "Christ never had any following among religious people," the implication being that true Christians were like George and himself, following ethical instincts as far and as freely as their minds would go. Hall apparently suggested a year of study in Turin, and Henry and George were seriously considering this.[14]

Like many progressive intellectuals, Mead was willing to do what most active politicians would not have dared to do: openly use the word *socialism* as the term to describe his new position. In fact, one way or another, progressives even of the most conservative sort often advocated measures that seemed socialistic in everything but name; most obviously, they did this in the popular campaigns for municipal ownership of natural monopolies such as electric, telephone, and water supply companies. If one excludes the foreign language socialism of recent American immigrants, with its violent language and intimations of class struggle, then socialism of some kind was a basic part of progressivism, and Mead was in line with dozens of other young people as he studied German institutions around 1890 and saw much that Americans could learn from them. As a thinker with minimal practical experience, he was impressed with the social ideals that were so noticeable in German society. He found "the opening toward all that is uplifting and satisfying in socialism," especially "in the introduction of the social idea into all forms of life." He told Castle that "the immediate necessity is that we should have a clear conception of what forms socialism is taking" in Europe, particularly at the local level: "how cities sweep their streets," how they "manage their gas works and street cars," as well as their poor, minor criminals, police, and even their houses of prostitution. Such ideas needed injection into America. For the moment Mead felt that he had to teach, but he did not intend to do so for long. He wanted a "more active life" and felt that physiological psychology had given him "exactly the right sort of a foundation for this work." His plan was to go to the University of Minnesota to teach and for Castle to join him, working as a lawyer. They would then get control of the *Minneapolis Tribune* and work for their new ideas.[15]

Mead's personality fit the popular perception of the academic. He was painfully shy, reticent, and withdrawn, and all too aware of it. He desperately wanted to involve himself with the larger democracy, in part because he could never really belong in it. He stressed democracy as an ideal form in large part because of his distance from true democracy; he stressed the social ideal because of his distance from true social life; he soon came to stress the unity of the thought and the act because he himself acted so little. His letters to Castle in the early 1890s were full of discussion about failure, as if he knew how hopeless it was for a student of Kant and Hegel to change the nature of American society. "I am perfectly willing to fail," he reiterated. "Life looks like such an insignificant affair that two or three & more years of utterly unsuccessful work would not seem to me in the slightest dampening." He also epitomized what became an all-too-common progressive belief: Set apart from the democracy yet wanting to be a part of it, aspiring yet fearful, his Protestant conscience told him that the effort was what really mattered. Good motives and honest effort were what God and one's conscience demanded; success, not to be expected anyway, became secondary. The very men and women who developed the notion of results into a philosophical antisystem were willing to settle often enough in practice for an honest effort that quite literally made no difference. Mead wrote: "I mean that I am willing to go into a reform movement which in my eyes may be a failure after all, simply for the sake of the work."

For Mead the laissez-faire philosophy of big business dominated American government. Yet, too many reformers looked only at the federal government in their efforts to change society. "The direction in which we must work seems to me very clear—city politics," he wrote. If we can get reform started in the city, if we can "remove the machine in large degree, educate the public politically, and start at the point where the new social duties and functions which our government can take in should make their entrance," then a pure federal government would have to follow. "If we can give American institutions the new blood of the social ideal, it can come in only at this unit of our political life, and from this starting point it will naturally spread." Start small in the city, and big results would follow in state and nation, he believed. He intended to follow the example of British Liberal John Morley and work toward "the practical application of morals to life" in education, in personal character, and in politics. He also wished to carry this impulse into his recent psychological work to encourage "a vigorous spreading of literature and methods that apply the psychology of moral development" to the study of the child. The effort would be its own reward: "I want a few years of activity that I can throw myself into, and I am perfectly willing to see no success—the subjective return is enough."[16]

By the fall of 1890, Mead's conversion into a progressive was essentially complete. He had begun as a teenager at Oberlin, devoted to his parents yet skeptical about the absolutes of their Protestant religion. He had experienced a period of profound self-doubt about religion and his own vocational future, had experimented with unrewarding jobs and new ideas, had managed several years in American and foreign graduate schools, had gained confidence in his own capabilities, and, finally, had evolved a sense that only in reform politics could he find meaningful work. He could teach; he could write for or run a newspaper; he could write books; he could conduct experiments in child psychology. He wanted above all to engage in the process of moral reform in this world; he would hope for results but would not count on them. He wanted, in effect, a Kingdom of God in America, and he found in some kind of democratic socialism the only theology he really needed. Seeking to tie his ideas to his actions, he developed a philosophy that did exactly that. All that he lacked was a suitable position that paid him enough to live on and presented him with an attentive audience.

He worked quickly to prepare himself for a professorship. He maintained contact by mail with William James, who sent him copies of his new book, *Principles of Psychology,* probably the most influential text ever published in the history of the field. He corresponded with James B. Angell and John Dewey. He sketched out his thesis, which was "largely a criticism from a Kantian or at least metaphysical standpoint of the sensational doctrine of space," and he planned to turn it in around April of 1892. But before he could get fully into gear to write, an offer came from the University of Michigan of an instructorship in philosophy. Mead responded quickly; he spent the summer of 1891 preparing course outlines in physiological psychology, the history of philosophy, Kant, and evolution. He wrote: "Kant will be a most admirable dessert—the drudgery will be the Phys. Psy." He regretted going back to teaching without his degree in hand, but expected to return in a year to take it then. He never did. Instead, he adapted well to a department that consisted of John Dewey and another instructor named Alfred Lloyd. He found Dewey "entertaining, clear, speculative, appreciative, and individual and western," and with him formed the great intellectual as well as personal friendship of his life. He wrote: "My thought assumed a definitely positive and cheerful aspect." He had come to some understanding "of the meaning of the faith that was in me—faith in the unity of meaning and life—and the absolute completeness of the statement which psychology can give." Mead found a great freedom at Ann Arbor, and was "able to study for the first time the real relation of the university to life—to see it in the idea organized—so that it can direct action," by which he meant the "direct relation to present activity." The medical and engineering departments of the state university already had close connections to democratic life and its needs, and he expected to play a role through "mothers clubs and psychological societies which will study the young child and the mental image." Thus, he could begin reform of the school "from beneath," and, of course, the school led directly to society.[17]

Dewey took Mead with him to the University of Chicago in 1894, and despite his lack of a Ph.D., Mead rose to associate professor in 1902 and professor in 1907. Unable to publish, he encouraged the prolific Dewey and contributed in immeasurable ways to what the world regards as Dewey's publications and influence. A progressive pragmatist in philosophy, he also worked as he had planned in Chicago. He was continuously involved with Jane Addams and Hull-House, worked quietly to improve the school system, and became an expert on labor problems.[18]

In this fashion, two of the most perceptive of the young progressives passed through the conversion experiences of their ancestors. Times had changed, but the process was old. But this time sainthood had to be earned on earth, and faith in one's own competence to do the job was not always available. The sharing of this experience with many others of their generation made it possible for figures such as Addams and Mead to become as important as they did.

Chapter 2

Two Visions of Democracy

B Y THE EARLY 1890s, the reform ideas and figures of the post-Civil War years were beginning to form coherent patterns. Creative individuals such as Jane Addams, George Mead, and Woodrow Wilson had articulated their religious needs, found conventional outlets inadequate to those needs, and groped toward new professions in the settlements and universities. Rural and urban poverty provided appropriate outlets for young people programmed for the righting of wrongs; they needed the poor fully as much as the poor needed them.

The normal stress of historians on political events has obscured the dual nature of creative life during the 1890s. In some ways the decade was one of the three or four most fecund in the history of American creativity, ranking with the 1770s, the 1830s, and the 1920s. Politics, on the other hand, was not an arena in which anything significant was taking place. Grover Cleveland and William McKinley are of a certain residual interest, largely because they occupied the most visible office in the land, but they left little for future generations to value. Bryan did somewhat better, exuberantly clasping change to his breast while hoping no one would notice that he did not understand most of what he was talking about. Several local politicians, such as Governor John P. Altgeld of Illinois, did much better, but history has been far kinder to Altgeld and those like him than voters were at the time. Except for a few vestiges of populism, politics was not a creative profession in the 1890s; until the inauguration of Theodore Roosevelt, politics provided little for the student of civilization to work with.

America remained dominated by patterns of religious thought. Some were explicit, giving rise to what became the social gospel movement. Most were implicit, shaping ideas that seemed to be about secular matters. Religion provided the central motivating force for adventurous thinking, but the subsequent secularization of modern culture has obscured the importance of religion in forming the minds even of the most secular thinkers. An intense study of two figures allows us to enter into this dualistic pattern of explicit and implicit religion in the 90s. George D. Herron was probably the most famous reform clergyman of his day and certainly the most publicized voice for reform within the church until his somewhat ignominious departure. John Dewey was a professor of philosophy, less well known than Herron but in the long run a far more significant figure.

II

Like an earlier generation of clergy, who had found Unitarianism cold and Congregationalism irrelevant and who became Transcendentalists and abolitionists, the socially conscious clergy of the 1890s felt that too many Christians divided their lives between righteousness on Sunday and rapaciousness during the remainder of the week. Some of the churches had become rich, yet their money never reached the poor. Many churches were located in inner-city areas, while their affluent congregations had fled to the suburbs; the picture of the well-dressed arriving in carriages on Sunday, with squalor often only a block away, distressed a large number of observers. The churches had not only become intellectually marginal in an age of Darwinism and agnosticism, they had become socially marginal in an age of increasing poverty, suffering, and an ever-widening breach between rich and poor.

A clergyman who perceived this situation found it hard to change. If he protested too loudly, he might be denied any important role in an established church. If he already had such a role, he could lose it. Clergymen rarely had independent financial resources, and church boards of trustees were often dominated by the very wealthy men who were obvious subjects of criticism because of the fruits of their business lives. Occasional clergy did speak up and survive, but they rarely spoke up in any fashion that seemed radical to their congregations. The most telling social commentary usually came from teachers and journalists. As the chorus of criticism grew louder, however, it became harder to ignore. It was especially hard to ignore if one had some awareness of primitive Christianity with its stress on poverty and other-worldliness, and had entered the ministry in order to help the less fortunate.[1]

The social gospel movement produced at least half a dozen significant leaders within or closely allied to the Protestant churches in the 1890s. None of these figures retains more than historical importance, but a figure like George Davis Herron stands out in the group both for his charismatic effect and for his extensive publications. He was born on 21 January 1862, the son of William and Isabella Herron, of Montezuma, Indiana. Written in 1891, Herron's own account of these years is virtually the only primary source, and he is insistent on the strictly religious nature of his home and upbringing. Through his father, "a humble man who believed the Bible and hated unrighteousness," he came "from an unbroken line of Christian ancestors, reaching back to the days of the Scottish Reformation." His mother, while pregnant with George, prayed constantly that her child would be a servant of God. "She received me as from God and gave me back to God as her free-will offering. She besought God to keep me upon the altar of a perfect sacrifice in the service of his Christ and her Redeemer." If Herron was correct, she never felt so exalted again, because she believed that nothing she had done or could do would be as important as giving God a child consecrated to his service.

Illness haunted the household. Herron was often thought to be near death, and his mother was an invalid most of the time. As a result, the child had an unusual upbringing, having few friends of his own age, and being entirely devoted to his father. "He taught me, very early, to read and selected my books, directed my thoughts. We were seldom apart, day or night. He drew out all there was in me and turned it Godward." At a young age, Herron worked his way through a history of the world, and through at least portions of George Bancroft's history of America, and emerged with the predictable vision of the past as the reign and plan of God. He wrote that he "could not form a conception from any other point of view. An accident, in the minutest detail of life, was a thing foreign to my comprehension. I was a slave, if I may so speak, to the will of God." Because he had so few friends in the real world, he felt that "God was my confidant. I never thought of myself as other than his child. I talked with him over my books and on my walks. He answered my prayers." For "imaginary playmates," Herron turned to books, and found especially "Joseph, Elijah and Daniel, Cromwell and John Wesley and Charles Sumner." Thus he "grew up in the company of God, with a daily deepening sense of a divine call which sooner or later I must obey."

The poverty of the family prevented continual schooling. Herron worked in a printing office, under what he later regarded as depraved and sinful conditions. At about age seventeen, in 1879, he decided to pursue a more formal education, so he entered the preparatory division of Ripon College in Wisconsin. After fewer than two years, ill health, poverty, and only mediocre success forced him to leave, and he worked briefly for a Ripon newspaper. Despite his rather dismal prospects, he convinced Mary V. Everhard, the red-haired daughter of a prominent physician who was also the mayor of Ripon, to marry him. The couple soon moved to St. Paul, Minnesota, apparently to further Herron's plans for a career in literature and journalism. About 1883, he experienced a profound religious crisis. After the classic manner of generations of American Protestants, he recalled later that he "groped in that horror of darkness which settles upon a soul when it knows that there is no sound thing in it, and that it merits nothing but eternal death and endless night." His anguished vision "laid such hold of me that all the eternities seemed overwrought with speechless pain." Only the mercy of God could save him, and he felt unworthy of that mercy. He wrote: "Jonathan Edwards' Enfield sermon was, at that time, the only thing real enough to answer my experience. But out of this horrible pit I cried unto the Lord and he heard me, lifting me up and planting my feet upon the rock of his salvation." Herron entered the Congregational ministry and took several small churches in the Midwest.

Meanwhile, like Jane Addams and other nascent progressives, Herron discovered the radical, socially conscious writers of Europe and America. F. D. Maurice, Henry George, and Giuseppe Mazzini were only three of the men who influenced Herron's thoughts during the next decade. He began to apply their teachings in his ministry almost immediately. In 1890 Herron gave the sermon that lifted him from obscurity and helped to make him famous throughout the area. To audiences nurtured on the quasireligious economics of Herbert Spencer and William Graham Sumner, he proclaimed that competition and inevitable progress were not tied together. He insisted that if progress meant anything at all, it meant moral progress, and that economic competition was always opposed to moral development.

In Herron's analysis the problems of economic life boiled down to Cain's question, "Am I my brother's keeper?" Herron thought that men were their brothers' keepers, and that capitalistic society assumed instead that selfishness and competition were appropriate moral guidelines. God's answer to Cain's question was the cross: by giving men Jesus to atone for their sins, God intended that men should follow his example by sacrificing themselves for others. Thus, the message of Jesus to every man, regardless of financial circumstances, was to sacrifice by doing service to others. The man who did not do this was not living a moral or Christian life. This message had peculiar relevance to the lives of the rich because they had so much more to sacrifice. If the wealthy men in America chose to follow Christ's example, then social problems would disappear. If they did not do so, they were immoral and ungodly.

Herron quickly found himself in demand as a speaker and possible pastor at more important churches, and he accepted a call from the Congregational Church of Burlington, Iowa. While at Burlington, Herron met three people who played key roles in the next decade of his life. Mrs. E. D. Rand, a member of his congregation, was the wealthy widow of a Burlington lumberman. She was financially, intellectually, and emotionally independent, and her support of Herron in whatever he wished to do became one of the more remarkable aspects of Herron's story. Her daughter Carrie was pretty and much attracted to Herron; the scandal surrounding their relationship, which began in Burlington, followed her the rest of her life. George Augustus Gates, a theologically unorthodox Congregationalist minister, had been called to the presidency of Iowa College in Grinnell in 1887. He first met Herron in 1892, when he asked the controversial new assistant minister to speak at the college. Gates knew of Herron's writings—and had himself been in trouble for radical views. He wanted to meet and know more of the man who so impressed midwestern evangelical circles. Herron came, spoke, and it seemed to Gates that "he did set the souls of our young men and women on fire with a high and holy passion, such as is not common in experiences of that sort." At that time Gates did not plan to have Herron come permanently to the college, but Mrs. Rand soon made it possible. A long-time friend of the college, she had already mentioned the possibility of endowing a professorship, but Herron's name had never before been mentioned as a possible appointee.

Gates consulted his faculty and they "unanimously" agreed that Herron should be offered a place on the faculty. Herron's radical theological and social views, well-known by 1893, were not impediments but attractions in the eyes of the school administration. He would come to promote the old college traditions, and in Gates' eyes, "Iowa College has always taught, teaches today, and, so God will, will always teach the actual applicability of the principles of Jesus Christ to every department of human life." On the face of it, no appointment could have been more welcome or appropriate to both Herron and the college.

Few people at the time gave thought to other motives for hiring Herron. It seems likely, however, in view of what was soon common knowledge on the campus, that the appointment had at least as much to do with Herron's growing affection for Carrie Rand as it did with Christian sociology. The mutual devotion of Herron and the Rands was as strong as their devotion to a just society. Yet outwardly, Herron voiced a view that was soon a hallmark of his public self-analysis. He struck the pose of a man chosen by God as if by an influx of grace, rendered powerless by himself, and made simply the instrument of divine will. He thus began a pattern of justification that made him at first a charismatic leader to young and evangelical reformers, and eventually, the devil himself to a large segment of more conventional society. "I believe God has sent me with this message of a new redemption through His Son. I must go as I am sent," he wrote. The chair of Christianity that Mrs. Rand had endowed "opens the way for me to speak to the church at large." He was not simply leaving one job and taking another: "I go to witness to the righteousness of society and the nation. I can do nothing else. I do not enter this open door because I expect to have an easier work. I go to toil as I never toiled. I go to suffer for the truth and name of Christ." Herron's appointment also brought the Rands and their money, time, and energy to the campus. Carrie soon became an instructor in social and physical culture, and then principal for women, while her mother apparently dominated the campus as a *de facto* dean and fundraiser.

In Iowa College the Rands and Herron had found an ideal institution for advancing their views. Founded before the Civil War, as a direct result of evangelical and reform activity, it was a small college designed after the New England manner, located in a town recalling New England as much as was possible in central Iowa. Strongly abolitionist, moralistic, and prohibitionist in the past, it had never had qualms about Christian morality, or the right of Protestants to legislate their views into practice. If Herron could not survive and prosper here, he would have great trouble doing so anywhere.

III

Herron's impact on the campus was enormous. His classes were popular, so frequently overflowing that for a while they had to take place in the chapel. Far from being silenced by conservative townspeople or nervous trustees, Herron received adulation on all sides when he began his work, and when he assigned radical reading matter to his classes, he received no strong opposition. The required reading for his courses included Richard Ely's *Labor Movement in America* and *Social Aspects of Christianity,* Washington Gladden's *Tools and the Man* and *Applied Christianity,* and Laurence Gronlund's *The Co-operative Commonwealth.* The college was so proud of its new Department of Applied Christianity that it published a three-page leaflet describing the course and its reading list. President Gates was enthusiastic, and proudly informed Richard Ely of Herron's campus popularity and the many outside requests for his speeches.

Unfortunately, the first flushes of success and enthusiasm for Herron and his message were the high points of his Iowa College career. In part the opposition to Herron that arose was the natural dislike of those more conservative and conventional than he for anything smacking of the social gospel or progressivism. Led by men like history professor L. F. Parker on the faculty, and the aged trustee Colonel John Meyer outside the faculty, the criticism of Herron for his radicalism slowly increased over the years until by the late 1890s it became very nearly a majority position even within the ranks of the college. As early as 1894, Charles A. Young, a Herron supporter, was fearful of the possible results of Herron's manner of speaking. As he wrote to Richard Ely, Young felt "very much concerned about Herron" and thought that "those of us who love him ought to counsel him to be guarded in his statements." In fact Herron seems to have become erratic in behavior, negligent toward his students, and extreme in his expression of opinions. Unlike other teaching progressives such as Richard Ely and Edward A. Ross, who were criticized for their views, Herron's problem was not merely a matter of academic freedom. President Gates, for example, remained personally fond of Herron and agreed with many of his views.

Herron's nonacademic activities soon all but preempted his time and energy. He joined the Kingdom Movement as soon as he arrived on campus, and through it achieved much influence over both Gates and Congregationalism. The movement had begun at a retreat in 1892 called by President Gates. It was an attempt to give formal expression to some of the reform impulses within Congregationalism. Through its journal, *The Kingdom,* with Gates as editor and Herron as associate editor, it soon influenced men in other Protestant denominations as well. Reformers as well known as Robert A. Woods, Washington Gladden, and John Bascom were for a while public supporters of the movement, as were people like Charles Noble and Josiah Strong, who later opposed Herron. The group met in a continuing seminar, during which Herron, Strong, or another writer would read from and discuss his next book; over several years what had been a small conference grew until it was called the School of the Kingdom. From all over the country, distinguished speakers and conferees came to exchange information on the application of Christianity to society. Iowa College became a symbolic center for the social gospel movement during the middle 1890s.

Herron's growing extremism, and Gates' open admiration for him, soon splintered the movement. Like Herron's teaching career, it peaked early, then faded, largely because many original supporters could not follow Herron's lead. When Noble looked back at the end of his career, he remained unhappy that "the movement was wrecked by the extreme socialistic tone which began to characterize Herron's utterances and to some extent those of President Gates." He could still remember Professor Parker taking him aside and indicating that the time had come to choose sides, since even Gates was beginning to sound like Herron. Parker felt that Herron expressed "an utterly needless antagonism to churches," that he "refused to recognize the extent to which the church is already applying the lessons of gospel living to our community life," and that his private life was becoming a "moral catastrophe." The quarrel destroyed the movement, but Noble insisted that even though it was "wrecked as a specific movement it left an influence which can never die."

Despite these premonitions the members of the Populist Party soon realized that they had a potential leader in Herron, whose religious and political views conformed to the pattern of agrarian discontent in the 1890s. Over the course of several years, as Herron moved slowly left and became more and more controversial, Iowa Populists pressed him to run for Congress or for governor on the Populist ticket. Herron was devoted to the views of the party, but he was never inclined toward direct political action, and poor health and marital problems made any vigorous, long-term career difficult. He did announce to the faithful his support of the religious socialism inherent in Populist planks. He thought the movement "permeated by a profound religious feeling," and hoped to see "out of these Western States the greatest religious movement since the Reformation. It will be a revival of faith in Christ closely akin to Primitive Christianity."

Far from finding any meaningful contrast between populism and progressivism, Herron moved almost imperceptibly into organizations and campaigns that marked the rise of both. Either as a member or frequent guest speaker, Herron associated himself with the National Christian Citizenship League, the Union Reform League, and most important, in 1899, with the National Social Reform Union. Through these groups he became well known to most active progressives and populists. Newspaper accounts of the Buffalo convention of the National Social Reform Union, 28 June–4 July, listed his name second only to its leader Governor Hazen Pingree of Michigan. Those following Herron's were an honor roll of political and social progressives that most accounts never really associate with religious movements at all: Richard Ely, Henry Demarest Lloyd, Eugene Debs, Graham Taylor, John P. Altgeld, Jane Addams, and John Commons.

Another delegate to the convention, Samuel M. Jones, the famous "Golden Rule" mayor of Toledo, had long admired Herron. He had tried to run his factory after Herron's principles, and a few weeks earlier in June 1899, had been the speaker at Iowa College graduation ceremonies. Herron campaigned vigorously for Jones during his unsuccessful third-party try for the governorship of Ohio. It was a first effort for Herron in democratic political campaigning. In 1900 Herron continued his activity by campaigning for Eugene V. Debs, and occasionally thereafter he participated in socialist candidates' campaigns. He thus managed to be a populist, a progressive, and a socialist, and was accepted readily as one of their own by each group, without having to change or moderate his views very much on any particular subject. The barriers historians erect among these groups, if Herron is at all typical, are a great deal higher than the facts warrant.

As his public career peaked, Herron's private life slowly disintegrated. He once confided to Upton Sinclair that his had been a loveless marriage to a dominating older woman, and he had found it all but impossible to continue. Even before he came to Grinnell, rumors circulated in Burlington that his relations to the Rand family were closer than was appropriate. He and they had a mutual influence on one another. Mrs. Rand apparently persuaded Herron to accept the call to Iowa College; having already put up the money for his salary, she all but took over aspects of college life when they arrived.

The students in those more "innocent" days apparently did not suspect that anything out of the ordinary was happening, but the relationship between the two families could hardly have escaped the eyes of a small college town. The Herron children called Carrie "Aunt Carrie," and one of them was named Caroline Rand Herron. The Rand house, directly across the street from the campus, had a room reserved for Herron so that he could rest between classes. His own home, however, was just down the street. By 1897 and 1898, most faculty took notice and at least a few were repeating the off-hand remark of one of Herron's children that "we don't have meat for dinner any more because Papa has his dinner at Aunt Carrie's." Even when the Rands and Herron traveled on European cruises together, people at first were not suspicious. The last of these trips apparently proved decisive for Carrie's and Herron's relationship, because when Herron returned, he asked his wife for a divorce, which was granted in March 1901. Two months later Herron married Carrie in a modern "socialist" ceremony that shocked as many people as the divorce itself.

In terms of Herron's career and influence, the scandal of his private life was anticlimactic. Herron's health had never been good even as a small boy, and the strain of public speaking was too much for him. He was often ill as an adult, and his lengthy trips had been in part attempts to restore his health. But the combination of travel for speaking engagements and then for his health caused him to neglect his duties at the college. Advanced students, and sometimes even President Gates, substituted for Herron in class, but inevitably, criticism arose because the highest-paid member of the faculty was not performing even the minimal duties expected of others. Enrollment in Herron's classes had dwindled rapidly after the first year, and after 1896 he was lucky to have twenty students register to study with him. At the same time, public controversy about his political views grew more intense, and pressure mounted on the Board of Trustees. Contributions from non-Rand sources diminished, and alumni threats of financial retaliation became louder. Herron was a topic at several meetings of the board, beginning in June 1896; this continued until his retirement on 13 October 1899 and the subsequent resignation of President Gates. Herron was soon expelled from the Congregational Church as well.

IV

Judged from afar, Herron did change radically during the 1890s. He began as a theologically conservative evangelical Congregationalist, a hero to many reform-oriented clergy and to many active in political and economic life, and became a left-wing socialist, despised by most of his former friends, expelled from his church, and denounced as a dangerous radical and practitioner of free love. At closer range, though, Herron changed little. Most of his later ideas were implicit in his early publications. Not his ideas, but public understanding and acceptance of them, did most of the changing.

In Herron's mind heaven and earth and God and man, were not separate but unified. God was human and man divine. Social and economic problems, properly understood, were religious problems, and religious problems always had social and economic implications. Christianity and sociology were thus synonymous. Within this framework Christ was central. In his character he represented the summation of the best that God thought man could become; he was the incarnation of God's will. The reformer, conscious of the discrepancy between what he sees in society around him and what he finds when he studies Christ and religious texts, tries to harmonize his society with his understanding of Christ's nature. "The search for some complete law of justice between man and man, the search for remedies for social ills, is essentially a search after the Christ."

Christ's chief message was that selfishness was evil, and self-sacrifice was central to any kind of religious life. Selfishness was a subject Herron returned to again and again; he saw it as a "separation from God and humanity. *It is the origin of evil. The man is a devil who is sufficient unto himself in whatsoever he is or does.*" The only way for a Christian to fight selfishness in himself and in society was to practice, at all times, self-sacrifice in imitation of Christ on the cross. "The most glorious career that love can conceive for its object is one of complete sacrifice in the service of the common life," Herron wrote. He then went on to insist that "no life is to be thought of as Christian that is not made sacred for the social service, and thus fully sacrificed in bearing away the sins of the world." Even detachment from self-interest would not be enough, for Christ's example showed that society would take its revenge, and might well demand martyrdom from anyone who was too insistent on changing its sinful ways. "For, I frankly acknowledge and declare, no man can practise this gospel without suffering loss and persecution through conflict with the opinions and customs of the world."

Unlike most Christian reformers, Herron apparently could not stop at simply trying to follow Christ's example. Instead, he seemed to identify himself so closely with Christ that, *mutatis mutandis,* his descriptions of Christ become very much like self-portraits, and his message to listeners was a demand that they follow George Herron as he took upon his head a crown of thorns and suffered martyrdom for the sins of the world. Once the observer is aware of Herron's own story and self-image, for example, it is hard not to find the author lurking behind Herron's evocation of Christ: "No one ever so hungered for human sympathy as Jesus," and yet "no one was so misunderstood—even down to the present day—by his own disciples, his own church." Christ yearned to express his affection and explain his message, Herron wrote, "yet his own mother understood him not; his brethren did not believe in him." All his companions were undependable, and those disciples who were loyal acted "mainly from selfish motives." The crowds "who had felt the healing power of his compassion, the authority of his words, the divineness of his being, and witnessed the miraculous and beneficent demonstrations of his power, shouted for his crucifixion." His nation rejected him, and so did his religion, "and thus, he presented himself, the willing victim of our sins, to be rent in soul, mangled in flesh, broken in heart, that he might show us the Father, show us ourselves, and lead us back to our Father's house." The message was that no one understood Herron, who was full of love for everyone. His friends betrayed him, and his church soon rejected him officially; his students left his classes; his country soon so pilloried him in its press that he felt he had to emigrate—the whole pattern of events and its similarity, as he saw it, to Christ's martyrdom, is too striking to ignore. Herron seemed virtually to have planned and then enacted his own martyrdom out of a twisted perception of Christ's life.

Indeed, whether or not by Herron's own suggestion, many of his followers noted his similarity to Christ. His personality did, in fact, give people the impression that he was divinely inspired. To one writer, the Reverend Mr. William T. Brown, Herron and Jesus seemed to share many qualities. Selfless love and devotion to social welfare dominated the lives of both men. Both were divinely inspired. Herron's life, Brown wrote, "has been one long crucifixion," and Herron was a selfless saint. Even his divorce and remarriage did not dim his qualities in Brown's eyes. Indeed, in his suffering, Herron's heroic silence and noble mien "exhibited qualities of character which are nothing less than divine."

Whatever his inspiration, Herron's message became clearer as the 1890s wore on: Capitalism was organized selfishness, and its only cure was socialism. Herron condemned competition and the profit motive in general terms, but by 1899, when his book *Between Caesar and Jesus* was published, no one could misinterpret his political views. Where he had earlier exhorted businessmen to follow Christian doctrine in their business lives, he now flatly proclaimed that this would be a contradiction in terms, "for business is now intrinsically evil, whatever good may come out of it." Throughout the course of history, he wrote, "private ownership of resources rests upon fraud, violence and force." Like Henry George, he insisted that "private ownership is social trusteeship, that is all: it is not private ownership in any real or right sense," and it therefore followed "that the public ownership of the sources and means of production is the sole answer to the social question, and the sole basis of spiritual liberty." Thus, if a man were to express his best qualities of character, to become the ethical being that Christ wished him to be, "the resources upon which the people in common depend must by the people in common be owned and administered." Only if the ownership of the earth is held in common could liberty and individuality take proper root. "Liberty as a human fact means communism in natural resources, democracy in production, equality in use, private property in consumption, and the responsibility of each for all and of all for each." To a generation that saw money as a sign of God's blessing, he said flatly that there could be "no such thing as a rich Christian."

Yet even though Herron became increasingly specific in his use of terms like "communism" and "socialism," and by his attendance at conventions and support of candidates and platforms identified himself with many progressive and socialist measures, his books remained empty of detail. He did not understand politics or economics, and he had no interest in studying them. What he had to say was mostly derivative exhortation from the previous generation of British, Italian, and American reformers. On this point Herron provides a lesson to students of progressivism. Like so many political progressives, Herron was basically an evangelist at heart; he wanted to convert his audiences. Herron once argued that "the spiritual alone is the real and eternal," and this thought, probably bootlegged from his occasional reading of Hegel, sets his whole reform outlook into perspective. If he could persuade one person to devote his life to altruism, if he could do this for every member of society, then the kingdom of heaven would reign on earth. The essence of Herron's politics thus was otherworldly since only the ideal was important. This attitude goes to the heart of the problem of why the progressives, by and large, were so rhetorical and so ineffectual. They seemed always to believe that personal regeneration would achieve social regeneration, and thus, that specific programs were merely way stations. A campaign for good men in office, by means of the initiative, referendum and recall, for example, was but a secular expression of this essentially religious and idealistic frame of mind.

Ultimately, Herron's reputation for radicalism rested on superficial analysis. His discontent with present conditions was great, and he thought society ought to be entirely changed, but he had no political program—certainly nothing resembling the very real grappling with issues that Eugene Debs attempted, and that was so characteristic of the New Deal in the 1930s. "We need no program of action save the words of our Lord, sending us, as he was sent by the Father, to please not ourselves, but give our lives as bread and meat to a hungry world," Herron wrote. The reconstruction of the world "will come through the acceptance of the mastership of Jesus on the part of the church that bears his name. *There are no social problems in fact. They exist only in the imagination of unbelief.*"

V

George Herron's path from religious to social reform involved no conversion of the sort that George Mead experienced. Herron merely continued to talk about new problems and insensibily moved into the arms of progressive socialism. Because his private life did not conform to contemporary standards, he had to give up his academic and church positions. After a brief period of public harassment, he gave up America as well and settled into life in a luxurious Italian villa—surely the best-kept radical in western Europe. But most progressive odysseys were less obvious in their religious origins and less political in their initial applications. The case of John Dewey provides a foil for the case of Herron. Dewey began as religiously as any young progressive, but by the 1890s, religion in any explicit sense had disappeared from his writings as well as from his private life, and he was rapidly becoming a leader in the fields of progressive philosophy and education.

In the fall of 1891, Dewey was just beginning to shape a consciously pragmatic philosophy. Dewey had already come a long way from his origins in small-town Vermont. His father was a Union army veteran; his mother, an evangelical Congregationalist. Dewey followed the typical progressive's pattern of maturation. He adapted easily to the demands of family and small town life while profiting from the presence of the nearby University of Vermont. Most of his early intellectual friends were clergymen, but his own turn of mind was more philosophical than theological, and Dewey opted for teaching rather than divinity school when he graduated from college. The future reformer of the American classroom disliked the experience, and after several false starts, he made it to Johns Hopkins University to undertake serious advanced study in philosophy. A professor at Vermont, Henry A. P. Torrey, had already introduced him to Scottish commonsense realism, Darwinism, and Kant; Dewey pursued these early interests in graduate school and wrote his dissertation on Kant. As much as any mind could parallel another, he had traversed the same landscape as had George Mead.[2]

Although the Hopkins was the best graduate school in America during the years (1882–84) Dewey was there, Dewey soon discovered that President Daniel Coit Gilman preferred scientific and medical studies and regarded philosophy as too closely dependent on religion in the American context to be worthy of heavy commitments on the university's part. No professor of philosophy held a permanent chair, and so Dewey had his choice of three untenured faculty with whom to work. Charles S. Peirce proved the least helpful. His personality was difficult; his teaching was inept; and his seminars tended to study logic as the method of physical science reduced to mathematical form, not philosophy as Dewey understood it. Only many years later did Dewey come to realize Peirce's genius through his publications. Relations between Dewey and G. Stanley Hall were more complex. Lacking a secure job for many years, Hall made himself expert in the new field of child psychology. Dewey studied physiological and experimental psychology with Hall and participated in an experimental seminar on pedagogy; he also conducted research in Hall's laboratory. But Hall had an unpleasant teaching personality; he was terrified of competition from his bright students, and Dewey was not the only scholar whom he drove from the Hopkins by his criticisms, which were all too often expressed in writing to President Gilman.

Dewey drew the most sustenance from George Sylvester Morris, a visiting professor from the University of Michigan. A "Christian spiritualist" throughout his life, Morris, like Dewey, adopted philosophical positions that seemed to help him answer his religious questions. He was often a confused man, better able to discuss what he opposed than what he favored, but in the study of Hegel, he had found a philosophy that seemed as satisfactory as any contemporary system could be. Morris did not study Hegel closely until after 1877, or about the time he began his work at Hopkins. He already had a secure reputation as the translator of a German book on the history of philosophy, and his chief interests when Dewey knew him were the history of English thought, Kant, and Hegel. He and Dewey analyzed Kant rather critically for his dichotomy between the phenomena and the noumena, and emphasized how Hegel had overcome this split with his all-encompassing synthesis. Morris' basic position was that a valid theory of knowledge led to a theistic conception of the universe, that no reality existed outside the spiritual, and that the more spiritual a thing was, the more real it was. Morris is today rightly regarded as a minor but historically important American idealist, most specifically a Neo-Hegelian.

In 1915, at the request of Morris' biographer, Dewey wrote his most important assessment of Morris. He found Morris' teaching "all the way through, an objective and ethical idealism." Morris managed the extraordinary feat of synthesizing Aristotle, Fichte, and Hegel. "The world, the world truly seen, was itself ideal," and he insisted on this ideal character of the world "as supporting and realizing itself in the energy of intelligence as the dominant element in creation." "That the struggle of intelligence to realize in man the supreme position which it occupies ontologically in the structure of the universe was a moral struggle, went without saying." Dewey then went on to discuss Morris in language that should make it quite clear that in Morris, and in Hegelian idealism, Dewey found the seeds of instrumentalism as well as the religious and moral solace already noted. Morris derived his method from Hegel, Dewey wrote. Morris seemed "at once strangely indifferent to and strangely preoccupied with the dialectic of Hegel." He had little interest in the technical aspects of Hegelianism, but from its study, he derived "an abiding sense of what he was wont to term the organic relationship of subject and object, intelligence and the world. This was the supreme instance of the union of opposites in a superior synthesis, and, as it were, vouched for the reality of the dialectic principle all along the line." To the best of Dewey's recollection, "the contrast was more moral and spiritual than physiological." Dewey felt quite sure that Morris' adherence to Hegel "was because Hegel had demonstrated to him, in a great variety of fields of experience, the supreme reality of this principle of a living unity maintaining itself through the medium of differences and distinctions."

Reasonably conclusive evidence thus illuminates the first key episode in Dewey's odyssey to instrumentalism. The evangelical New England Congregationalism that so deeply affected both men was not something they could simply abandon while they pursued professional careers. The impact of any number of influences, from German biblical criticism to Darwinism, made retention of the old Congregationalism all but impossible in men of such intelligence, but it helped define their psychological yearnings, and thus, their preferences in philosophical systems. Only with these considerations in mind can anyone comprehend Dewey's remarkably emotional description of this period in his key autobiographical essay "From Absolutism to Experimentalism." Using language that is both religious and psychological more than it is philosophical, he described how one of the most eminent philosophical minds in America, and one of the country's leading progressives, came to adopt a system:

> There were, however, also "subjective" reasons for the appeal that Hegel's thought made to me; it supplied a demand for unification that was doubtless an intense emotional craving, and yet was a hunger that only an intellectualized subject-matter could satisfy. It is more than difficult, it is impossible, to recover that early mood. But the sense of divisions and separations that were, I suppose, borne in upon me as a consequence of a heritage of New England culture, divisions by way of isolation of self from the world, of soul from body, of nature from God, brought a painful oppression,—or, rather, they were an inward laceration. My earlier philosophic study had been an intellectual gymnastic. Hegel's synthesis of subject and object, matter and spirit, the divine and the human, was, however, no mere intellectual formula; it operated as an immense release, a liberation. Hegel's treatment of human culture, of institutions and the arts, involved the same dissolution of hard-and-fast dividing walls, and had a special attraction for me.

VI

Dewey followed Morris to the University of Michigan for a decade that was crucial in his development from traditional Congregationalism to instrumentalist progressivism. When he arrived, he was an ill-paid instructor with no reputation, little self-confidence, and a philosophical position clearly derived from his teachers. He was quickly promoted to assistant professor and given a raise in salary. His classroom duties led him to the publication of several books that established him as a promising younger scholar. He met Harriet Alice Chipman, a small-town Michigan girl slightly older than himself, when they lived in the same boarding house. She was already interested in philosophy and took three advanced courses with Dewey before she graduated in 1886. They were married on 28 July 1886, and their first child was born the next year. In early 1888 Dewey was offered the chair of philosophy at the University of Minnesota. Although he was happy at Michigan the advance in salary and status was too great to refuse; that fall he took up his new duties in Minneapolis. The next March, Morris died unexpectedly, and Dewey was the obvious choice of both faculty and graduate students to replace him. He returned, this time as full professor and department chairman at $2,200. Thus, his career at Michigan neatly falls into two parts, 1884–88, and 1889–94. The second five years were those of early maturity, showing original new ideas and fewer references to the religious and philosophical influences of his apprenticeship. He also brought to the department three of the more talented younger men in the field: George Mead; James Haydon Tufts, later his collaborator on an important book; and Alfred Henry Lloyd, a student of James and Royce who became an idiosyncratic ally of the pragmatists. Dewey was apparently quite happy, and when he left in 1894 for the new University of Chicago he did so not from discontent or mistreatment, but because Chicago offered him more contact with graduate students and advanced research, and because he would there have the opportunity to study young children and pedagogy under something resembling laboratory conditions.

Dewey received attention during the late 1880s for his work in psychology, culminating in his text of 1887 called *Psychology*. This work was not distinguished, however, and to the student of progressivism, the religion that Dewey practiced in this period is far more significant. Continuing his youthful interests, he was active in the Student Christian Association group, which sponsored regular meetings where faculty members spoke, encouraged Bible classes, and even published its own monthly bulletin. Dewey's surviving talks to this group provide specific evidence of the link between his overtly religious past and the covertly religious underpinnings of his mature work in education and philosophy. Dewey also joined the First Congregational Church in November 1884, faithfully attending Sunday services and even teaching a Bible class on church history. His year at Minnesota marked a decline in conventional religious activity. He revived some of his activities after his return—most obviously in 1892, when he was the only professor at a three-day Bible Institute—but when he finally arrived in Chicago in 1894 he did not join the local Congregational Church, nor did he continue the student religious speaking that had been so important at Michigan. Thus, Dewey's conventional religiosity faded between 1890 and 1894, and his impulses in this direction were displaced onto political, social, and educational concerns.

Even to list the titles of Dewey's talks will indicate his concern for religion, and its appropriate fusion with philosophy, social thought, and politics: "The Obligation to Knowledge of God" (November 1884), "The Place of Religious Emotion" (November 1886), "Christianity and Democracy" (March 1892), and "The Relation of Philosophy to Theology" (January 1893). When other little-known works of this period are added, like "The Ethics of Democracy" (1888), what had previously appeared to be a time of often arid psychological and philosophical investigation in Dewey's career suddenly becomes alive with his growing interest in social and political problems, and their solution within a religious framework.

Read in conjunction with Dewey's more technical work, we see that his early religious ideas developed into his concern for progressive ethics: the new ideas about how to conduct life so common to many progressive leaders, whether in political theory, social and industrial relations, or personal growth, as well as in philosophy. As his overt participation in religious activities declined, his concern for a moral universe became central to his work. A word like "dualism," for example, which seemed applicable only in a philosophical context—as in his Hegelian critiques of Kant—suddenly became political, applied to the class divisions between aristocracy and democracy, and fully as wicked as when it seemed to describe the separation of the "is" and the "ought." A word like "democracy," properly understood, became far more than a form of government; it became a form of social interaction that evolved into a political structure having divine sanction because it fitted so well the community for appropriate, healthy growth.

The most valuable of these short pieces is "Christianity and Democracy," originally given at the Sunday morning services of the Students'

Christian Association on 27 March 1892, first published a year later, and largely ignored by students of Dewey until recently. Dewey's religion here is clearly not his mother's evangelicalism, but rather, a liberal conception in the tradition of a Theodore Parker. "Every religion is an expression of the social relations of the community," Dewey said: "Its rites, its cult, are a recognition of the sacred and divine significance of these relationships." Over time these relationships became condensed in symbols and dogma; their origins forgotten, they seemed to become ends in themselves. They then decayed, but meanwhile, society evolved and developed new symbols and ideas, and so religious feelings were always being renewed, and feelings of despair were appropriate only to those who had lost contact with the most vital life around them. Dewey then took up several problems associated with Christianity, and summed up the first part of his paper: "Christianity is revelation, and revelation means effective discovery, the actual ascertaining or guaranteeing to man of the truth of his life and the reality of the Universe."

At this point the significance of democracy appeared for Dewey. He made the crucial link between Christianity, human intelligence (or "reason"), and the social matrix, which indicated the religious origins of his mature philosophy. "The kingdom of God, as Christ said, is within us, or among us. The revelation is, and can be, only in intelligence." Dewey thought it strange that people could call themselves Christian teachers, and yet condemn the use of reason in relation to Christian truth. "Christianity as revelation is not only to, it is *in* man's thought and reason. Beyond all other means of appropriating truth, beyond all other organs of apprehension, is man's own action." Any person trying to interpret the universe does so "in terms of his own action at the given time." If Jesus Christ "made an absolute, detailed, and explicit statement upon all the facts of life, that statement would not have had meaning—it would not have been revelation—until men began to realize in their own action the truth he declared—until they themselves began to *live* it." Only in actions did man have a means for perceiving truth, and a man's action "is found in his social relationships—the way in which he connects with his fellows. It is man's social organization, the state in which he is expressing himself, which always has and always must set the form and sound the key-note to the understanding of Christianity."

Thus, "the significance of democracy as revelation" was that democracy "enables us to get truths in a natural, everyday, and practical sense." Democracy for Dewey was not mere governmental machinery, but "a spiritual fact," the "means by which the revelation of truth is carried on." It was only in democracy that "the community of ideas and interest through community of action, that the incarnation of God in man (man, that is to say, as organ of universal truth) becomes a living, present thing, having its ordinary and natural sense." The obvious conclusion, now that Dewey was moving from the abstractness of Hegel to the concreteness of poverty, was a demand for social action, and he could close with this exhortation: "Can anyone ask for better or more inspiring work? Surely to fuse into one the social and religious motive, to break down the barriers of Pharisaism and self-assertion which isolate religious thought and conduct from the common life of man, to realize the state as one Commonwealth of truth—surely, this is a cause worth battling for."

Within the context of Dewey's own life, such a sermon or homily seems quite right, the logical outcome of an intelligent mind as its religious and philosophical training confronted the realities of the depression of the early 1890s and its human agonies. Indeed, taken on its own terms and within its proper biographical context, Dewey's future career becomes that of a religious prophet. As he told the students in January 1893: "The next religious prophet who will have a permanent and real influence on men's lives will be the man who succeeds in pointing out the religious meaning of democracy, the ultimate religious value to be found in the normal flux of life itself. It is the question of doing what Jesus did for his time." It was the same question that Jane Addams and George Herron were asking; it was the key question asked by most progressives as they entered maturity.

VII

In his letter recommending Dewey to President William Rainey Harper of the University of Chicago, James H. Tufts recounted Dewey's extensive publications and impressive classroom and administrative record, and stressed that Dewey was a modest and simple man who made friends easily. Dewey was also "a man of a religious nature, is a church member, and believes in working with the church. He is, moreover, actively interested in practical ethical activity, and is a valued friend at the Hull House of this city." Tufts thus neatly included in his comments both the religious origins and the ethical and political results of Dewey's maturing views. Displaced onto the real world, Dewey's strong ethical drive soon found outlets in efforts to reform the Chicago public schools, to bring the university into closer contact with Chicago institutions, and to work with Jane Addams. Dewey was a member of the first board of trustees at Hull-

House, and he worked at the job. In the process he had much contact with the radical workers and social thinkers who spoke and attended meetings there, and he found a special attraction in the ideas of Henry George. So great was his admiration for Jane Addams that he named his youngest child after her.

Officially, Dewey was the head professor in the departments of philosophy and pedagogy until he resigned in 1904. In philosophy he continued what he called his "experimental idealism," a transitional phase between his Hegelianism and the instrumentalism of his later years. He continued to give much the same course work and to administer his department in much the same way as he had at Michigan; his important publications did not come until after 1904. He concentrated most of his energies and ethical fervor on pedagogy. His prior work in psychology and education seemed to anticipate the establishment of an experimental school in which to test his new ideas in a scientific fashion similar to that in any laboratory in physics or chemistry. At Chicago, with its new president and lack of outmoded traditions, he seemed to have the ideal environment. Unfortunately, as Dewey found to his intense irritation and frustration, while President Harper was sincerely enthusiastic about pedagogy, his support was more moral than financial. Other departments of the university had come to the bank first, and little remained for Dewey or his school. Except for token seed money and a few institutional facilities, the school had to rely largely on tuition and gifts for its continued support, and Dewey found himself again and again defending his budgetary requests to Harper, even when he felt them ridiculously inadequate and almost indefensibly low. Dewey also demonstrated clearly that he did not have the appropriate temperament for an administrator. While he could creditably chair a small university department of philosophy, he was out of his depth running what became a large establishment devoted to young students, independent and sometimes uncooperative teachers, critical parents, and an occasionally uncomprehending university bureaucracy.

In one of his articles during the 1880s, Dewey had argued that psychology really *was* the philosophic method, because consciousness was the whole of reality, and psychology was the science of consciousness. As head of the philosophy department, Dewey brought together one of the most remarkable groups of men ever to serve together in an American philosophy department, probably the only real philosophical school ever to exist in America, and one that persisted in showing Dewey's influence long after he went to Columbia. The department seemed to develop directly out of Dewey's early preoccupation with psychology as a philosophic method, catalyzed into instrumentalism by William James' *Principles of Psychology*

(1890), Dewey's own concern with his growing family and their education, the Laboratory School, and his intimate associations with Hull-House. James Tufts was already on the faculty and helped to bring in Dewey. George Herbert Mead and James Rowland Angell took positions in philosophy and psychology; Edward Scribner Ames and Addison Moore took their places shortly thereafter. Other figures were involved for brief periods or through other departments. Central to this school of thought was the concept of activity, which they regarded as being psychological and ethical as well as biological. Activity implied an organism, and an organism implied growth, and a growing organism implied a social milieu in which to grow. Emersonian ideas such as "organism" combined with Darwinian and Hegelian ideas such as "evolution" and "process" into what appeared to be an entirely new and modern intellectual stance, with absolutes gone and values relative to the growing organism and its socially generated problems. Yet however clearly a historian can discern the origins of Dewey's ideas and the ideas of the Chicago school of philosophy, the school was remarkably new and coherent in its emphasis and was recognized as such by the best men in the field. As William James wrote, shortly before Dewey's departure: "Chicago University has during the past six months given birth to the fruit of its ten years of gestation under John Dewey. The result is wonderful—a real School, and real Thought. Important thought, too! Did you ever hear of such a city or such a University?" At Harvard, "we have thought, but no school. At Yale a school, but no thought. Chicago has both."

Yet in the mid-1890s, the fame of the school and Dewey's major philosophical publications still lay in the future. At the time he was more concerned with education and his new Laboratory School. When Dewey arrived in Chicago, it was already the scene of considerable educational controversy. Colonel Francis W. Parker was the focal point for a concerted attempt to reform the education of young children and the training of their teachers. At the Chicago Normal School and its successor, the Chicago Institute: Academic and Pedagogic, Parker developed theories about organic growth and returning children to nature that often bore a striking resemblance to the ideas Dewey was developing. In addition he had interested the wealthy heiress Anita McCormick (Mrs. Emmons) Blaine in his theories, and she had put up the money for Parker's newest school. Dewey quickly became a public supporter of Parker, and Dewey's laboratory school, inevitably became connected in many minds with Parker's earlier efforts. Dewey tended to emphasize society whereas Parker emphasized nature, and Dewey's school was to be a laboratory for the testing of psychological hypotheses rather than a training school for future teachers, but the efforts of the two men did run parallel, and each respected the other. Parker even shared many elements of Dewey's religious and Hegelian background, although at a noticeably lower intellectual level.

Dewey began his school with great enthusiasm, and quickly built up the student body and a small but loyal group of parents and friends willing to donate funds. The old dualism that separated everyday life from learning, and dictated that children adjust to a preconceived curriculum, disappeared. Instead, Dewey instituted the solving of problems as the key to children's educational growth, and insisted that moral and educational values could only be generated in the process of solving the problems posed by modern society as the child actually encountered them. Just as the old reflex arc artificially separated the parts of a response to a stimulus, so the old school had separated learning from experience; Dewey saw learning as inseparable from experience and arranged the curriculum accordingly. Children tried to approach history through primitive social and economic tasks, obtaining through a knowledge of weaving and carpentry, for example, a knowledge of how societies evolved. They learned how to cook properly, and by examining and analyzing food came to understand the role of proteins, fats, and carbohydrates. Teachers were always present and prepared to organize choices subtly, but the burden of student growth was suddenly on student shoulders, and the rigid structures of the past became far more fluid, flexible, and relative to each individual child, his rate of growth, and his interests.

For progressives, the Chicago experiments at the Dewey school represented what a school could and should be to a child. It was a reality and an ideal, a pragmatic realization of the energy and the moral significance that Dewey had once devoted to the church. The school became the key institution for the nurturing and the saving of souls for democracy. Democracy itself became a kind of heaven on earth, where salvation would be possible for all those who were brought up organically, according to their true natures, so that growth could always continue and always reinvigorate society with its insights. The school would be the agent of nature as it benevolently shaped souls; the assumption was that all could be saved.

This religious and biographical background explains why Dewey's writings at the end of the nineteenth century have so much morality at the heart of their rhetoric. This was no secular skeptic, working in the spirit of science regardless of consequences; this was a man of deep faith finding a new institution—the school—and a new congregation—the citizens of democracy. Dewey's key essay during this period is "Ethical Principles Underlying Education," published in 1897. Its message is that

MINISTERS OF REFORM

"the moral responsibility of the school, and of those who conduct it, is to society," and thus, it is necessary to discuss "the entire structure and the specific workings of the school system from the standpoint of its moral position and moral function to society." The child who attends such a school "is an organic whole, intellectually, socially, and morally, as well as physically. The ethical aim which determines the work of the school must accordingly be interpreted in the most comprehensive and organic spirit." The school must also never lose track of the fact that the child will grow up to live "in the United States, a democratic and progressive society."

Dewey shifted smoothly from the ethical to the pedagogical to the social reform level of discourse. "Apart from the thought of participation in social life, the school has no end or aim," he said. "We get no moral ideas, no moral standards for school life excepting as we so interpret in social terms." The terms, of course, were Dewey's liberal democratic ones, and at times he seemed to equate the school with a kind of universal Hull-House for the young, responsible for training them to be the young saints of modern democracy: "Interest in the community welfare, an interest which is intellectual and practical, as well as emotional—an interest, that is to say, in perceiving whatever makes for social order and progress, and for carrying these principles into execution—is the ultimate ethical habit to which all the special school habits must be related if they are to be animated by the breath of moral life."

Dewey's closing paragraphs thus have a peculiarly religious, even late Victorian, ring to them. They are Victorian because the religiosity has an emphasis on moral principles as it did in so many other late Victorians who were losing their conventional faith. "What we need in education more than anything else is genuine, not merely nominal faith in the existence of moral principles which are capable of effective application." These moral principles need to be "brought down to the ground through their statement in social and psychological terms." "The one thing needful," Dewey said, turning the language of Charles Dickens' baleful educator on its head, is for people to realize "that moral principles are real in the same sense in which other forces are real; that they are inherent in community life, and in the running machinery of the individual." Once reformers have won recognition of this fact, "we shall have secured the only condition which is finally necessary in order to get from our educational system all the effectiveness there is in it." Any teacher who works while holding this faith "will find every subject, every method of instruction, every incident of school life pregnant with ethical life."

Or, as he said even more graphically in "My Pedagogic Creed," of 1897:

"In this way the teacher always is the prophet of the true God and the usherer in of the true kingdom of God."

VIII

The Dewey who achieved his mature reputation at Columbia seemed a different man from the shy Sunday-school teacher at Michigan and the school innovator at Chicago. He became known for his emphasis on science and reason, for his lack of any supernatural religious belief, and for his involvement in a long series of liberal causes. He became, in short, a man who did not look like a typically moralistic progressive. Because he lived so long, and because he could comment on so many later issues, from the New Deal to Leon Trotsky, he seemed more a figure from the 1920s and 1930s than one from the 1890s. But Dewey was a progressive to the bone, and his progressivism left a mark on his thought that perhaps even he never recognized. He was, in a basic sense, never the skeptic he seemed to be. He was a man of deep and abiding faith; he believed in reason, in democracy, and in organic growth, just as once he had believed in the tenets of Congregationalism. He could not question these concepts even when they yielded him disastrous results, as in his support of American entry into World War I. The movement for progressive schools often went awry, and Dewey himself later criticized some of his overly enthusiastic followers. Reinhold Niebuhr and others questioned Dewey's ability to understand evil and the irrational, and pointed out how his principles seemed able to deal only with societies that were already basically sound. Dewey could criticize overzealous followers; but he could not change his basic beliefs. He tended to ignore his critics rather than answer them, and to elaborate his beliefs rather than question them. He remains in history as perhaps the greatest progressive intelligence, but the emphasis must be on the word "progressive"; even a man with his great gifts remained essentially within the intellectual world in which he came to maturity. His religion was implicit, but it was always there.

Chapter 3

A Hull-House of the Mind

THE COURSE of John Dewey's career has much to tell later generations about the way a climate of creativity like progressivism can shape professional concerns. The religious pressures of life in small-town New England made certain questions seem more pressing than others; the need for employment made certain choices essential; the restricted vocational options channeled these needs for answers and for money into professional graduate study; the most attractive employment opportunities took him to the Middle West; life there provided a relatively unstructured intellectual and professional climate that encouraged new insights; universities furnished the departments within which to refine these insights, schools in which to test them, and journals through which to publicize them. Had Dewey been born two decades earlier, such a path would have been far harder to follow and the questions asked and the answers given would have been different. Had he been born two decades later, the pioneer work would have been done, perhaps along other lines, and a far different, less original career could have resulted. Both the people and the ideas were rooted in the places, problems, and opportunities of their times, and to forget this is to do violence to the history of creative innovation.

Many specialists in the history and sociology of knowledge prefer to ignore the context within which creativity occurs. They stress the internal history of science, art, or law, as if each discipline alone provided the precedents, the impetus, and the opportunities for creation. Thus, students find themselves confronted by the disciplines they must choose among and adjust to as if knowledge moved forward by immaculate conception. Biological breakthroughs produce further biological breakthroughs; legal decisions generate further legal decisions. The pressures of mothers on their sons before those sons enter college is ignored; the models provided for young women at the age of twenty-one go unmentioned. The relative degree of prestige involved in choosing among a life in the military, the ministry, or in chemical engineering seems too tawdry even to mention. People seem to don white coats and proceed to their appointed destinies unbothered by diseases, money, religious obsessions, the media, or accidents of birth. Any study of a dominant climate will produce a wealth of evidence that the daily concerns of life have significant impact even in the most "professional" and "scientific" parts of a culture.

Progressivism provides countless examples of this interplay. Some areas of American culture, like physics or mathematics, have no noticeable progressive era in their histories. For others, such as biology or psychology, the progressive influence is visible but not basically significant. But in most of the social sciences, progressivism played a crucial role. It contributed to the establishment of certain disciplines, like sociology; in others, like history or economics, progressivism inspired a number of important leaders in the profession to ask and provide answers to significant questions in such ways as to change the course of seemingly "internal" professional controversies. Some of these concerns subsequently proved so significant that they still provoke professionals to extensive new research and publication. Often, the footnotes to such contributions display a complete innocence of the fact that the basic terms of the argument were established almost a century earlier.

The institution that best symbolized the "present reality" for young progressive intellectuals during the 1890s was Hull-House. Ignored by many historians of the professions until recently, Hull-House was the great catalyst to progressive social science. It, and the other social settlements, reified the problems of modern American civilization for a generation of disturbed young Protestant intellectuals at the same time that it provided them with a definite, personal alternative for action. Hull-House brought the problems of the city to the one place where they could be confronted, classified, and organized. It provided room, board, and congenial friends. In time it generated ideas. The source of those ideas lay deep in the rural Protestant past, and the means to articulate them lay chiefly in the new colleges and universities. But without the inspiration of Hull-House, of Jane Addams and her many friends and associates, social science as it developed in America would not have taken the pattern it did. Jane Addams was in no significant aspect an intellectual or social scientist herself, but her influence lurks everywhere, all the greater for having been unrecognized. In fact books such as *Democracy and Social Ethics* (1902) and scores of articles began to flow from Hull-House and its visitors soon after its founding; for subsequent generations they reached significant synthesis with the publication in 1910 of *Twenty Years at Hull-House*. Much of this autobiography consisted of personal detail and factual reportage, but concealed beneath the anecdotal style were the sort of concepts that provided professional social scientists with organizing hypotheses. Three examples demonstrate this clearly.

On one occasion, a shipping clerk whom Addams had known for some time had lost his job and come to the settlement for help. Still believing that good Americans should take whatever jobs were available before applying for unemployment assistance, she advised him that jobs were available on the drainage canals. The man replied that he had always worked indoors and did not think he could survive heavy outdoor work in the bitter Chicago winter, but nevertheless, he disappeared and took one of the suggested jobs. He survived for two days, caught pneumonia, and died within a week. Addams, of course, blamed herself for this unfortunate result and always tried to keep track of the two young children who survived their father. The moral that she drew from the experience was that "life cannot be administered by definite rules and regulations; that wisdom to deal with a man's difficulties comes only through some knowledge of his life and habits as a whole; and that to treat an isolated episode is almost sure to invite blundering." Addams had discovered that the emphasis on people as interchangeable economic units, so common in intellectual circles during the Gilded Age, did not fit the realities of their biology or their place in society. People could not be manipulated without regard for their past, their experiences, and their social context. Forcing someone to "struggle for existence" was an inappropriate application of a biological analogy.

In a similar vein, Addams told the story of the young German girl Marcella, who spent her life working and devotedly bringing home her wages to her domineering mother. The family had many mouths to feed, and her mother had European notions of morality and dress; Marcella was badly dressed even though she sold silk in a fashionable department store and was constantly surrounded by fine clothes. The crisis came when

Marcella needed a new dress for a ball, and her mother refused to allow her to spend her own money for one. Despairing of injustice, Marcella wrapped an appropriate piece of cloth for a customer who carelessly dropped the parcel. Marcella quickly seized the item, only to be arrested by a vigilant store detective. The case went to court, well watered by the tears of a mother who could not understand how such a well-raised daughter could have become a common criminal. Jane Addams understood very well. Nineteenth-century morality might insist that some people were born wicked and sinful, and that only rigid legal and social customs held society together, but progressives were coming increasingly to another view. For them people were all born much the same, and notions like a biological propensity to criminal behavior were as outmoded as notions of original sin. The idea of environmental determinism was entering social science. Marcella stole because she was poor and tempted beyond her ability to withstand it. Give her more money and greater opportunity, and she would become a responsible middle class citizen.

Even closer to the progressive heart was a third lesson. During the summer of 1902, a typhoid epidemic devastated the ward in which Hull-House was located, and two residents made a detailed investigation of the causes and possible cures for the disease. In the process they came across a small, comfortable house still inhabited by one of the old residents from the days before the waves of Italian immigrants had arrived. Proud and aloof, the woman refused to have anything to do with her new neighbors or any of the efforts at community improvement undertaken by the settlement workers. She lived only for her two daughters, who attended eastern colleges. That summer they returned, caught typhoid, and one died. No matter how much distance the woman kept from her neighbors, she could not keep the diseased water out of her home. The social philosophy of the Gilded Age had glorified the forgotten, middle-class men and women who lived apart and minded their own business and refused to shoulder responsibility for the faults and miseries of others. For Jane Addams and the progressives, however, "the entire disaster affords, perhaps, a fair illustration of the futility of the individual conscience which would isolate a family from the rest of the community and its interests." Such a community and such interests were at the core of Dewey's instrumentalism and his notions about primary schooling. Summed up in the word "democracy," such communal awareness formed the core not only of a social concept like progressivism, but also of the emerging social sciences as they tried to differentiate themselves from outmoded philosophical and moral categories to find a professional place under the American sun.[1]

MINISTERS OF REFORM

II

Hull-House was able to play its catalytic role in the history of the social sciences because it managed to appeal to old and new tendencies in American social thought. On one hand, Hull-House appealed to the tradition in American reform of the aroused Christian conscience. Its founder came out of Lincoln Republicanism and Quaker abolitionism; outside of the settlement, she continued this family emphasis most obviously with her active participation in the founding of the National Association for the Advancement of Colored People (1909). She was a living example of applied Christian reform thought. On the other hand, to the newly emerging disciplines within the social sciences, she was also the person who brought empirical, scientific data to the eyes of professors eager to prove the rigor of their calling. Her impoverished immigrants from Europe, with their needs and problems, provided the children of the Anglo-Saxon middle classes with experiences otherwise unknown to them, and this occurred within a manageable setting. Hull-House cut both ways: it satisfied the amateur conscience wanting to do good and the professional need to research well. It stood halfway between the Protestant churches and the University of Chicago.

Organized social science had begun in America only after the Civil War. In 1865 the American Social Science Association was founded as an institutional umbrella that covered what remained of the Christian reform endeavor of the pre-war years. Its key figure was Franklin B. Sanborn, the New Hampshire Unitarian abolitionist who had assisted John Brown in planning the raid on Harper's Ferry and had been the friend of most of the men of letters in the Boston area. Already secretary of the Massachusetts State Board of Charities, he was soon active as well in the National Conference of Charities, the National Prison Association, the Clarke School for the Deaf, and the Massachusetts Infant Asylum. For Sanborn and most of his associates, social science was always connected with reform. In practice the focus was even more narrow. Members of the association were overwhelmingly a white-collar, male, Protestant elite from the professions, the world of letters, or comparable careers. For at least fifteen years, ties with institutions of higher learning were infrequent, and the profession of social scientist did not properly exist: the men had no social mission beyond a vague idea of reform, no systematized body of knowledge, no legally sanctioned authority or role, no code of ethics, and no sense of themselves as a separate group. In the early 1880s, the same forces that had opened

Johns Hopkins University in 1876 were working both in universities and in the larger culture to end this amateur status for social science. Men trained in Germany had returned with their prestigious degrees and were, in turn, training the first native generation of professional students of social problems. Insecure over their future in a democracy, eager to improve society, yet also eager to prove themselves men of science, these new scholars slowly began to challenge Sanborn and his amateurs for dominance. New professional groups began to split from the parent organization, weakening it progressively until it died quietly in 1909.[2]

The first social science to emerge out of these conflicting crosscurrents was economics. Early in the 1880s, a number of the better-trained students of political economy, ready to lead their developing discipline away from its origins in moral philosophy, began making suggestions about establishing some kind of national organization. They held numerous meetings in Boston, New York, and Washington. The groups included many figures who were firmly conservative in outlook and active figures in the business world. Much of this activity was disturbing to future progressives like Richard Ely. They had to face a situation in which political economy had long been all but synonymous with doctrines of natural law and laissez-faire. The devotion of a figure like William Graham Sumner of Yale to theories of Darwinism as applied to economics, for example, has long been something of a cliché. Sumner was a much more complex figure than critics have believed, but he did strongly oppose most governmental intervention in the economy and he helped give prestige to the commonly held notions of businessmen that they should be left to their own devices unhindered by the state. Fresh from Germany and full of the precedents of the prestigious *Verein für Sozialpolitik,* Ely and a few friends were eager to legitimize their policy disagreements with Sumner and to further the professionalization of the study of the economy. At the same time, Ely was especially candid about his desire to bring Christian ethics to the new discipline. As he wrote Joseph Labadie in mid-1885, he did not "believe it possible to accomplish anything of lasting value without the aid of true Christianity."[3]

Eager to eliminate the "Sumner crowd" from respectable economics, Ely corresponded with those friends and former students who largely shared his devotion to Christian ethics and German professionalism. Henry Carter Adams, Edmund J. James, Simon Patten, John Bates Clark, Albert Shaw, and Woodrow Wilson were among the men he consulted, all of whom were in general agreement with Ely's stance toward reform. Only Edwin Robert Anderson Seligman, a Jew destined for the leadership of the profession in the early twentieth century, was both an Ely confidant and outside the pale of evangelicalism. He was also skeptical about the motives of Ely's group and much more concerned with the extension of science to economics than he was with applying ethical judgments to politics. Ely could be persuasive, and for the moment, he was willing to deemphasize public expressions of contempt for Sumner and any insistence on Christian reform goals. But by the time his organization finally gathered at Saratoga Springs, New York, in 1885, more than twenty of the fifty people in attendance were either practicing clergy or had been in the recent past. Ely's bitter enemy at Johns Hopkins, Simon Newcomb, could sneer with some justice that Ely wanted the American Economic Association "to be a sort of church, requiring for admission to its full communion a renunciation of ancient errors, and an adhesion to the supposed new creed."[4]

Newcomb had every right to complain. Ely had written his statement explaining the purposes of the organization shortly before the meeting. The document stood as a living challenge to laissez-faire. "We regard the state as an educational and ethical agency whose positive aid is an indispensable condition of human progress," he had written. The document took notice of individual initiative but insisted that "laissez faire is unsafe in politics and unsound in morals." It rejected the permanence of the natural laws assumed by the conservatives, insisted that the science of political economy was still evolving from a primitive state, and urged an "impartial study of actual conditions of economic life for the satisfactory accomplishment of that development." According to Ely, "church, state, and science" had to unite in examining changing economic conditions and in developing new ideas to meet the challenges that resulted.[5]

The majority of the members of the association, while sharing many of Ely's biases, were disturbed by his evangelical devotion to reform. To them such a stance smacked of the pulpit and the press and was unbecoming to a new profession eager for respectability and fearful that wealthy businessmen would penalize them for radical views by putting pressure on university boards of trustees. They preferred an implicit religion of progressive reform to any explicit commitment to a social gospel. Ely, to their minds, was much too close to friends like George D. Herron and the American Institute of Christian Sociology in which Ely, Herron, and John Commons cooperated during the early 1890s. For a while Ely seemed oblivious to the rising sentiment against his form of progressivism, but his regular attendance at the Chautauqua summer sessions and his enormous output of popular articles finally did him in. Without consulting key colleagues, he chose Chautauqua as the site of the 1892 annual meeting of the AEA. He succeeded, but this blatant attempt to retain the unity between religion and social science caused his downfall. His friends deserted him, and he found that he had to resign the key post of secretary of the association. His replacement, according to a compromise solution, was his loyal student Edward Alsworth Ross, who was fully as firm a progressive as Ely, but the new president of the association was Charles Dunbar of Harvard, no friend to innovating ideas, Christian or otherwise. The profession of economics had emerged only by denying many of the forces that had contributed to its formation.[6]

Ely was bitter. He withdrew for some years from national activity. At the same time, he found himself involved, as did several other advocates of Protestant reform, in serious cases of academic freedom. Ely retained his job by denying his right to be a socialist and a professor at the same time, a somewhat inglorious victory that historians have occasionally misunderstood as a victory for toleration. Ely learned from his experiences and turned away from popular and openly religious work toward more acceptable, professional, scientifically neutral studies. The explicit religion became implicit within professionalism and a vaguer progressivism, but Ely never abandoned his original attitudes and goals. He remained in close touch with Hull-House, for example, and was closely involved with a Milwaukee settlement. Students followed his example and were active in settlements in Cincinnati, Toledo, Minneapolis, and Pittsburgh. As years passed, and a progressive climate of opinion began to dominate the country, Ely found himself less controversial. His home became something of a progressive hotel. When the Wisconsin Idea became a national inspiration, Ely was host to, among others, Albert Shaw and Woodrow Wilson, his old students from Johns Hopkins; new friends like Jacob Riis and Theodore Roosevelt; and, of course, Jane Addams.[7]

III

The progressive economists were important in their own time but retain mainly a historical interest. In history and political science, the progressives made a greater contribution. The progressive historians were major figures of controversy. They established the framework for most disciplinary controversy for the next two generations; scholars still find books written by progressive historians useful for presenting the progressive point of view to students who have come to accept more contemporary models.[8]

The two disciplines, long separate in most academic institutions, were all but indistinguishable until after the turn of the twentieth century. History seemed to be past politics; political science was current political practice; and the line between vague. A figure like historian Herbert Baxter Adams at Johns Hopkins could run a seminar that produced as many political scientists as historians; students and professors moved easily from past to present according to the demands of dissertation directors and the requests of publishers. Woodrow Wilson could publish a study of congressional government that made him a nationally known figure in political science only to follow it with studies of the Civil War and a wide-ranging American history. Charles A. Beard managed to be president of the major scholarly associations of both historians and political scientists during his career. For scholars trained later, such facility was highly exceptional. A given book might prove useful to a neighboring discipline; a methodology from another discipline might be discreetly borrowed; but professionals could rarely bring themselves to admit outsiders to their *inner sancta* once the institutional lines had hardened. The persons involved in these two disciplines did their best work in history, which not only was somewhat older and better established as a scholarly discipline, but also had a long tradition of amateur excellence outside the university.

Political science in America had no William Prescott or Francis Parkman to look back upon, and its students found instead what models they could, chiefly in England and Germany. Although a number of future political scientists came out of the seminars of Adams and Richard Ely at Johns Hopkins, the true center of political science was Columbia, where John W. Burgess had begun serious graduate teaching in 1880. No progressive, Burgess did bring German training and considerable administrative ability to his work; the framework he established allowed Columbia to dominate the new field until a more progressive figure like Charles Merriam established a more influential center at the University of Chicago in the 1920s. Columbia also tended to stress the importance of a year abroad for its students, which Johns Hopkins never did. This bias toward an older tradition helped give depth and authority to a field often uncertain of its mission. Among the students attending the early courses were Theodore Roosevelt, William A. Dunning, and Herbert L. Osgood. Soon to join the faculty were former students like Seligman and Goodnow and one distinguished new appointee in the person of John Bassett Moore, Professor of International Law and Diplomacy. Not all of these figures became progressives; occasionally someone was an unwavering conservative. But the environment provided an excellent place for serious students to discover what was wrong with American administrative practices and to begin the sort of analysis from which progressive reform could proceed.[9]

The great trailblazer for the progressive historians was Frederick Jackson Turner. Descended from puritan ancestors, Turner had the usual background of Sunday school, Republican politics, and Civil War memories in his childhood. At a loss for a satisfying career, he tried journalism briefly, but Professor William F. Allen at the University of Wisconsin so excited Turner with the study of history that he was hooked for life. Turner learned "scientific" history from Allen, a history that stressed empirical data compiled from primary sources, the evolution of whole societies, and the process by which social change occurred. Turner desperately wanted to remain at Wisconsin, even as an overworked assistant, but the realities of professional advancement dictated a shift to Johns Hopkins. There, Turner made a lifelong friend of Woodrow Wilson—"homely, solemn, young, glum, but with the fire in his face and eye that means that its possessor is not of the common crowd." He studied with Adams, Ely, and Albion Small. When Turner returned to Wisconsin, he had made contact with some of the key figures of progressive social science, and the men promoted each other's careers for years. Wilson was the most enthusiastic publicizer of Turner's ideas in the days when few people noticed them; Ely came to Wisconsin in large part because Turner paved the way and was able to survive a nasty battle over academic freedom in part because of Turner's support. Small, as a Chicago sociologist, was instrumental in the concerted effort of William R. Harper to win Turner for Chicago. The Hopkins network, while hardly coextensive with progressive social science, worked again and again to raise spirits as well as salaries across the nation and to spread the new ideas that were to reform it.[10]

Historians of Herbert Baxter Adams' generation seemed obsessed with the German origins of American institutions: the germs of the American town meeting, of the legal system, etc. They stressed scientific research and the use of primary sources to write history, but they demanded from their students confirmation of their Teutonic, Anglo-Saxon, and Darwinian presuppositions. Turner, however, was inordinately proud of his own family and its slow migration west, each generation coming closer to Wisconsin and shedding the remaining vestiges of its European origins. Like Woodrow Wilson before him, he managed to compromise with Adams, taking the cherished degree without alienating the powerful, temperamental professor. But Turner was too intelligent simply to swallow whole the platitudes of scientific history; besides, he had prejudices of his own to address. What was important about the America he knew was not its European origins or its institutions but the process of democratic living that all people experienced as they conquered the frontier and established new cities, new universities, and new ideas. Adams' view of history as past politics was far too narrow. History, Turner declared in 1890, "is past literature, it is past politics, it is past religion, it is past economics," and all of these were essential to society's attempts "to understand itself." Every age had to write history to enhance self-knowledge, and since society was always in flux, its history should always be in flux. Turner's history, like Dewey's philosophy, was a relativistic effort to use the intellectual materials at hand in order to fashion a better future for all peoples. It refused to accept the European past as determinative of the American future. It rejected the intellectual preconceptions of the East and emphasized increasingly the contributions of movement toward the West and life there as the core of the American past. The phrase, the "New History," had not yet come into use, but by 1890 Turner had thought out its essential premises.[11]

Turner was never the kind of historian who wrote concrete scientific interpretations of anything. Instead, he produced influential "think" pieces that offered frames of reference to two generations of scholars who sought to demonstrate and illuminate the democratic experience in America. In 1893 Turner informed the annual meeting of the American Historical Association that "the existence of an area of free land, its continuous recession, and the advance of American settlement westward explain American development." Of course, these did no such thing, but like most of Turner's insights, the vision had an elemental grandeur. It swept away the detritus of German institutionalism, and in its place put an explanatory model that was simplistic, chauvinistic, and obsessed with space and motion. To a country still discovering itself, Turner's vision produced a shock of recognition: the frontier, democracy, individualism, pioneering—surely these really were the essence of American experience. What was best about America was what was least like Europe.[12]

Before Turner left Wisconsin for Harvard in 1910, he managed to train a generation of historians in the study of the minutiae of western democratic life. He was also a key progressive force within the university, not only training figures like Paul Reinsch and Charles McCarthy, who helped make the university an integral part of progressive state administrations, but also supporting President Charles R. Van Hise in his perennial battles against conservative regents and legislators. Within the context of progressive history, his greatest contribution may have been the training of shy, young Carl Becker. Never really a western historian, Becker had in his early years an odd dual career. He taught European history while publishing chiefly in American history. His dissertation, *The History of Political Parties in the Province of New York: 1760–1776* (1909), stressed social and economic rather than the political and institutional factors, and added the influential emphasis to historical thinking that the revolution was as much over who should rule at home as it was over home rule. His most probing essays about historical methodology, the relativity of historical data, and the impossibility of a truly objective historical science did not appear until after the end of the Progressive Era. They were nonetheless progressive for all the delay and were among the more significant methodological contributions of the interwar period. Meanwhile, the emphasis on economic and social factors in colonial politics set the frame of reference for much of the research of the next two generations.[13]

In one of those intriguing shifts that intellectual history sometimes observes, Becker moved briefly to Columbia in 1898 and worked there with the other most influential college professor of history, James Harvey Robinson. Robinson presented a sharp contrast to Turner. In his early work, he had proved to be one of the more passionate devotees of primary sources, producing substantial translations of key European historical texts and scholarly articles that quickly made their mark in the field. But far beyond Turner, Robinson was a committed progressive who wanted to make history immediately relevant to state legislators as well as to ordinary citizens making daily decisions. Where Turner's progressivism made him a scholar's scholar, Robinson's made him a pedagogue's scholar. Like Dewey, Robinson poured himself into school work, speaking often to teachers and producing textbooks which were professionally insignificant. He produced popular essays that his colleagues felt cheapened the whole profession. Thus, Robinson's influence has been hidden from historical discourse. Conveyed chiefly during his classroom lectures, it was nevertheless a real influence that eminent scholars recognized. Not only Carl Becker, but also historians of the stature of Preserved Smith and Carlton J. H. Hayes have praised the quality of Robinson's mind and its impact on anyone in the field during the Progressive Era.[14]

Robinson differed from Turner in emphasis and in personal style. Both men were interested in explaining the present in terms of its immediate past, but Robinson carried this further when he insisted on explaining the past in terms of the present. For him, as for Becker in his maturity, the past was a pointless collection of facts, much like marbles. Historians, being in large part shaped by their own needs and interests, of necessity sorted out the marbles and placed them in an order that satisfied those needs. History was a prose narration, usually in chronological order, of what they found. In this sense Robinson proved to be a more forceful analyst of the past than Turner. Where Turner was a born professorial bureaucrat, conciliating people and charming them into changing their ideas and habits in small ways, Robinson was a born skeptic, abrasive in personal style, sarcastic in

MINISTERS OF REFORM

his contempt for any past conservatisms and impatient at university bureaucracies. Where Turner left Wisconsin only to go to what seemed to be the pinnacle of the profession at Harvard, Robinson resigned from Columbia out of a loathing of deans, grades, degrees, and the restrictions on his freedom of expression exerted by an authoritarian university president and his excessively patriotic board of trustees. In effect Becker retained the personal style of Turner but toughened his intellectual stance with the rigor of Robinson. Perhaps symbolically, he also shifted fields in Robinson's direction as time went on. His greatest contribution to knowledge proved to be a study of the French philosophers who in many ways had resembled Robinson: *The Heavenly City of the Eighteenth Century Philosophers* (1932).[15]

Robinson united the key elements of his historiographical approach in a volume of essays entitled *The New History* (1912). Quoting such progressive social scientists as Dewey, Mead, and William I. Thomas, he restated the case for a relevant history that could be used even by mechanics and artisans. He attacked the popular emphasis on the flashy, violent, or extraordinary event that might capture student or public attention, but which was basically a distraction from the more important historical problems. He also attacked the tendency of most of his scientific historical colleagues to stress the inert facts of government—when kings were born and died, when crusades occurred, or inquisitions began—with little effort to tie these data to larger, current concerns. "Man is more than a warrior, a subject, or a princely ruler," he insisted; "the State is by no means his sole interest."

Buried beneath this insistence on the relevance of his profession lay a justification of his own particular orientation and its social utility. The "history of thought should play a very important part" in solving the contemporary social problems of mankind, he insisted, "for social changes must be accompanied by emotional readjustments and determined by intellectual guidance. The history of thought is one of the most potent means of dissolving the bonds of prejudice and the restraints of routine." Like all progressives Robinson believed in progress. To him intellectual history of the kind he taught "promotes that intellectual liberty upon which progress fundamentally depends." As a discipline history "seems to justify the mystic confidence in the future suggested in Maeterlinck's *Our Social Duty.*" Robinson, in effect, defined himself as a Relevant Progressive, worthy to stand next to Dewey and Mead. "At last, perhaps, the long-disputed sin against the Holy Ghost has been found," he began his conclusion: "it may be the refusal to cooperate with the vital principle of betterment." History condemned all attempts to impede progress "as a hopeless and wicked anachronism." "History," of course, did nothing of the kind,

76

and by its nature could not possibly do so. But to believe in this fashion in the mission of history, economics, or sociology was typically progressive. Regardless of how skeptical he thought himself to be, James Harvey Robinson was descended from the Pilgrim pastor Dr. John Robinson and all the intervening generations of staunchly Calvinistic Presbyterians. He had to believe in something, so he believed that somehow salvation would come from "science," "primary sources," "relevance," "democracy," and "progress." He, too, had his heavenly city; Carl Becker understood.[16]

Robinson's great kindred spirit at Columbia was Charles A. Beard. The descendant of Quakers and Methodists whose Scots-Irish ancestors had been in America for 200 years, Beard grew up in a Whig and Republican tradition that underscored the value of owning property, the virtue of farming, and the positive use of the government to achieve moral goals. He carried on the family traditions by assuming that the wage system was the modern version of Negro slavery, and that any decent Christian should work to end it. The depression of the early 1890s had the same effect on him that it had on many other progressives. Like them he visited Hull-House to immerse himself in the realities of modern America and read books like John Ruskin's *Unto This Last* to supply himself with an intellectual rationale for helping the laboring classes. In 1898 he went to Oxford, met the young American socialist Walter Vrooman, and became the primary force behind the establishment of Ruskin College, the workingman's college that he hoped would pioneer adult education throughout England. For two years he spoke constantly in the Manchester area, playing "the industrial counterpart of the frontier preacher, delivering lectures to temperance reform societies, co-operative associations, and units of the Independent Labor Party."[17]

In 1902 Beard abandoned his efforts to reform British education and entered Columbia to take his M.A. and Ph.D. degrees. He joined the faculty in 1904; by 1907 he was an adjunct professor of politics in the department of public law, and by 1915, a professor of politics. He rebelled almost instinctively against the formalism of senior colleagues like John Burgess, and quickly became a key member of a circle of faculty and students that was developing new pragmatic and socialistic approaches to both the study of American government and the best methods of achieving reform. Inspired by figures like E. R. A. Seligman, Algie Simons, and J. Allen Smith, Beard applied his past experiences in reform to current problems, stressing economic determinism that may well have gone back, as he insisted, to the blunt combination of self-interest and moral reform so apparent in Conscience Whiggery. Populism, Hull-House, and British Labour doctrine had simply reinforced and modernized family traits. Books from outsiders like Arthur R. Bentley's *The Process of Government* (1908) or from American thinkers like James Madison of *The Federalist #10* provided him with a suitable framework within the confines of political science. Even as he was collaborating with James Robinson on one of the key textbooks of the new history, he was also developing a new pragmatic political science that remains the clearest expression of progressive thought in that discipline.[18]

Beard delivered "Politics" as part of a series of lectures defining the disciplines at Columbia. Implicit in the talk was his scorn of Burgess and formalist political science, his dislike of Hegelian idealism applied to the state, and his refusal to take mechanical ideas of law and politics seriously. In their place he demanded the study of history, economics, and sociology, and the placing of human beings in all their complexity in the contexts in which they lived. Although he could sneer at "the high priests of the mathematical and the exact," he could also point proudly to the piles of new data that were discrediting the old, grand theories of political action. Always distrustful of ideas as motivating forces, he viewed the political process in the same way that Dewey had viewed the philosophical process: "Political philosophy is the product of the surrounding political system rather than of pure reason." Despite the fact that he was himself a moralist whose writings praised and condemned others with gusto, he also insisted that scholars "should not praise or condemn people or acts, but rather, understand and explain." It was a characteristic progressive pose—the tough-minded, scientific skeptic insisting on detachment, factual research about the actual process of human activity and insisting, too, on his own disinterest in the results.

Beard approved of the achievements of progressivism: the short ballots, the civil service, improved administration in government, and the growing popular recognition of the need for more positive governmental interventions in the economy. He welcomed the declining prestige of natural rights and laissez-faire ideas. Modern scholars, at least, had learned to leave the laws of nature to theologians and had come to "treat politics as a branch of sociology." He saw a close connection between the realms of scientific research and legislative reform: scientific scholars worked disinterestedly, merely seeking "the truth concerning special problems simply in the spirit of science," but, of course, that data could live its own life in service to the nation. He closed "Politics" with what sounded like the Columbia edition of one of his old Manchester sermons. A professor should tell his students: "Observe these facts, consider these varying explanations, ponder upon these theories, study the most impartial records of political operations, look to the future as well as the past, and as a citizen of this great nation build this discipline of the mind into the thought and action of after life." Anyone who planned to participate in politics needed to understand that "the wisdom that comes from a wide and deep and sympathetic study of the political experiences of men is the true foundation of that invisible government, described by Ruskin. It wears no outward trappings of law, diplomacy or war, but is exercised by all energetic and intelligent persons, each in his own sphere, regulating the inner will and secret ways of the people, essentially forming its character and preparing its fate."[19]

Always a man to practice what he preached, during the next two decades Beard proved to be an indefatigable advisor to and participant in various administrative agencies and commissions on the East Coast. But his greatest contributions to progressivism were his contributions to the early history of the country. Especially in *An Economic Interpretation of the Constitution of the United States* (1913) Beard set out to take the myths of idealism out of American legal and political history. Posing as the dispassionate scientist only interested in quantifiable facts, he insisted that the types of property that the Founding Fathers possessed had determined their votes on the Constitution. Abstract principles had nothing to do with it. Since real, self-interested human beings had written the document, the implication was clearly that equally real, modern progressives could change it. Times, interests, and power shifts had created a new situation which demanded a revised Constitution. Two generations of modern scholars and the invention of the computer have been required to prove definitively that most of Beard's conclusions were wrong. In so doing they demonstrated that he had set the terms of professional debate until about 1960. Few other progressive creations enjoyed such durability.[20]

IV

The last of the progressive social sciences to achieve a professional identity was sociology. Economics might have its problems with moral philosophy, and political science with history, but sociology served as a wastebasket for expanding academic institutions, gathering in topics unwanted by other departments and receiving a correspondingly meager amount of money. Often, the establishment of a department of sociology depended on the whims of a university president, the willingness of a key donor to supply funding, or the presence of a powerful professor eager to guide a new discipline. At Harvard, for example, the influence of the social gospel and the settlement houses was significant throughout the 1890s. Francis G. Peabody taught his famous courses in "drainage, drunkenness and divorce" in a department of social ethics. John Graham Brooks,

Edward Cummings, and William James Ashley brought the world of Unitarian social reform and Toynbee Hall to the campus, but they remained in the economics department. Between 1900 and 1902, a series of deaths and departures eliminated anything resembling sociology, and the discipline had to wait thirty years for the arrival of Pitirim Sorokin and the founding of a true department of sociology, which quickly absorbed the remaining vestiges of social ethics.[21]

For a brief period, Columbia seemed likely to produce a sociology department to rival its program in political science. On the surface Franklin H. Giddings seemed an excellent choice for a new chair; the son of a Congregationalist minister who had studied engineering, he was a capable journalist. Converted to an interest in sociology by reading Herbert Spencer, he entered academia when he replaced Woodrow Wilson at Bryn Mawr. He was soon dividing his time between Bryn Mawr and Columbia, and in 1894 he was the obvious candidate to head the department of sociology at Columbia. Full of a positivistic desire to improve society through scientific research, he seemed on the surface a typical academic progressive, even to the point of being a director of the University Settlement and a significant figure on the social-work scene in New York City. In time he served on the Board of Education and was a leader of the League to Enforce Peace. But there was less to Giddings than met the eye. With great ambitions as a theorist and quantifier, like so many of that early generation of social scientists, he satisfied himself with urging others into scientific research without ever indulging in it himself. His theoretical ideas filled many pages without producing much that was new or influential. Worst of all, he was so dogmatic, conceited, and prejudiced that he offended most of his colleagues at Columbia, and they quietly made sure that sociology remained insignificant institutionally as long as Giddings remained there. Like all too many progressives, he was rabidly anti-Semitic. Since two of the most significant figures on the Columbia scene were E. R. A. Seligman and Franz Boas, the result could only be a stalemate. Over time Columbia produced a number of excellent sociologists, but the record suggests that they emerged in spite of Giddings rather than because of him.[22]

V

Of all the social sciences, sociology was the most progressive and the least secure in its professional status. More than one historian of the discipline has noted that the first generation of sociologists with official positions in the field resembled nothing so much as a private club for Protestant clergy interested in mitigating the impact of industrialism on America. Lewis Coser has remarked, for example, that of the early presidents of the American Sociological Society, "Giddings, Thomas and Vincent had been born in clerical homes, while Sumner, Small, Vincent, Hayes, Weatherly, Lichtenberger, Gillin and Gillett pursued careers in the Protestant ministry before they became sociologists." Not all clerical sociologists became progressives, but the vast majority seemed to be oriented toward settlement work, reform Darwinist ideas about government activity, and the gas-and-water socialism common among municipal reformers around the turn of the century. A 1927 survey discovered that 61 of 258 practicing sociologists for the preceding period had been in the ministry, and 18 others had attended divinity school at some point in their lives. Most came from rural or small-town backgrounds, and the evidence strongly indicated that had the ministry not been steadily loo ing economic and social status, these individuals would have chosen it as their profession. The urge to help people shaped many of the early concerns of professors in sociology, and they often wavered indecisively between a desire to be active and relevant and a desire to be professional and detached.[23]

The University of Chicago proved to be the major home for innovative, institutionalized sociology. As a new, wealthy, and expanding institution, it could afford to give a home to a new discipline, and professors in that discipline felt themselves the equal of other professors within the university—hardly the case in older institutions where more traditional fields held the prestige and the funding. Just as the accidents of hiring, promotion, and death tended to keep sociology from success and expansion elsewhere, they worked at Chicago to imbed the new field firmly in the curriculum. President William Rainey Harper, the vigorous Baptist scholar working for the firmly Baptist interests of John D. Rockefeller, at first tried to hire Herbert Baxter Adams and Richard Ely for his new department, but they proved to be too secure where they were. Harper then turned to yet a third figure who had studied and taught at Johns Hopkins for a chairman —Albion Small. Among the lesser figures attached to the department over the next two decades were Charles R. Henderson, a Baptist minister who taught the administration of charity; Charles Zueblin, a Yale ministerial student trained in Old Testament studies who kept close ties to the Northwestern University settlement house; and George Vincent, the son of a Methodist bishop who was also the founder of Chautauqua. The periphery of sociology also included Frederick Starr, who taught anthropology, which did not become a separate department until 1929, and Marion

Talbot, who taught "sanitary science" until the formation of a home economics department in 1904.[24]

In this intensely clerical, earnest environment, Small proved to be an ideal administrator. The son of a Baptist minister from Maine, he had attended Newton Theological Seminary before beginning a career as a teacher of history at Colby College. He spent most of the 1880s teaching, took a Ph.D. at Johns Hopkins in 1889, and was president of Colby when Harper persuaded him to come to Chicago. Already a veteran of some of the earliest college-level courses in sociology, Small entered into administration with zest. He founded the *American Journal of Sociology,* became president of the American Sociological Society, and recruited figures such as William I. Thomas, Robert Park, and Ernest Burgess, who formed the core of the department during its two decades of unchallenged preeminence (1910–1930). Having studied at Berlin and Leipzig for two years and married the daughter of a Junker general, Small was also fluent in German and the perfect vehicle for transporting German sociological ideas to the newer American discipline.[25]

Small's major contribution to sociology was to domesticate the conflict theory that Gustav Ratzenhofer and Ludwig Gumplowicz had developed in Europe. Small also stressed social process and the satisfaction of interests in much of his theoretical work. His language was redolent with Darwinism: "The social process is the incessant evolution of persons through the evolution of institutions, which evolve completer persons, who evolve completer institutions, and so on beyond any limit that we can fix," he argued in his major text. To a progressive the key problem was how to guide this evolution toward social goals that satisfied ethical demands for a just society. Small answered that sociology was "a moral philosophy conscious of its task." From the beginning of his career, his writings displayed a Christian imagery that undermined their demand for acceptance as scientifically objective. The "tendency of sociology must be toward an approximation of the ideal of social life contained in the Gospels," he wrote even before arriving in Chicago. The sociological worldview, he argued, in 1903, "turns out to be the theater of a plan of salvation more sublime than the imagination of religious creed-maker ever conceived. The potencies which God has put in men are finding themselves in human experience." Almost a decade later, he insisted that sociology was "really assuming the same prophetic role in social science which tradition credits to Moses in the training of his nation." Sociology "is the holiest sacrament open to men. It is the holiest because it is the wholest career within the terms of human life." Indeed, all of social science "is the indicated field for those 'works' without which the apostle of 'salvation by faith' declared that faith is dead." He was even capable of defining his key methodological contribution, the social process, as "we are members one of another," purposely echoing the language of Ephesians 5:25. In a letter at the end of his life, written to a friend who was compiling a memoir, he brought together his Christian, sociological, and reform activities in a manner that could have come from Jane Addams. "The one impulse that has remained constant in spirit . . . has been the conviction that experience will never let itself be interpreted as an affair of aggregated monads, but that some-how we live, move and have our being as members one of another," he wrote. "Some day that will be the unquestioned *a priori* of everybody." He hoped that if his name lived that it would "have a tag attached with the memorandum 'he had something to do with laying the individualistic superstition.' "[26]

Small's colleague, William Isaac Thomas, was the most significant social scientist to function within the progressive frame of reference. Born in rural Virginia in a family devoted to farming and Methodist preaching, Thomas seemingly grew up in an isolated and unstimulating environment. Yet he majored in literature and classics at the University of Tennessee, and taught literature and languages during his early professional life. He attended the universities of Berlin and Göttingen, spent several years on the Oberlin faculty, and went to the University of Chicago when its reputation for innovative social science reached him. Already the holder of the first Ph.D. ever granted by the University of Tennessee, he took another at Chicago in sociology and anthropology, joined the faculty, and rose to full professor by 1910. In some ways Small's greatest contribution was to give Thomas an institutional home for his astonishing breadth of interests.[27]

Thomas took inspiration not only from literature, but from biological science, psychology, ethnology, and his own restless explorations of the cities of Europe and America. He scorned most of the books and disserta-tions written in sociology and proved to be the person most responsible for making them irrelevant to the history of significant knowledge. When pressed later in life to express his greatest intellectual debts, he mentioned Herbert Spencer, though not always sympathetically; German folk psy-chologists Moritz Lazarus and Hajim Steinthal; American behaviorist psy-chologist John B. Watson; his collaborator, the Polish philosopher Florian Znaniecki; the founder of American anthropology, Franz Boas; and Ameri-can sociologist Charles H. Cooley. Despite the assumptions of several historians of sociology, he admitted no influence at all from John Dewey; but he did acknowledge, as did so many others, the inspiration of George Herbert Mead. Such a motley list of influences was perhaps only possible in the early years of a discipline; the formalities of professional training soon made it difficult for such a wide-ranging intelligence to survive the training process. Even with Thomas, the results were long uneven. His early work at times bogged down in biological and biochemical explanatory models, particularly in his efforts to explain sexual differences while other publications seem to be little more than accumulations of known anthropological data.[28]

But Thomas' inquisitive mind and exuberant personality also fastened on the same kinds of problems that Jane Addams had. To a progressive intellectual in Chicago, one of the most obvious social problems was the arrival of huge immigrant communities, ignorant both of the English language and the mores of a large city. Already facing severe social pressures in Europe, Poles, Italians, and Jews faced something like a complete breakdown of their communities under the new pressures. This social disorganization seemed to be directly connected to the outbursts of violence, the unemployment, the prostitution, and juvenile delinquency that was so prevalent in immigrant wards. In 1908 Thomas spoke of his interests to Helen Culver, heiress of the man who had endowed Hull-House, and she agreed to support the study of a key immigrant group with $50,000. Fascinated by the unpredictable and often violent nature of Polish social maladjustment, Thomas chose Poles as the ideal group to study. He found the available scholarly sources and articles valueless, so like a good ethnographer he set about collecting whatever written folk material he could. He found a weekly journal full of peasant letters covering several decades; he advertised for family letters and bought what was available; he purchased newspaper archives; he commissioned autobiographical accounts; he examined parish and court records; he even traveled to Poland in search of data. On at least one occasion, he was almost hit by a package in transit from a window to a pile of garbage and upon examining the missile, discovered a key collection of letters. The mixture of random sample and commissioned primary source bothered the next two generations of social scientists; Thomas positively invited skepticism about his methodology and sampling techniques. By instinct more an anthropologist than a sociologist, Thomas nevertheless insisted that these documents were the perfect sources for such a study and shrugged off his critics.[29]

The Polish Peasant in Europe and America, written with his friend Florian Znaniecki, proved to be the masterpiece of progressive social science. The five large volumes of the first edition, 1918–20, were not without flaws, and professional critics have pointed out the oddities of the source-gathering techniques, the way the conceptual apparatus and the data often seemed unconnected, the frequent vagueness of key concepts like "attitude" and "value," as well as an overall impression of two authors more gifted at microanalysis of individuals, groups, and circumstances than at the larger conceptual picture of social disorganization. But to dwell on the flaws of a masterpiece should be only a prelude to a celebration of its virtues and innovations. The book introduced the study of social attitudes to a profession that soon became devoted to their study. It introduced many subjective sources, from private letters to parish records, to a discipline that still preferred armchair grand theorizing on the one hand and a desiccated, pointless objectivity on the other. It introduced numerous key terms and emphases: the "four wishes," the "definition of the situation," and "social attitude," to name but a few. It showed a generation of moralists yearning for professional status and acceptance how they could contribute to the solution of social problems without compromising their precious objectivity; indeed, the work made older notions of science quite irrelevant and substituted a new sense of detached empiricism for the older sense of the neutral and the typical. *The Polish Peasant* made what became known as scientific sociology possible in America. In the process it stimulated work in culture and personality studies, social stratification studies, work in attitude change, and explorations of field theories of behavior.

Like all works of social science, the work was embedded in the frame of reference that led to its conception. In a contemporary article, Thomas already managed to sound like James Harvey Robinson when he wrote: "We must first understand the past from the present. We must view the present as behavior. We must establish by scientific procedure the laws of behavior, and then the past will have its meaning and make its contribution." Whether Thomas was conscious of the influence of Robinson, Mead, Dewey, or Addams, or whether they in turn were feeling his influence, matters little. The significant point is that progressives in many disciplines seemed to be pursuing parallel lines of thought at roughly similar times. Thomas wished to establish *"a system of laws of social becoming"* that would enable scholars to focus on the process that was so important to pragmatic philosophy and on the social control that progressives felt was essential in order to Americanize new immigrants. Like Cooley, but with more emphasis on the individual, Thomas dwelled at length on the importance of primary groups, and the way in which individuals formed their personalities through interaction with these groups. The process of thinking, for Thomas as much as for Mead and Dewey, was the solving of problems. Ethics were situational, relative to the time, the person, the place, and the problem. Sociologists in Thomas' world should speak of probabilities and not laws, and they should speak "as if" they knew certain things when, of course, their knowledge was never total, and their materials were never perfect. Indeed, although Thomas never mentioned William James, that great catalyst of progressive thinking surely lurked behind such key concepts as "if men define situations as real, they are real in their consequences."[30]

The books received immediate recognition, and scholars have never been in doubt about the stature of Thomas' masterpiece. Unfortunately for his career and for sociology, the university turned out to have other ways of measuring scholarly eminence. An urbane and convivial man, Thomas had long irritated the Baptist administrators at John D. Rockefeller's monument. He may have been the son of a preacher and a man who shared many of the values and concerns of Chicago's progressive intellectuals, but Thomas was also outspoken on matters of sex, divorce, and religion; his views were broad and tolerant in ways that the university could not approve. Despite his two doctorates and frequent publications, promotions came slowly. William Rainey Harper died, and in 1918 the president at the University of Chicago was the aged, conservative, and intolerant Harry Pratt Judson. In that year, as the first two volumes were going through the university press, the Federal Bureau of Investigation arrested Thomas for violation of the Mann Act. The circumstances surrounding the arrest have always been unclear, and the charges were dropped, but the suspicion seems to be that Thomas' wife, who was active in various peace movements, had excited the enmity of various self-styled patriots, who then managed to engineer the scandal as a means of discrediting her as well as her outspoken husband. Thomas steadfastly refused to discuss what had happened in any detail, even in the face of a public scandalous assault. The president and the board of trustees moved immediately to rid the university of such an embarrassing figure, and since modern concepts of tenure and the separation of public and private morality had not yet taken root in Chicago, Thomas saw his career ended. He continued to publish important books and to participate in national sociological affairs, but he never again held an influential academic position. At least one of his major publications even appeared under two other names, to avoid embarrassment to the publisher, and was not publicly labeled as Thomas' work until after World War II.[31]

With the departure of Thomas, Chicago sociology might have collapsed. Small was rapidly lapsing into the numbing reminiscence that marked his classes before his retirement, and no new figure on the regular staff matched Thomas in stature or promise. But the department survived to prosper as never before because Small and Thomas had attracted a regular visitor in Robert E. Park. Working for minimal wages, and sometimes for none at all, Park lived on a small inheritance while he taught the material he loved and also discussed it with Thomas and younger colleagues, like Ernest W. Burgess. With the parsimony that so frequently marked Harper's successors, the university dragged its heels about formalizing the relationship, and not until Park was already the major inspirational force in the 1920s did it condescend to make him a full professor. The department Park dominated proved to be the major force in sociology until it declined in originality in the 1930s and lost preeminence to Harvard and Columbia.

Park's life and career effectively brought together many of the major themes of progressive thinking about society. Several analysts of his career have chosen the imagery of Quakerism to describe his early life. Frequently depressed, restless, and energetic, he was a Seeker looking for a meaningful vocation that would give his life meaning and utility. As a child in Minnesota, he had lived in a world of American boosterism softened only by his mother's attempts to encourage an interest in literature and the arts. The town of Red Wing was full of Civil War memories; the ideals of Lincoln served as a secular replacement for the ideals of the puritan New England in which so many of the townspeople had their family roots. Park quickly developed an aversion to the world of business that dominated the life of his blustering father, but not for years could he find a satisfying replacement. He made his first serious attempt at the University of Michigan, where he took six courses with the young John Dewey, thus developing the interests in communication and democracy that marked his entire later career. Society for both progressive philosophers and sociologists was a functioning organism that had to be analyzed while it functioned; it could never be the static, reified abstraction that it seemed to be in the minds of other kinds of social thinkers.[32]

Like many progressives, Park was attracted to journalism; he approached the field with a great faith in its importance to a functioning democracy. Inspired by Dewey and Dewey's friend Franklin Ford, Park saw in journalism a means of uniting a democratic community around certain key ideals. In pursuit of this chimera, Park spent most of the 1890s as a reporter for various newspapers, chiefly in the Midwest, but he soon realized that he was burning himself out and had no satisfying future in the field. He, therefore, decided to go to Harvard to study the philosophical aspects of the impact of journalism in a democracy. Park worked with several of the great names in philosophy, but was most impressed with William James. James read to Park's class an early version of his essay, "A Certain Blindness in Human Beings." Its plea for an understanding of the complexities of knowledge and the uniqueness of the individual made so deep an impression on Park that in later years it seemed to be almost all that he could remember of his stay in Cambridge. James was never typical of much of anything, however, and certainly not of philosophers; despite his example Park soon left Harvard for Germany and what seemed to be the place where significant sociological knowledge of the human community originated. There he wandered from Berlin to Strassburg to Heidelberg in search of the best people and ideas to help him satisfy his conscience and his need for a vocation. He found most helpful the ideas of the Russian social thinker Bogdan A. Kistiakovskii and the lectures of Georg Simmel. Kistiakovskii had been a student of Wilhelm Windelband, and so Park determined to go to their source, and thus, he left Berlin to work with Windelband in Strassburg, and then followed him to Heidelberg where he finally completed a Ph.D. in 1904. The technical issues of his educational progress were largely irrelevant to his place as a progressive; what remained of his education was a fascination for the cultures and ideas of Europe that enabled Park to put his American pragmatism into a comparative framework and thus transcend the parochialisms of all too many of the sociologists of the era.[33]

As many new "Drs." have since discovered, the title did not come with a job, let alone with psychological satisfaction. Although Park had returned for a brief period as assistant in philosophy at Harvard even before completing his dissertation, he was soon at loose ends. He became interested in the Congo Reform Association and its efforts to publicize the depressing conditions in the Belgian Congo. Park contributed several articles to the cause, but still found nothing worthy of a serious vocational commitment. In the process of researching and writing, however, he met Negro leader Booker T. Washington. Always eager for white support and in need of a ghostwriter for his rapidly expanding journalistic contributions, Washington convinced Park that he would find a situation worthy of him at Tuskegee. He was right. Park spent seven years working for Washington, much of the time in residence in Alabama. While his family remained in Massachusetts complaining about his neglect, he wrote substantial portions of books and articles published under Washington's name and became the single white scholar most familiar with black life. He had found his Hull-House.[34]

In Park's eyes a sociologist was "a kind of super-reporter," and so in a sense, he had long been a sociologist. But he had turned down earlier requests from Albion Small to lecture at the University of Chicago, and only when he met W. I. Thomas at Tuskegee did a connection form that led to his extraordinary career as a professor. Park was tiring of working for Washington and welcomed the chance to have yet one more attempt to organize his ideas. He and Thomas were almost the same age and shared many of the same interests and biases. Their friendship proved stimulating and endured long after the scandal that ended Thomas' connection with the university. Park, too, was convinced that the city and its problems were the most productive focus for social investigation in his generation, and he was insatiably curious about every aspect of urban life. He was loud in his contempt for progressives as "do-gooders," but deeply progressive himself in his sense of the city as an organism in process and his faith that reportage about the city provided the information essential for social control and social reform. His wife was a devoted follower of Theodore Roosevelt and the Progressive Party, but Park held back from such outward displays.[35]

In effect the Chicago sociology department spent its greatest period using the methods of W. I. Thomas to answer questions about the city that Park asked in his first major scholarly article written as a sociologist. Demonstrating a wide range of reading not only of sociologists such as Charles Cooley and William Graham Sumner, but also in philosophers of history like Oswald Spengler and biological scientists like Jacques Loeb, Park asked an astonishing number of questions about the processes of city life. Always psychologically an outsider himself, he emphasized the "marginal men" who seemed to be unable to conform to normal Protestant expectations of behavior. For him the city was a "state of mind" that could be analyzed like any other philosophical idea. Fascinated by anthropology, he wanted sociologists to do for the city what Franz Boas and Robert Lowie were doing in their studies of the American Indian. Only novelists like Emile Zola had even come close to the sort of examination he wanted, and, of course, the insights of a novelist were hard to use in formulating either concepts or laws. Just as Jane Addams had compiled the Hull-House maps and papers, so the more professional sociologists should get control over the city and in the process get control over their own lives. Individual need, social reform, and professional scholarship had finally merged.[36]

Chapter 4

Innovative Nostalgia in Literature and Painting

T HE KEY TERMS for any discussion of the life of the artist in the Progressive Era are "innovative" and "nostalgia." It is always difficult to use terms that cover more than one art at a time, but those two words used together help focus attention on the dilemma of the progressive artist around the turn of the twentieth century. He was full of new ideas about religious and political life, sure that they would help usher in an America of great promise. He was, however, an artist more than a politician, and thus naturally he wanted the arts to participate in this great adventure. At the same time, like progressives in other disciplines, he had a nostalgia for the past: for the small town, for religious certainty, for the morality of his parents, and for the America that always seemed to exist about the time he was born. This tension between the old and the new helped generate art, but it also made that art confusing and contradictory. Most often, the new ideas found themselves in works of art that were based on traditional forms, as if the comfort of the old forms took the sting out of the threat of the new ideas. Many artists could never solve this dilemma and their works have sunk from sight, the progressive ideas not compelling enough to carry the nostalgic forms. But with the greatest artists, innovation and nostalgia fused to create works, chiefly in music and architecture, which remain the chief monuments of progressive creativity.

Both artists and audiences struggled with great handicaps. Three centuries of Anglo-Saxon Protestantism had left a legacy of suspicion about all art, as if art might insensibly lure people's minds away from the pure doctrines about God. Art had a place within orthodox Protestantism, but that place was rigidly prescribed: hymns praising God, paintings commemorating appropriate ancestors, and novels inculcating virtue in the behavior of teen-age girls were all acceptable. But art never could exist for its own sake, and artists were well advised to go out of their way to lead virtuous lives. Like preachers, artists had to be role models to the impressionable young. Their lives had to appear to be morality plays, and their creations had to be sermons in stone. Occasional precursors of modern aesthetic ideas, such as those of James Whistler or Henry James, arose to challenge such attitudes, but the public reception was so chilly that such artists often emigrated to Europe seeking a more sympathetic environment.

The ancient difficulties of assimilating the genuinely new further complicated the situation. The taste even of the most educated and sensitive people lagged a generation or two behind the products of the most creative people. The atmosphere of home, church, and school had been full of convention rather than originality. Almost by definition the most "cultivated" people cultivated art forms that were safely dead. The well-educated progressive grew up reading stories full of virtue in magazines like *Our Young Folks.* These led naturally into the key religious texts, such as the Bible and *Pilgrim's Progress.* The shelves in the living room often contained the novels of Jane Austen, William Thackeray, and Charles Dickens; the poetry of Lord Tennyson, Robert Browning, and Matthew Arnold; and the essays of Ralph Waldo Emerson, Thomas Carlyle, and John Ruskin. All these works taught moral lessons; all were linear in their development; all seemed replete with Protestant assumptions on every subject from sex to salvation. Above all, a progressive knew where he stood with such books, and rereading an old book, seeing a three-volume novel in paint, or looking at a Gothic church helped give a spurious sense of permanence to a life increasingly in flux. If progressive artists were to address their own times at all, they had to speak through these established forms, in effect conservatizing their art to transmit their message. The result, all too often, was a clichéd novel about political corruption or a technologically modern building dressed out in Beaux-Arts classicism. A look at urban planning provides a concrete example of what musicians and painters also faced.

When Cleveland Mayor Tom Johnson wanted help in planning a new city, he turned to Daniel Burnham, the much-honored impresario of the Chicago Columbian Exhibition of 1893. Quite aside from Burnham's experience, the choice had much to be said for it, as if one progressive creator had turned to another for technical guidance. Burnham had done much to work out a plan for the City Beautiful in Washington, and like most urban reformers, Tom Johnson wanted to remake his city so that it would remake the citizens who grew up in it. Burnham was always sensitive to the kind of organic considerations that received emphasis in the works of John Ruskin and William Morris, and he took full advantage of any possibilities for remaking waterfronts, building parks, and otherwise adapting to any inherent possibilities for the beautiful in the landscape. He also made a good, progressive effort to dispose of the Skid Row area and to focus the spiritual city on a new civic center. Yet despite these aims and the subsequent approval of Mayor Johnson, Burnham was himself no more progressive than was his plan.

Born in 1846, Burnham was, for one thing, a bit too old. But his family was comfortably Republican and middle class, largely Anglo-Saxon in heritage and Swedenborgian in religion, and thus compatible with the heritages of most other progressives. As an artist, Burnham was not original enough to be a genuine innovator such as Frank Lloyd Wright or Charles Ives. He was a successful entrepreneur with enormous ability, capable of recognizing true talent when he saw it, but he never could free himself from his respect for academic precedent, for the taste of the wealthy men who normally hired architects, and for the one great urban model that captured his imagination: the remaking of Paris under Baron Georges Haussmann in the 1860s. For him, the Place de la Concorde was the way a city should look, and the Beaux-Arts classicism taught in Paris represented the way he thought great buildings should look. Progressive politicians, such as Tom Johnson or Burnham's friend Theodore Roosevelt, saw no incompatibilities between this historical revivalism and their own social goals, and, of course, they were largely correct. Surface compatibilities seemed to be all that mattered, and Burnham went on to his well-known efforts to reorganize San Francisco, the Philippines, and Chicago. The true progressive leader in the field of architecture and city planning, the young Wright, kept his distance from "Uncle Dan," eschewing Beaux-Arts classicism in most of his mature work. Instead, Wright developed an organic, progressive aesthetic that won little public approval until long after Burnham, Roosevelt, and Johnson were dead. When it came to the arts, progressives could not recognize their own.[1]

Thus, progressive creativity in the arts presented several paradoxes. The taste of progressives in nonartistic fields singled out work that appeared appropriately reformist and dressed it in styles that were comfortably old. Conservative artists who felt sympathy with reform goals may well have thought of themselves as progressives in the arts, but they were not creating art that was genuinely progressive. As in most ages, those artists who were both progressive in their instincts and in their creative impulses produced work that was too new for immediate acceptance. Daniel Burnham was acceptable to Theodore Roosevelt because he tried to create an environment that was beautiful and organic even when it properly belonged to an earlier day. Frank Lloyd Wright was not acceptable because he seemed too unprecedented, too willing to take chances to make the necessary adjustments to business and political reality. Somewhat paradoxically, when Wright's work was complete, anyone who examined it could see that he, too, looked backward and forward. Like the work of Burnham, his innovations had much that was nostalgic about them. Understanding them was the task of the next generation, and so like most truly original artists, he had to wait before public taste caught up with his work.

II

The most visible of the arts during the Progressive Era was prose writing, an area where their genuine achievement was small. Indeed, the whole literary generation that came of age around the turn of the twentieth century had a doomed air about it. The 1880s had been one of the great harvest periods of American prose, as a generation of realists had produced their major works. Within a few years of 1885, Henry James published *The Bostonians* and *The Princess Casamassima;* Mark Twain, *The Adventures of Huckleberry Finn;* and William Dean Howells, *The Rise of Silas Lapham* and *Indian Summer.* These were only the most memorable of many works that helped put American literature on the cultural map of the world. Yet during the 1890s, Twain lapsed into literary fragments, James turned to writing unsuccessful plays, and Howells devoted himself increasingly to social issues. A new generation, led by Stephen Crane, Frank Norris, and Harold Frederic seemed about to take over, yet all three died young.

The five most talented writers who emerged during the Progressive Era showed little sign that they were progressives. Willa Cather worked on *McClure's Magazine* for several years and was the ghostwriter of McClure's *My Autobiography,* but her work gave little evidence of progressive commitment; she was merely doing available journalistic jobs, and her true interests and abilities lay elsewhere. Ellen Glasgow tried her hand at a pair of naturalistic urban novels before shifting to explorations of the social history of Virginia. Although she occasionally dabbled with progressive themes in her early work, her true métier was ironic social comedy, which was alien to the earnest progressive mentality. Edith Wharton, the best of the lot, tried once in *The Fruit of the Tree* (1907) to write a progressive novel, but its many flaws only underscored the fact that she lacked the middle class, evangelical psychology necessary for a commitment to social reform. Her studies of the status of women had political implications for progressives if they wanted to perceive them, but few gave any indication that they did. Of the men, both Theodore Dreiser and Jack London lived in worlds that progressives knew only when on guided tours with Jane Addams. The naturalistic determinism of their work, with its stress on biological or chemical causes, could not possibly appeal to the average genteel progressive, as prim as he was about sex and as sure as he was about the unique value of the individual and the efficacy of an informed conscience.

The progressive novel as a genre has never received the study it deserves. Early in the period, populists produced volumes like *Caesar's Column* (1890), the somewhat absurd dystopian romance by Ignatius Donnelly, and *A Spoil of Office* (1892), one of Hamlin Garland's weaker attempts at an extended narration. Populists, as a rule, were not skilled in literary expression, and the only memorable works to emerge from them were a few of the single-tax short stories that Garland collected in *Main-Travelled Roads* (1893). Urban progressivism was more productive. Upton Sinclair's *The Jungle* (1906) was the best-known book of the genre, but like many writers of the period, Sinclair was a journalist, and any analysis of his major book more properly belongs to the study of muckraking journalism. In many ways Robert Herrick represented the opposite pole: a distant relative of Nathaniel Hawthorne and George Herbert Palmer, Herrick was a son of puritan New England and Harvard; he retained an air of respectability and even snobbery that made him a sharp contrast to the faintly raffish and socialistic Sinclair. Yet both men were doctrinaire moralists, convinced that the acceptance of their views would liberate a society headed down a dangerous path. They strove to present a factual view of real life in their key works, yet were attracted to naturalistic imagery, to circumstantial detail, and to characters moved by physical stimuli. At times each man pursued strange religious visions, and each was capable of degenerating into romantic sentimentality.

Considered on strictly aesthetic grounds, Herrick was the best of the progressive novelists, achieving almost the stature of Howells in his three most significant works. *The Common Lot* (1904) was a study of a social-climbing architect who abandoned his ethical standards only to see one of his buildings collapse, thereby causing many deaths and leading to his own spiritual rebirth. The theme of sin and redemption, subtly handled, made the book an important example of progressive realism. *The Memoirs of an American Citizen* (1905) was Herrick's one venture into the mind of the "enemy," a businessman-narrator who worked his way through life with convenient moral blinders, producing the sort of world that progressives wanted to change. *Together* (1908) caused a major scandal. Long, diffuse, and sometimes forced in its plotting, it analyzed the "new woman," marriage, and divorce in such a way as to imply that the author approved of her behavior—which he did. Less successful as art than as a document analyzing women's status, the book was an impressive record of progressive intentions and of the moral earnestness with which progressives explored sin.

Three other writers wrote on progressive themes. Brand Whitlock, reform journalist, lawyer, and mayor of Toledo for four terms, wrote light social comedies and a political novel of some merit before he produced *The Turn of the Balance* (1907), a study of criminal psychology and police brutality based in part on actual trials and genuine prison letters. His characters all too often were caricatures of poor laborers or vicious cops, and the plot got a bit out of control, but the book was the most passionate work in the genre. Winston Churchill, born into a New Hampshire family that could trace its ancestry to Plymouth Colony, turned rather facile notions of virtue and vice to good use in his journalism and early romances. After producing historical dramas about the revolution and the Civil War, however, he met Theodore Roosevelt, became a guest at the White House, and turned his attention to current politics. Elected to the New Hampshire legislature, he quickly wrote two novels based on his experiences there. *Coniston* (1906) and *Mr. Crewe's Career* (1908) dealt with the issues of political bossism and the influence of railroads on state politics. Roosevelt, Upton Sinclair, and Albert J. Beveridge all sent approving remarks. Churchill meanwhile became the progressive candidate for governor. He campaigned in favor of primaries, restrictions on lobbying activities and the issuing of railroad passes, and other related reform measures. Defeated for governor, Churchill then wrote *The Inside of the Cup* (1913), a controversial study of the social gospel in an Episcopal church.

David Graham Phillips completed the group. Having roomed at college with Albert Beveridge and studied urban corruption in Cincinnati and New York, Phillips concentrated on writing novels when he discovered that the public bought his early work in great quantities. Despite melodramatic plots, extraordinary coincidences, turgid language, and unconvincing characterizations, Phillips covered the whole field of progressive exposé: local politics in *The Conflict* (1911), state politics in *George Helm*

(1912), national politics in *The Plum Tree* (1905), the insurance scandals in *Light-Fingered Gentry* (1907), and marriage in *Old Wives for New* (1908). The least talented of the progressive novelists, he became notorious for his series of articles entitled "The Treason of the Senate" (1906), a muckraking survey of the ties between wealthy senators and big business. These articles, combined with work like *The Jungle,* so irritated President Roosevelt that he delivered his famous speech on muckraking, turning his back on those journalists who had been among his most enthusiastic supporters. Upset at the president's reaction, Phillips turned increasingly to novels on the theme of the "woman problem." When he was murdered in 1911, he left for posthumous publication his major effort in this area, *Susan Lenox: Her Fall and Rise* (1917). Overrated by a few critics but ignored by most, the book effectively ended the genre of the progressive novel. Brand Whitlock lived on to produce his best book, the neglected *J. Hardin & Son* (1923), and Upton Sinclair to produce millions of words on a variety of topics, but "progressivism" as a viable category had no relevance for the novel after 1917.[2]

III

The progressive poets tended to be younger or to mature later as artists. The great harvest period for progressive poetry, at least in terms of book publication, lasted only from 1912 until 1917. Of the three significant writers, Edgar Lee Masters was the only one born in the progressive generation; Carl Sandburg and Vachel Lindsay were of the second progressive generation, a situation that changed some of their formative experiences. Since Masters wrote decent poetry only in his maturity, and Lindsay and Sandburg were so much younger, all three men had models unavailable to most other progressives. As a model, Lincoln was there for everyone, and all three men could use William Jennings Bryan as well—especially the "Boy Bryan" of the 1896 convention—as well as Governor John Peter Altgeld of Illinois, the German immigrant reform leader brave enough to free the Haymarket anarchists. A nonpolitical leader such as Salvation Army General William Booth could also serve as a role model. These circumstances, though, made the world of the progressive poem "past" in a way that the progressive novel usually was not. Where the novelists looked forward to reform, the poets as often looked backward to great reformers and were vague about precisely what a reader could do now. The narrow prejudices of the small town that Masters exposed were as much in the past in the Chicago of his present as the campaign of 1896 was for Lindsay when he wrote about it two decades later. All three progressive poets knew the city, especially Chicago, even if they seemed often to dwell pyschologically in small towns, like Springfield, Illinois.

Thus, the tension between new reform content and older sentimental and realistic forms, so evident in progressive novels, was even more complex in the poetry, enhancing its value for posterity. Much of the progressivism in the poetry was as nostalgic as the evocation of the small-town morality. What was truly progressive in the poetry was not the politics or the new sociological views that appeared in the novels, but the new use of the language, the new rhythms and modes of presentation, and the new subject matter. These innovations, at their best, made progressive poetry akin to, as well as a pathbreaker for, modernism.

Progressive poets operated in an English-language world that was in rapid flux. Since Thomas Hardy's early verse, English poets had begun to dispose of poetic diction and traditional meter and to change what was permissible subject matter for poetry. Friends were comrades, steeds charged, and the brave were gallant in high-Victorian diction; Hardy was the first significant British poet to employ a new range of plain speech and to deal with the depressing subject matter that later became the hallmark of a modernist poem like T. S. Eliot's *The Waste Land.* What Hardy began, the Great War completed. All the old words, like all the old moral verities, died in the savage irony that suffused the language during the war.

Speaking broadly, England lacked a progressive phase in its arts. Even in politics, the picture was not validly parallel to the progressive scene in America. Having no dominant evangelical tradition, Britain channeled what moral reform currents it experienced into peripheral institutions like Toynbee Hall. Its social reforms sometimes had a superficial resemblance to progressive crusades, as with women's suffrage or reform of the upper house of Parliament, but the British effort as a whole more resembled the New Deal than the New Freedom. British poetry was thus free to pass directly into the modernism of T. S. Eliot, Ezra Pound, and William Butler Yeats. America, however, always had to contend with its more dominantly Protestant past; it also never experienced the Great War the way Europe did. But what America did have was the precedent of Walt Whitman. Almost alone Whitman altered the diction of poets and the meters of poetry during the 1850s. He brought the city and democracy into verse, and with his writings about the Civil War, he moved unpleasant realities to the center of significant verse. His evocation of Lincoln's death was for progressives a supreme statement of what their lives encompassed.

MINISTERS OF REFORM

Progressive poetry was thus something of an American way station on the road to modernism—a modernism already available to a few readers who had discovered Edwin Arlington Robinson and Robert Frost, but one not yet widely recognized or popular.

The shift from Victorian to progressive and then to modernist forms of discourse appeared most obviously in the symbolism of nature and the role of the poet. To Victorians, nature was always central to a poetic vision of the world—usually it was a symbol of something divine that connoted optimism. By 1920, April was the cruelest month, and drought the outward and visible sign of an inward and spiritual inadequacy. To the Victorians the poet was a sage: Arnold, Tennyson, and Browning were guides for living, role models, the great preachers of the day. To the modernists, preaching was in bad taste, and morals were the object of mocking irony. The idea of the public as a congregation, indeed, the notion of a large audience of any kind, was gone. Poets preferred to speak mainly to each other.

Those Americans who were important in the earliest phases of modernism got little response from progressive America. Pound and Eliot moved to England at the first opportunity, and lesser artists like Hilda Doolittle spent considerable time there. At home the new voices were Masters, Sandburg, and Lindsay, who cultivated the tradition of Whitman and fused it to their progressive political views. Displaying some of the older Victorian tendencies in their handling of nature and their attempts at preaching morals, they also found much to criticize in American society. At their best they established new ways of doing this.[3]

All the progressive poets had characteristics or experiences that put them on the fringes of progressivism and yet kept them from being representative. Masters' mother had come from an old New England family and her father had been a Methodist minister, but Masters' father and paternal grandfather were Democrats hostile both to Calvinism and to Abraham Lincoln. Masters could long remember William Herndon, a friend of his father and the law partner of Lincoln, and out of this complex of conflicting experiences, Masters emerged with an obsessive concern for Lincoln as man and myth that ultimately produced one of the least friendly biographies in the literature. All his life, Masters felt divided about his heritage. On the one hand, he read his Bible "through and through"; on the other, he was always skeptical of it as revelation. He loved his prohibitionist mother yet admired his tolerant, moderately drinking father. Never happy with either his education or his career, Masters practiced law but was always preoccupied with poetry, ever eager to abandon bill collection for Greek verse. He attended the 1896 Democratic convention and was devoted ever after to Altgeld and Bryan. He remained on the periphery of Chicago reform activity, but his energies went into his writing. Friendly with Theodore Roosevelt as well as with Sandburg and Lindsay, he combined his poetry and his politics in his admiration for Walt Whitman: "What had enthralled me with Whitman," he wrote, was "his conception of America as the field of a new art and music in which the people would be celebrated instead of kings" and where "the liberty of Jefferson should be sung until it permeated the entire popular heart." He looked forward "through Whitman to a republic in which equality and fraternity should bind all hearts with a culture of that profound nature which enabled an Athenian audience to sit in the Theater of Bacchus and follow with appreciative delight a tragedy by Sophocles."[4]

If Masters' Copperhead animosity toward Lincoln and the Republican Party set him apart from other progressives, Carl Sandburg's family circumstances separated him in a different way. The family was devoutly Lutheran and Republican, which fit into the progressive ethos, but it was also Swedish immigrant and working class, which did not. Sandburg learned from his father that "Republicans are good men and Democrats are either bad men, or good men gone wrong, or sort of dumb," yet by the time he reached adulthood, he was a socialist and an active Social Democrat in Milwaukee. He met his wife Lillian Steichen, sister of the photographer Edward Steichen, at party headquarters. For two years, Sandburg served as secretary to Milwaukee Mayor Emil Seidel. A journalist by profession, he soon moved to Chicago, where he published poems in *Poetry* and *The Little Review* even while continuing his reform efforts. Sandburg's early work, frequently on the subject of Chicago, won the support of early modernist critics like Amy Lowell, and his children's stories won the praise of Frank Lloyd Wright. Sandburg never forgot the memorial constructed in his hometown commemorating one of the Lincoln-Douglas debates. He had walked by it daily for years. He admired Bryan and read Coin Harvey and thought about socialism. He also evolved his concept of Lincoln as a democractic folk hero, an American for all seasons. This vision dominated his life, and the six-volume biography that he wrote in the two decades after the death of political progressivism was both the greatest contribution to the popular conception of Lincoln and the greatest contribution of the progressive poets to American literature.[5]

Carl Sandburg became known for his method of reading his poems to the accompaniment of a guitar, but the true performer, as well as the best poet, among the progressives was Vachel Lindsay. With his gospel rhythms, his whoops, his drums, and his hobo persona, Lindsay was at times more a presence than a poet. English critics like William Butler Yeats praised his work so vigorously in large part because Lindsay represented so well Yeats' conceptions of American originality in the arts. Despite the snorts of Ezra Pound, Lindsay brought the tradition of Whitman into the twentieth century, and all but shouted Victorian verse out of its comfortable place in middlebrow minds and journals.

Coming from a devoutly fundamentalist home, Lindsay never doubted the truths of the Disciples of Christ, yet paradoxically, he was ecumenical and syncretistic as an adult. He seemed to value Buddha as much as Jesus, and was always eager to learn of new forms of religious experience. In a sense he was almost too Christian to be a good progressive, since he displayed an uncharacteristic pessimism about progress and human nature. Although he rarely appears in conventional accounts of larger political and social-reform movements, he corresponded with progressives like Upton Sinclair and Brand Whitlock, adapted ideas from reformers as different as "Golden Rule" Jones and William Allen White, and used the ultimate expression of progressive city planning, Frank Lloyd Wright's Broadacre City, as the model for his own utopian community. Long a friend and admirer of Jane Addams, he also had so great a reputation by 1915 that Secretary of the Interior Franklin K. Lane introduced him to members of President Woodrow Wilson's cabinet. That audience, which included his hero, Secretary of State Bryan, heard among other works a poem celebrating the opening of the Panama Canal. Despite Wilson's support of World War I, about which Lindsay had grave misgivings, Lindsay was a progressive to the end, devoutly supporting the attempts to get America to join in the League of Nations. For him religion and politics were the same.

Although his father was a Southern Democrat, Lindsay identified with his mother, growing up as he did in Springfield, Illinois, in the heart of Lincoln country. As a child, he knew by heart such Union songs as "The Battle Hymn of the Republic" and "John Brown's Body." Their rhythms and ideas permeated his mature poems. He needed only to look out the window to see Governor Altgeld walking slowly to his office every morning, and he insensibly conflated Lincoln and Altgeld into one mythological figure who gave depth to his political analyses. Even more than most progressives, however, Lindsay had trouble finding a career. Indeed, until he became a professional reader of his own verse, he never did find one, and he supported himself on family money and handouts. He attended, but never graduated from, Hiram College, and then went to the Chicago Art Institute to fulfill his ambition to become a Christian cartoonist. He then went to New York, where he worked with William Merritt Chase and Robert Henri. Henri, in particular, became a friend and counselor, and was a key influence in persuading Lindsay that his talent was greater with words than with pictures. Nevertheless, his friendships with fellow students like Rockwell Kent, George Luks, and George Bellows helped establish points of connection between progressive art and progressive poetry, and Lindsay's regular performances for four local New York City settlements demonstrated his concern for the larger community of artistic interest within the reform movement.

Lindsay first achieved limited public attention in 1911, when his elegiac memorial to Governor Altgeld, "The Eagle That Is Forgotten," was printed in an Illinois paper. It attracted approval throughout the Midwest. Other poems of the period, like "Why I Voted the Socialist Ticket" and the prohibitionist "King Arthur's Men Have Come Again" were less popular, but nevertheless equally obvious in their progressive leanings. Lindsay finally achieved critical recognition when Harriet Monroe accepted "General William Booth Enters into Heaven" for the January 1913 issue of *Poetry*. "From Lincoln's own country a poet of Lincoln's own breed," she commented. The compulsively rhythmic composition had markings indicating that it should be sung to the tune of "The Blood of the Lamb," with instruments including a "bass drum beaten slowly," banjos, "sweet flute music," and tambourines. Hull-House itself could hardly have produced such a motley crew of convicts, lepers, drug addicts, and saints as those catalogued in Lindsay's verse, as they followed their indomitable leader toward his meeting with the Lord. Progressives rarely could enter into the spirit of the ghetto the way a true hobo like Lindsay did, and in no other work of art did progressive artists so successfully produce a religious vision of a classless society, where everyone stood equal at the gates of heaven.

Two other poems completed the case for Lindsay's being the greatest progressive poet. "Abraham Lincoln Walks at Midnight" depicted Lincoln as a Christ figure, pacing near the Springfield, Illinois, courthouse in distress at the carnage of the European war. Just as Lincoln suffered over the Civil War, so he suffered again over the miseries of Europe, unable to sleep until peace was restored and the world had some kind of workers' league of nations to prevent such things from happening again. One of a series of six poems that Lindsay wrote opposing war, "Lincoln" underscored his allegiance to the kind of pacifism usually associated with Jane Addams. The end of the war allowed Lindsay one final burst of creativity before he lost his ability and began to fritter away his life on other subjects. In the summer of 1919, he recalled the political campaign of 1896, so important to Midwestern progressives, and chanted of "Bryan, Bryan, Bryan, Candidate for president who sketched a silver Zion." The issues were clear between silver and gold, good and bad, Democratic reform and Republican reaction, and the sixteen-year-old Lindsay could never forget. Altgeld was still in office when out of Nebraska came the "Boy Bryan," "In a coat like a deacon, in a black Stetson hat/ He scourged the elephant plutocrats/ With barbed wire from the Platte." Drunk on words, Lindsay had the desert producing even a "prodactyl" as well as an eagle to greet the new leader. "Prairie avenger, mountain lion," Bryan was "Smashing Plymouth Rock with his boulders from the West." Standing there with his best girl, Lindsay heard Bryan make his most famous speech: *"You shall not crucify mankind/ Upon a cross of gold."* But it failed. Mark Hanna bought the election and William McKinley sat in the White House. It all seemed so long ago. "Where is McKinley, Mark Hanna's McKinley,/ His slave, his echo, his suit of clothes?" Hanna and McKinley, Ben Tillman, and Pierpont Morgan, even "Roosevelt, the young dude cowboy,/ Who hated Bryan, then aped his way," all were gone. Even Bryan was gone, at least the Boy Bryan of 1896:

> Where is that boy, that Heaven-born Bryan,
> That Homer Bryan, who sang from the West?
> Gone to join the shadows with Altgeld the Eagle,
> Where the kings and the slaves and the troubadours rest.

By late 1919 the poem was all too appropriate, a dirge for a lost progressivism, penned and ready for slaughter in the White House and the Senate. Neither Lindsay nor the country would ever find it again.[6]

IV

Like the progressive writers, the progressive painters often seemed radical because of their subject matter, their various acts of rebellion against institutional conservatism, and their outspoken allegiance to political ideas that were outside the mainstream. The dislike that most creative figures felt for the clichés of their calling led to a revolt against notions of "correct drawing," "composition," and "technique" as taught by such hallowed schools as the National Academy of Design in New York and its spiritual ancestors, the Royal Academy in London and the Ecole des Beaux-Arts in Paris. "Correct" institutional art critics accepted any number of older styles. The establishment insisted that works of art depict scenes of relevance to the moral life of the contemporary community. Paintings should either represent a significant historical event or tell a moral story in realistic forms that most sensible people could grasp easily. Progressive painters rebelled against the conventions of the day in several ways. They accepted the realistic shapes of conventional work, and often tried to paint stories and morals of their own, but they also experimented with dark palettes and new theories about color and introduced as subjects lower classes of people and contemporary urban themes into their work. Finding their inspiration in artists such as Daumier and Doré, and especially in Rembrandt, they gloried in depicting the common, ordinary democratic life of their time as much as the progressive writers had. Vague in their political allegiance, they established their credentials as progressives more by rediscovering indigenous American topics than by campaigning for reform candidates. They invoked the ideals of Emerson and Whitman, and in the process, continued the pioneering realism of Thomas Eakins and Winslow Homer.

Progressive painting actually began in the world of Philadelphia journalism in the early 1890s. During that decade the city was a key center for periodical publication and the art work that still accompanied many articles: the illustrations, etchings, and engravings that continued to have a market until the development of more modern photographic techniques around the turn of the century. Artists frequently raced to the scene of important stories to capture the details of an accident or a profile in a quick sketch that they could turn into an acceptable periodical illustration. This training in visual memory and facile skills of illustration was exceptionally useful and did much to form the artistic assumptions of the men involved. They worked repeatedly on subjects that were relevant, contemporary, democratic, and representational, saying in pictures what a writer was saying in words in the next column. Newspapers also encouraged cartoon art, which combined the traditions of Western anecdote and the tall tale with work done by artists with European precedents: the scathing political cartoons of Thomas Nast, as well as the decorative illustrational work of Walter Crane, Kate Greenaway, and William Morris.

A group of five friends became the core of the progressive achievement in art. Hardly uniform in personality or style, they nevertheless shaped a creative environment of democratic rebellion that attracted other artists and gained them considerable publicity for their work, as well as the uncomprehending hostility of the establishment. Edward Davis, art director of the *Philadelphia Inquirer* and father of painter Stuart Davis, was one of several employers who helped to produce an environment in which good art could also yield a decent income. Probably the most admired man in the group was William Glackens, whose memory and facility with quick sketches were the envy of even those artists who later became superior painters. George Luks, the bibulous *bon-vivant,* and young Everett Shinn,

who half a century later recalled the period most vividly, were key participants. But the two figures who came to have the most influence on American art were Robert Henri, the elder statesman of the five, and John Sloan,
the most gifted painter to emerge from the group.

Henri's childhood was rough even by American standards. Despite a
background that included church attendance in many Protestant sects and
considerable interest in Swedenborgianism, Henri was the son of a professional gambler and real-estate speculator. His father had taken his family
over much of the West, only to find himself involved in murder charges
which resulted not only in the precipitous flight of the entire family, but
also in the adoption of assumed names. Despite his unsettled upbringing,
Henri nevertheless managed to attend the Academy of the Fine Arts in
Philadelphia and to spend much of his early adulthood in Europe. In
Philadelphia Henri enrolled at the academy a bare eight months after
Thomas Eakins had been expelled for having too cavalier an attitude
toward the nude in his classes, but he nevertheless attended a school in
which Eakins' influence remained great. Particularly through Thomas Anshutz, the Eakins tradition of scientific study of anatomy, perspective, and
photography helped instruct Henri in techniques that always remained
with him. In his mid-twenties, Henri enrolled in the Académie Julian in
Paris, where he worked until he qualified for the more rigorous program
of the fabled Ecole des Beaux-Arts, the same school in which Eakins had
acquired his chief principles of instruction. Henri emerged with two pre-
cepts: that painting should always deal with real people and real events,
and that an artist had to impose on his subjects his own moral or religious
point of view. Great art was the product of this interplay between the real
and the ideal. Henri returned to a Philadelphia almost barren of genuine
interest in contemporary art to become the teacher, as well as the friend,
of the younger progressives. John Sloan later referred to him as the "catalyst" and "emancipator" of American art.

Henri became one of the most respected teachers in the history of
American painting. Many of his pedagogical ideas came as much from
Emerson, Whitman, and Ruskin as from his own art teachers. He brought
them together in a volume, *The Art Spirit,* compiled by one of his students
and published in 1923, shortly before his death. Above all the book
stressed that life and art could not be separated, and that no artist could
produce anything beautiful that was dissociated from human feeling. The
strongest motive for an artist was life, and even in a still-life, he wrote that
"you will find the appearance of interweaving human forms, the forms we
unconsciously look for." All artists used these human forms to communicate with each other, "teaching the world the idea of life." Such art always

had "a subtle, unconscious, refining influence" on the community; even a landscape that seemed barren of ideas must "express some mood of nature as felt by the artist." Henri insisted that the modernist idea of art for art's sake held no interest for him; he was "interested in life." Like a good Emersonian, he saw all the laws of the universe in the specific object before him, and to his French-trained eye, no object was more significant than the nude: "There is nothing in all the world more beautiful or significant of the laws of the universe than the nude human body." Anyone who objected on religious and moral grounds made not only a moral error but an aesthetic one as well. Significance for Henri and his companions was never in the object painted, because the object was never the subject; the subject was the intention of the artist, and the nude or the scene was merely the vehicle of expression. Thus, an ugly tramp could become the subject of a beautiful work of art, for "the beauty of a work of art is in the work itself." Henri dwelt on the implicit politics of his aesthetic views less than most socially aware artists, but when he did, he sounded like John Dewey in a smock. America could achieve greatness in art only "by the art spirit entering into the very life of the people." Art should "make every life productive of light—a spiritual influence. It is to enter government and the whole material existence as the essential influence, and it alone will keep government straight, end wars and strife; do away with material greed."[7]

Henri spent roughly seven years in Europe spread out over a somewhat longer period that also saw him in Philadelphia for lengthy stays. Glackens and Luks both managed to get to Cuba to report the Spanish-American War pictorially, and Glackens was in Paris with Henri for a year and one-half. John Sloan, by contrast, never left America and seemed somewhat afraid both of European women and European artistic ideas. Six years younger than Henri, Sloan grew up in a poor, religious family. His father, an artistically talented failure, spent his time repairing furniture, decorating china, and painting without much success. As a boy, Sloan discovered the work of Hogarth, Walter Crane, and Tenniel and as a clerk in a bookstore, he spent hours admiring the Doré illustrations in the books he was supposed to be selling. With no interest in college and no money to pay for it, Sloan began working for the *Inquirer,* and for a long time, thought of himself as a newspaperman who only painted on the side. Henri introduced him to the lithographs of Daumier, the drawings of Forain, and the aquatints of Goya, and with his encouragement, Sloan spent more and more of his time painting. By the turn of the twentieth century, he was probably the most significant newspaper illustrator in the country. With the sophistication of half-tone photography, however, the newspaper market evaporated and Sloan followed his friends to New York. By 1904 the Philadelphia Five were located chiefly in that city, which became the center of magazine illustration in the country. Sloan's interest in painting inevitably grew out of his assignments to illustrate the stories and journalism then appearing: the realistic short stories of writers like Brand Whitlock, appearing in journals like the socially conscious *McClure's*. But progressive writing and painting did not "cause" or "influence" each other; instead, they came out of similar environments and were responses to similar currents in American life.

The progressive painters also found themselves in the midst of a long tradition of rebellion. New York galleries had enjoyed a reputation as the most depressing places in the city. They "vied in coldness and aloofness with the museums," according to Guy Pène du Bois. "They were hung in horrible red velvet and a pall of stuffy silence," and if one dared enter "one was invariably attended" by "an excessively well-mannered gentleman in afternoon clothes who seemed incapable of looking straight at anything without looking down his nose." Art in such environments was intended solely for "the captivation of tycoons, and everyone else was an intruder." As for the National Academy, its standards reflected smug mediocrity; Sloan snorted that its standards of teaching were such that "you checked your brain at the door." Against this unresponsive environment, many artists had been brandishing new ideas for a generation before the five arrived. William M. Chase began the rebellion in 1877, and there was a revolt of "the 10" which Childe Hassam and J. Alden Weir had led as recently as 1898. Henri and his friends began their own movement in 1901 with a show at the Allen Galleries. Other shows followed. The artists in each show changed slightly, but essentially the same group staged a show at the Colonial Club in 1903 and the National Arts Club in 1904. In 1907 a particularly obtuse National Academy jury refused to accept the selections supported by Henri for the annual exhibition, and with Sloan in the vanguard, the group determined to stage their own version of the famous French *Salon des Refusés.*

The common denominator of the resulting exhibition was not, as casual observers have assumed, a devotion to democratic art depicting urchins playing in jolly fashion around their neighborhood ashcans. Certainly Henri, Sloan, and Luks created such pictures on occasion, and Glackens had done so until his passion for Renoir and color permanently altered his artistic interests. But Everett Shinn was a member of the group largely because of past friendships, and the three new collaborators—Arthur B. Davies, Ernest Lawson, and Maurice Prendergast—joined the rebellion only because they opposed the National Academy and not because they ever had a democratic thought or peered aesthetically at their local trash.

Rather, the common denominator was rebellion against institutional and accepted forms of art. Even when applied to Henri and Sloan, later terms like "the Ashcan School" were inaccurate for their creative achievement and cheapened the aesthetic values these men shared. A far better way to remember the rebels is to recall the official title of the participants in the 1910 show, the Society of Independent Artists—which included seven of the Eight of 1908, plus a number of others. These artists were independents. Many were friends, and a few friends shared aesthetic goals, but in no sense were the progressives of the group typical of the rebels, nor was the act of rebellion synonymous with a proletarian view of art.[8]

At their most progressive, the rebels tied their ideas of social equality and the worth of the common people to pictures of slums, chambermaids, and boxers performing in bars. James Huneker, a sympathetic if erratic art critic, used the term "darkest Henri" to describe the dim palette that was common. Sloan reveled in those palettes for years and claimed to have distinguished twelve distinct tones of grey in the New York atmosphere. Before 1895 Henri had been something of an Impressionist in his approach to art, and he was quite capable of painting colorful landscapes and cheerful portraits of pretty women in ways not easily distinguishable from the work of John Singer Sargent and James Whistler. But social consciousness did not limit their art to depressing urban scenes, nor were all realistic urban painters part of these rebellious exhibitions. Sloan was the most consistent user of a limited palette, but even he changed radically after 1913, and much of his later work bore little resemblance to the work of his progressive years. Glackens especially liked to take sketching trips through Nova Scotia and Newfoundland, and after his palette brightened again, he became far better known as a colorist and painter of still lifes and nudes than as a painter of ashcans. Some artists who were genuine progressives found themselves ignored for some of the important exhibitions of progressive art. Jerome Myers, for example, was a friend of Sloan's whose work was far more progressive by any sensible definition than that of Prendergast or Davies, and he was also a rebel against the domination of the Academy. Yet he found himself ignored, apparently for no better reason than that Henri found his work too sentimental.[9]

George Wesley Bellows was the key progressive artist left out of the independents' exhibitions before 1910. Born in 1882, he was too young to be in the first progressive generation; he had not yet matured enough as an artist. In many ways, though, his career illuminates the progressive experience both in art and in the larger world of politics. Growing up in a home "surrounded by Methodists and Republicans," he absorbed all the political, religious, and Civil War attitudes of Columbus, Ohio. His earliest ambitions were in athletics, and his conservative family firmly disapproved of any career not devoted to religion or American capitalism. After withdrawing from Ohio State, Bellows went to New York to study with William Merritt Chase at the New York School of Art. There he found Robert Henri the most perceptive teacher, and under Henri's tutelage, he explored the urban scene for subjects to paint and read writers like Emerson and Whitman for appropriate ideas. The school was a center of creativity: other students included Edward Hopper, Rockwell Kent, and Glenn O. Coleman. For a brief period John Sloan took Henri's place, and through Sloan, Bellows met the other members of the Philadelphia five. When Henri established his own school early in 1909, Bellows lived in the same building—as did a young man named Eugene O'Neill, whose successes with women were at the time more impressive to Bellows than his talent in the theater. Bellows quickly absorbed the subjects and style of the older progressives, his athletic interests producing some of their most impressive works. By 1910 Bellows was so much a part of progressive painting that he not only showed in the exhibition that year, he soon found himself treated as if he had been one of the charter members of the five.

Bellows is also politically significant. Despite his middle-class background and his respectable marriage into Upper Montclair, New Jersey, society, he was simultaneously a sexual prude, a progressive artist, a socialist voter, and an anarchist by emotional attachment to Emma Goldman. Once again, with Bellows, as with Henri and Sloan, conventional political differences between progressivism, anarchism, and socialism became meaningless in practice. Bellows shared with Sloan an affection for socialism, and with Henri an admiration for Goldman and her brand of anarchism. On at least one occasion in 1916, the three men and a number of friends demonstrated in Goldman's favor. They gave a dinner in her honor at the Brevoort the night before her trial, and when she was released, they escorted her in triumph to a celebration at Carnegie Hall. Both Henri and Bellows taught at Goldman's Ferrer School and had no difficulty reconciling their own more conventional lives and ideas with this unconventional, unprogressive figure. One can get an indication of the depth of this commitment from the tale of one visitor to the Bellows household. Leon Kroll found daughter Anne serving tea for four. When asked who the three invisible guests were, Anne replied in all seriousness: "God and Rembrandt and Emma Goldman," the three names most often mentioned in that home.[10]

The political combinations and recombinations among the progressive painters make conventional ideological analysis a study in contortion.

With the exception of Henri, whose family were Southern Democrats, the Philadelphia five were Republicans, so much so that the slang term in the group for cockroaches was "Democrats." No one in the group seemed to have much interest in politics until the Theodore Roosevelt years, and even then, the chief evidence of political activity was Everett Shinn's quaint scheme for a Theodore III Club that would stimulate his puzzle business by using the 1908 election and the possibility that Roosevelt might run for a third term. But by then the group was slowly shifting toward the Democrats. Sloan voted for Bryan without enthusiasm and claimed not to belong to any party, although he had a great aversion to William Howard Taft and thought the Republican Party had become an unappealing spectacle. Despite his admiration for Emma Goldman, Bellows remained similar in his views to Woodrow Wilson through all the difficult years of Wilson's presidency. Some of his violently antiwar cartoons actually appeared after both he and Wilson had shifted suddenly to a belligerent stance. Sloan's diary provides the most reliable chronicle of his own shifting views. By December 1908 he was clearly thinking about socialism. A Christmas visit from Thomas Anshutz cheered him up when he learned that his old teacher also intended to vote Socialist in the future. Piet Vlag of the Rand School was interested in Sloan's work, and the relationship between the two men continued intermittently for both artistic and political reasons. Sloan refused to draw specifically for Socialist Party goals even as he freely offered his political support, although he occasionally sent pseudonymous contributions to *The Call.* He read the party platform on which Eugene Debs had run in 1908 and could not "understand why the workers of the country were so disinterested or intimidated as not to vote en masse for these principles." By Christmas 1909 he was writing about "that great Socialist Jesus Christ. He was a Revolutionist." and castigating both his "hyper religious sisters" and institutional Christianity for polluting the Christian message. Four days later he and his wife joined the Socialist Party.[11]

As Sloan's diary indicates, the creativity of the New York progressive artists arose as much out of personal contacts as from any set ideology or agreement on a specific reform platform. Just as it was possible for George Bellows' daughter blithely to serve tea to God, Rembrandt, and Goldman, likewise a magazine like *The Masses* could have any number of different viewpoints that clashed from page to page and could appeal to a wildly varying audience. Explicitly revolutionary in tone, the magazine could nevertheless win financial backing from a wealthy progressive like Amos Pinchot and approval from so typical a progressive as Frederic C. Howe. Known for its radical socialist bias and its support of pacifism in the early

MINISTERS OF REFORM

years of World War I, *The Masses* was, in fact, far more important as a vehicle for American artistic creativity than as a vehicle for ideological dissent. The work of artists like John Sloan and cartoonist Art Young won it more notice than the editorials of Max Eastman or the book reviews of Floyd Dell. Indeed, the artists and writers disagreed about any number of issues, which in time led to the exodus of several important artists, including Sloan. Only personal friendships could have brought together such a disparate group, and only New York had an environment that encouraged such diversity. The journal was one of those rare anomalies in history that transcended its period and often encompassed opposing trends. In Sloan and Young, it had progressive artists of the first generation; in George Bellows and Boardman Robinson, it found progressive artists of the second generation; in Stuart Davis, it had an artist who became one of the great American modernists; in Robert Minor, it had probably the most important artist of American collectivism of the 1930s, a Communist Party stalwart who was hostile to most of the values of Sloan or Davis. Only the course of time made some of these distinctions evident, but the coexistence of diverse attitudes in one journal even for a brief season indicates the fluctuating nature of American creativity between 1912 and 1917, and perhaps inadvertently provides a reason why more dogmatic and ideologically uniform publications were so much less important to history.

The friendship between Piet Vlag and John Sloan was the primary relationship around which the magazine formed. It was an attempt to reach a wider public than *The Call* could reach. Politics and political goals were always important, but artistic radicalism was central from the first. Art Young was involved at an early stage, and he commented that the editors wanted "a magazine which would have the bold tone and high quality" of *Simplicissimus, Jugend,* Steinlen's *Gil Blas,* and *L'Assiette au Beurre,* "all of which were inspiring to the world's rising young artists." As much the product of midwestern Protestantism and Republicanism as any of the other progressives, Young had also been a student with Henri at the Académie Julian in Paris. Having once parodied the populists and criticized Governor Altgeld for pardoning the anarchists, Young had slowly come to admire Eugene Debs and to accept socialist doctrine. When Vlag left and his journal seemed about to fall apart, Young suggested the charismatic Max Eastman to be the next editor. Eastman's success encouraged them all. Young "felt as many a Crusader must have felt long ago as he set forth to rescue the Holy Land from the infidels." Young thus managed to combine a progressive background, European training, and socialist ideas with the choice of an editor who had never heard of the Philadelphia five or the New York eight—indeed, who freely admitted that he did not know anything about art at all—and whose hostility to modernism in art became legendary in critical circles. It could only have happened on *The Masses.* [12]

The progressive painters never shied from political radicalism in their personal lives. Henri taught at the Ferrer School out of his admiration for Goldman; his students even included Leon Trotsky during the brief months of his American exile. Bellows shared that admiration for Goldman and her school and was one of several painters willing to dine with I.W.W. leader Bill Haywood. He also felt great admiration for radical dancer Isadora Duncan, and was one of a number of young men whom Duncan asked to father her child. John and Dolly Sloan helped Socialist Party efforts to aid strikers during some of the more vicious labor encounters of the period. But in contrast to Art Young, who was a propagandist at heart, progressive artists as a rule disliked any explicit connection between their political allegiances and their art. Sloan insisted that the artist had no duty whatever to society and clearly stated that he was no cultural democrat, however democratic he might be in his social and political views. For Young and Henri, art reflected life and the relationship of people to nature. As Sloan once noted laconically: "When propaganda enters into my drawings, it's politics not art—art being merely an expression of what I think of what I see."[13]

However clear this distinction was in Sloan's mind, it was not always obvious with regard to his or anyone else's daily conduct on *The Masses.* Max Eastman and Floyd Dell often created subjects for illustration, usually of a political nature and designed to make a radical point. Sloan and the other artists produced works on these themes, but they disliked the fact that the literary editors added captions to their work. The artists could veto these captions, but apparently found printing drawings with no captions at all difficult. Sloan found this practice irritating, while younger artists like Stuart Davis and Glenn Coleman found it infuriating. Art Young, Boardman Robinson, and Robert Minor were among those inclined to go along with the literary editors, while Bellows sat on the fence, agreeing with Sloan but not wishing to aggravate the controversy.

Floyd Dell later spoke of the "smouldering grudge" that resulted between artists and literary people, although the true division seems to have been between those who believed primarily in using art as propaganda and those who believed primarily in using art as personal expression. According to Dell, the key spokesmen were both artists, "fat, genial Art Young" arguing for propaganda, and Sloan, "a very vigorous and combative personality" who was himself "hotly propagandist" in everything but his art, defending "the extreme artistic-freedom point of view." Young thus supported Eastman's policies, praised his work as editor, and insisted that the Sloan group "want to run pictures of ashcans and girls hitching up their skirts in Horatio Street—regardless of ideas—and without title." The rest of us "believe that such pictures belong better in exclusive magazines." As with so many other issues, progressives could disagree on this as well, each side certain of the righteousness of its position. Art as propaganda was in great vogue in this period. It was given legitimacy by the writings of Leo Tolstoy and the artistic precedents of Hogarth and Daumier, but in the long course of art history has only rarely obtained substantive and continuing support from the best critics.[14]

The issue that killed progressivism in painting as in literature was modernism: a far more creative climate was already revolutionizing the way Europeans understood the arts by stressing form over content, the lack of responsibility of the artist to society, the simultaneity of past and present time, and the expression of three dimensions in two. The isolation of America before World War I being as great as it was, the progressives hardly realized what was happening to them until it was over. Indeed, they found that they themselves had initiated their own displacement from the cutting edge of creativity in America. The Armory Show of 1913 separated innovation from nostaligia in American painting and in scarcely more than a month made the aesthetics of Protestantism seem irrelevant.

The first plans for the Armory Show were laid late in 1911. On the surface the show was to be yet another harvest of innovative American art such as the half dozen or so other shows in recent years that had publicized the work of independents and served to loosen the clammy grip of the Academy on popular taste and critical ideas. Several painters had formed still another society and had begun recruiting support for a show of American art with only a token representation of new French work. Calling themselves the Association of American Painters and Sculptors, the group included a motley variety of artists devoted to many styles. They attempted to name impressionist J. Alden Weir as president, but after due consideration, Weir decided that the tone and intention of the group was too hostile to the establishment, and he declined. Robert Henri received some support for the position, but he seemed too closely identified with progressive realism to serve. Someone less a part of a well-known dissenting group would be more appropriate. The nod soon went to Arthur B. Davies, a painter of mysterious and ethereal works whose style was unique and whose private life and friendships were largely a mystery. He was the most respected painter among the rebels, but mild in manner and well acquainted with rich art patrons whose support could be vital. The choice of Davies unintentionally helped to shift the focus of the show from

American to European innovation even though the organizers had intended no such thing.

The first board of trustees included Henri and Jerome Myers, and the various committees included Bellows, Luks, Glackens, and Sloan. But Davies worked with great skill to keep the realists out of power, and on the key committee that chose the best American art, he included only Glackens, by then long involved in his colorist, Renoir phase, and a realist only by virtue of his past styles and friendships. The important figures were Davies, Walt Kuhn, and Walter Pach, all of whom were enthusiastic about European modernism. They toured European exhibitions like the Cologne *Sonderbund* and came away full of admiration for Van Gogh, Cézanne, Munch, Picasso, and a spectrum of new artists of diverse styles. With Kuhn and Davies doing most of the traveling, and Paris-based Walter Pach providing contacts with modernist art dealers, the men accomplished an extraordinary coup in gathering an admirable body of works then unknown to all but a few Americans. Henri in particular was displeased by the results and soon dissociated himself from the show; all the realists were disturbed and none emerged untouched by the experience.

Henri's opposition has probably been overdone in certain of the accounts of the Armory Show. At least some of his opposition had more to do with the usurpation of his position as the leading teacher of the independents than with aesthetic theory. He could be quite sympathetic to the modernist art of Max Weber, for example, a former student of Henri Matisse who was as modern as one could get. But Henri was irretrievably mimetic rather than musical in his sense of form, and the absence of the natural and the human in the works of the Cubists and the Fauvists seemed almost as much a crime against art to him as it seemed to the conservatives. Both he and Sloan had been exhorting artists to paint people and nature from fresh memory; no matter how much they might talk about the role of ideas and forms in their work, their art always depended on what they could see and remember. Sloan died insisting that he did not "believe in art for art's sake," and that "a concern with the abstract beauty of forms, the objective quality of lines, planes, and colors, is not sufficient to create art." He exhorted his students: "Get your impulse for making a picture from some incident in nature." No modernist could accept such a statement.[15]

The Armory Show is a legend in art history; unlike many legends, it deserves to be. It received considerable publicity, much respectable opinion being favorable but with much of the yellow press ranging from the amused to the scurrilous. The National Academy was never heard from again. Their critical apologists ranted some, but slowly sank into postures of irritated grumbling against contemporary tendencies in art. When the show went to Chicago, its American content dwindled; by the time it arrived in Boston, it was entirely European. The change was symbolic. American independents, led by progressives, had initiated their own obsolescence, and they had the choice of adjusting or becoming prematurely old. Sloan and Bellows responded positively even if their own art seemed unable to thrive on the new impulses. Only Stuart Davis proved able to adapt and become a genuinely important modernist. Ironically, even Arthur Davies had trouble adapting to the new ideas, and he never recovered his position of leadership in the world of art. Maurice Prendergast, always his own man, serenely continued to paint as always, endorsed by the Armory Show but not ruined by it. The only group to emerge from the show with any kind of psychological security was that surrounding Alfred Stieglitz: Max Weber, John Marin, Alfred Maurer, and Marsden Hartley became the core of the modernist Americans, and they created an entirely different creative climate than the progressives. As Myers noted later: "Davies had unlocked the door to foreign art and thrown the key away. Our land of opportunity was thrown wide open to foreign art, unrestricted and triumphant; more than ever before, our great country had become a colony; more than ever before, we had become provincials."

Perhaps the country as a whole had an inkling of what Myers meant, but if so the result was cautious opposition and amused dismissal more than insight into provinciality. In addition to the bad doggerel and joking cartoons, the Show produced any number of examples of intelligent lay opinion. The last president ever to dare comment in detail on an exhibition of art spoke for progressives as well as any man possibly could. Theodore Roosevelt cautiously congratulated the Armory Show exhibitors for their enterprise, but he did not really approve. For him, the European modernists were "extremists" whose work implied "death and not life, and retrogression instead of development." Roosevelt thought Americans erred "in treating most of these pictures seriously." He found little to endorse in the Cubists, the Futurists, and the "Near-Impressionists." "The Cubists are entitled to the serious attention of all who find enjoyment in the colored pictures of the Sunday newspapers." As for the famous Marcel Duchamp painting, "Nude Descending a Staircase," which became the focus of the Armory Show, Roosevelt called it "A naked man going down stairs," and he claimed to have "a really good Navajo rug" in his bathroom "which, on any proper interpretation of the Cubist theory, is a far more satisfactory and decorative picture." As for the Futurists, he thought they could as easily be called "past-ists" because of their resemblance "to the later work of the paleolithic artists of the French and Spanish caves." He found the sculpture equally mystifying and disliked the praise that certain pieces received: "Why a deformed pelvis should be called 'sincere,' or a tibia of giraffe-like length 'precious,' is a question of pathological rather than artistic significance." He thought such works stood in sculpture "where nonsense rhymes stand in literature and the sketches of Aubrey Beardsley in pictorial art." Nature had always provided Roosevelt with his implicit standards. Like Henri and Sloan, he wanted healthy, patriotic Americans to retain their sanity by using traditional forms to suggest moderately unsettling new ideas. He could not tolerate any art that distorted man or nature, and he retained a suspicion that such art was somehow immoral.[16]

The innovative nostalgia implicit in progressive creativity thus never quite resulted in work of the first rank in literature or painting. Both audiences and artists were too aware of Victorian traditions of interpretation: nature was benign, evolving, and intelligible, people were basically decent and comprehensible, and art should contribute in some way to the solving of both personal and political problems. Modernism had its appeal. A progressive poet could play with the themes of urban life; a progressive painter could involve himself with left-wing social ideas. Artists, almost in spite of themselves, increasingly wrote or painted for each other. Perhaps the real problem was lack of genius. For all their virtues, progressive artists and writers proved unable to create works of permanent value. Yet they played an important historical role in changing people's tastes and in awakening them to the possibilities of a new aesthetic. For this, their art deserves some attention.

Chapter 5

Innovative Nostalgia in Music and Architecture

IN TWO REMAINING AREAS of the arts, music and architecture, progressivism expressed its influence more clearly than in writing and painting. The innovations in these fields proved so striking that they burst the older forms. By fertilizing innovation rather than smothering it, the nostalgia of the progressives led to works that not only paved the way for modernism, but were often of higher quality and more lasting value than anything produced a generation later.

The leading progressives in these fields, Charles Ives and Frank Lloyd Wright, fit the general picture of the progressive personality. Both grew up under the cultural influence of New England, and both displayed the influence of its Protestant traditions in their works. Both seemed incongruous in their chosen professions. Ives all too often appeared to be a cranky Yankee modernist, cut off from the rest of his culture. He had no audience in his own day and no impact on the next generation of composers. He emerged as a genetic mutation, without predecessors or progeny, only to be rediscovered after World War II by a country suddenly eager to find national greatness in its arts. Wright emerged as a populist of similar idiosyncrasy, somehow creating theories and structures that had little architectural precedent. He spawned much of architectural modernism, only to deny having done so with great passion. Both men seemed to delight in contributing to their own legends and in refusing to correct the public record.

Both men in fact lived easily in America. During most of their early creative periods, they did not feel alienated from their societies. Ives and Wright had definable roots in American culture and achieved distinguished places in that culture. Ives achieved his place mostly through his business career, to which he was sincerely devoted, while Wright achieved early success within architecture. Two factors that do not always seem relevant to the conventional histories of music and architecture are essential to understanding these careers. Ives and Wright claimed that they drew their chief resources less from their chosen fields than from American religious and philosophical literature. It is not always easy to establish the links between an influence in one field and creative achievement in another. Both overstated this claim, perhaps out of a need to seem original and "American" far beyond the actual evidence in their work. They were also characteristically progressive in that they innovated while thinking about the past. Unlike the modernists, who were intensely interested in pure works of art and intensely disapproving of social convention, they were deeply involved with family ideals, and persistently nostalgic. Their central aesthetic concern was with nature, and they remained ever conscious that nature was an expression of God. The art that resulted was always, in some way, programmatic: It had a meaning, and even a moral, in the eyes of its creators and that meaning was a realization of something in the past. Such a programmatic dependence on nature proved foreign to the modernists of the 1920s, with their concern for purity of form, sound, and color. The creative leaders of the 1920s in America commonly ignored the work of both Ives and Wright.

The best way to understand the work of Ives and Wright in its progressive cultural context is to concentrate on their memories. Their innovative nostalgia was the direct result of their upbringing. In Ives' case his recollections of early childhood, of his father, of the religious atmosphere of his youth, of Transcendentalism and its heroes, and of Yale College all provided the raw material of his vision. His career in life insurance and his Wilsonian political views helped relate him to the rest of the culture and gave an appropriate context for the consideration of his musical achievements.

In his *Memos,* written after his musical career had ended, Ives acknowledged his debt to his father, crediting George Ives with his early training and interest as well as much of the originality in his music. George's "personality, character, and open-mindedness, and his remarkable understanding of the ways of a boy's heart and mind," combined with "a remarkable talent for music and for the nature of music and sound, and also a philosophy of music that was unusual," provided Ives with a childhood that was extraordinarily stimulating musically. George Ives was himself a professional musician who started musical lessons for his son at the age of five; acquainted him with the best of classical music; insisted that he learn harmony, counterpoint, and musical history; and, above all, in Ives' own words, "kept my interest and encouraged open-mindedness in all matters that needed it in any way."

George was convinced that most men never used their full faculties. He was fond, for example, of having the family sing a tune like *The Swanee River* in the key of E flat, while he played the accompaniment in the key of C. Ives wrote: "This was to stretch our ears and strengthen our musical minds, so that they could learn to use and translate things that might be used and translated . . . more than they had been." George Ives probably was not thinking of what modern composers have come to call polytonality; he simply had a musical idea and a desire for his students to have disciplined and independent musical sensibilities.[1]

Ives' father demonstrated his own independence in many musically productive ways. Once, Charles remembered, the family saw George standing in the rain, without hat or coat, listening in the garden to the ringing of the church bell next door. He ran indoors and tried to duplicate the note on the piano. He failed, and then ran outside to listen and try again. "I've heard a chord I've never heard before—it comes over and over, but I can't seem to catch it," he explained. For most of that night he stayed up, trying to find the chord on the piano, but he failed. Unable to locate the chord, he decided that the half-tones that the piano produced were too far apart, and that he needed quarter-tones to reproduce the sound of the bell accurately. Soon afterwards, he began work on an instrument that could produce quarter-tones. The instrument was designed to stretch twenty-four or more violin strings, tuned to match the sounds in his mind. He even tried to get the family to sing the quarter-tones. He tried similar experiments with glasses and bells.[2]

Long after his father's death in 1894, the memory of him obsessed Ives. As his friend, the pianist John Kirkpatrick recalled many years later, Ives seemed to live "almost in a state of Chinese ancestor worship. He talked of his father as if he were still living, as if he were still a member of the household—it was that immediate. He spoke with great reverence and with great intellectual interest in his father's curiosity about everything." Those memories were an integral part of Ives' character and his music, and played an important role in the maturation of his creativity.[3]

Next to his father, Charles Ives best remembered the religious atmosphere of his youth. Historically, Connecticut had long been a hospitable climate for revivalism: Jonathan Edwards, Timothy Dwight, and Lyman Beecher played central roles in the state's history. Charles grew up in a family that took this religious atmosphere for granted, and he remembered the revivals and church services that were so much a part of evangelical Christianity. Throughout his life he was a conventional churchgoer, undoctrinaire but secure in his faith. He served much of his musical apprenticeship in Protestant churches of various denominations—Congregational, Presbyterian, Episcopalian—and composed several pieces that are overtly religious, or that use religious material in obvious and specific ways.[4]

Religion shaped Ives' mature life and work in indirect ways as well. While an organist at the Central Presbyterian Church in New York, from 1899 to 1902, despite some misgivings about exploiting a captive audience, he bewildered his congregation with his unconventional musical experiments. "I am sure that various members of the congregation were in a state of continual quandary whether Charlie was committing sacrilegious sins by introducing popular and perhaps ribald melodies into the offertory," one of his close friends recalled years later. "I am sure that he was under grave suspicion, but the melodies were so disguised that the suspicious members of the congregation could never be sure enough to take action." Ives' friend was "certain that he did interweave them, not from a sense of humor or tantalization, but because that was the way in which his musical mind found certain satisfaction." On the surface, such experiments could look like simple foolery, but Ives was serious, and his experiments had direct musical consequences.

Likewise, conventional church music was important in shaping his talent. The *Fourth Symphony,* one of Ives' most complicated experimental works, has as the chief inspiration of its final movement, "a Communion service, especially the memory of one, years ago, in the old Reading Camp Meetings," according to Ives. In the middle of the movement, "there is something suggesting a slow, out-of-doors march, which has for its theme, in part, the remembrance of the way the hymn, *Nearer My God to Thee,* sounded in some old Camp Meeting services." The piece went into one of Ives' organ compositions, and he played it at the Central Presbyterian Church in New York in 1901. "The rest of the movement gradually grew out of this."[5] A childhood memory, an adult experiment, and a mature symphonic composition: Ives' memory affected his music in this fashion in countless ways.

Despite his religious ties, however, Ives shared the progressive feeling that institutions fettered a man and so conventionalized his responses as to block his communion with God, with Nature, and even with the divinity in his own soul. Ives' terminology was not always precise, and his sense of the ironic could obscure his literal meaning, but one discussion about Emerson was clear. "But every thinking man knows that the church part of the church always has been dead," Ives wrote. He believed that the essence of religion was a state of mind, and he had no illusions about the stuffiness of churches. "Many of the sincerest followers of Christ never heard of Him," he continued, assuming that the essence of Christianity was inherent in the way a person thought and acted and not in any institutional affiliation or creed. Being true to ourselves, Ives insisted, "*is* God" and "the faintest thought of immortality *is* God, and that God *is* 'miracle.'" Ever since the days of Emerson and Theodore Parker, New Englanders had been translating such ideas into reform activities, and in doing so himself, Ives became a member in good standing both of the tradition of Transcendentalism and of the cultural atmosphere of progressivism. For him "most of the forward movements of life in general and of pioneers in most of the great activities, have been [the work of] essentially religious-minded men."[6]

Ives' attitude toward religion was thus closely connected with his love of Transcendentalism and its major prophets. The ideas of the Transcendentalists were the most important influences on his writings, their names adorn various movements of the Concord Sonata, and their attitudes toward art and religion especially provided him with a frame of reference that he never abandoned.

Perhaps the most striking similarity between Ives and this New England tradition lies in the subject of hero worship. Most obviously, Ives idolized Emerson, the great prophet who helped reveal the secrets of Nature to democratic man. To Ives, Emerson was "an invader of the unknown— America's deepest explorer of the spiritual immensities—a seer painting his discoveries in masses and with any color that may lie at hand—cosmic, religious, human, even sensuous." Emerson was constitutionally unable to embrace any precise doctrines, and in his intentional vagueness Ives found great solace and a whisper of Divinity, for "vagueness is at times an indication of nearness to a perfect truth." Emerson instead made his own way, wringing "the neck of any law that would become exclusive and arrogant, whether a definite one of metaphysics or an indefinite one of mechanics." He hacked "his way up and down, as near as he [could] to the absolute, the oneness of all nature, both human and spiritual, and to God's benevolence."[7] In effect Emerson became a divine messenger, bringing God's plans to man and showing men the path to God.

Also implicit in Transcendental musical thought was a distinction between music and sound that was unknown in modern usage. For Henry David Thoreau, music did not mean the audible notes that reached the ear. Music, Thoreau remarked, "is the sound of the universal laws promulgated," and thus did not really need to be played or heard at all. Ives agreed that sound and music were not necessarily the same. Once, frustrated that he could not write music or get performers to play or sing it as he imagined it, he exclaimed: "My God! What has sound got to do with music! . . . Why can't music go out in the same way it comes in to a man, without having to crawl over a fence of sounds, thoraxes, catguts, wire, wood, and brass?" It is not the composer's fault that man has only ten fingers, or that his voice is inadequate to sing a divine theme. "That music must be heard is not essential—what it *sounds* like may not be what it *is.*" The point was reasonably clear: the sounds a composer wrote down or that a performer produced were intended only to represent and evoke in the listener some mystical idea or vision that illuminated the basic structure of Nature and the universe, and thus, of necessity, each real work of art had its own organic form, a form that could easily violate established musical rules and human abilities. Ives, therefore, could be more than a little eccentric when he discussed his attitude. To his way of thinking, "a sharp is a kind of underlying sign of, or senses and reflects or encourages, an upward movement, tonal and more perhaps spiritual, at a thing somewhat more of courage and aspiration-towards than the flat carries. . . ." The flat was "more relaxing, subservient, looking more for rest [and] submission, etc. —often used as symbols as such when they're not needed as the signs of tonality in the usual way."[8] Seen in relation to this philosophy of music, all of Ives' compositions became program music, that is, they were always *about* something usually connected in some way with an ideal, the world in Ives' head, Nature, or some related concept. The notes on the page were merely poor representations to help someone else share the vision. The innovations of form that resulted were not the product of pure experimentation; they were, rather, the innovations of nostalgia, as the composer tried to recapture the ideals of Transcendentalism and its leaders.

Yale College also had an important place in Ives' memory. When Ives enrolled there in 1894, the school did not have the academic prestige it later acquired. It had, instead, social prestige, and the philistine values that often accompanied it. The teaching was stiff, formal, and dull, and the faculty often showed a perceptible distaste for the students. Even professors who could be quite winning in private seemed to become distant and humorless before a class. The students, in turn, seemed not to take the faculty seriously; they tended to be exuberant, healthy-minded, athletic, and self-confident. Outside observers found the school conformist and patriotic, unlikely to question the values that dominated the comfortable homes of most students. The divisions in the town, Henry Seidel Canby recalled, were not so much those between "town" and "gown," but between "town, gown and sweater."

Canby, who entered Yale two years later than Ives, wrote in 1936 that he and his fellow students "were strenuous without thought to ask the reason why," and had little interest in the humanistic values of their liberal education. "No, the goal was prestige, social preferment, a senior society which would be a springboard to Success in Life." Music scarcely existed in any formal, pedagogical sense, and the histories that have been written about Yale and about New Haven during this period all but ignore it. "To be musical and indulge in music privately was a sure sign of freakishness, and as bad as private drinking or the reading of poetry in seclusion," Canby recalled. "The banjo, the mandolin, and the guitar were respectable, since skillful players of these instruments could 'make' the musical clubs and so gain social recognition; but proficiency on the violin was a sure sign of something wrong, as was skill on the piano not confined to 'beating the box,' and also the singing of 'classic' music."[9] Ives, in short, had entered a school that could give him little encouragement or room to grow, and that supplied him with friends who were, at best, uncomprehending of his values and desires. He managed to survive, at least outwardly, happy and popular and a member of a senior society, but he also convinced himself that music could give him neither a career nor a position in society.

Ives' one hope for instruction and assistance was the newly appointed Battell Professor of the Theory of Music, Horatio Parker. Ives always retained a certain respect for Parker. Very much a moralist, Parker intended that his art have abiding values and not simply gratify the senses. Like Ives, he approved of the Transcendentalists, but his musical training made his art seem foreign to the Transcendentalism Ives admired. Parker had studied in Boston under George W. Chadwick and in Germany under Joseph Rheinberger—both strong-minded, conservative teachers. He brought this training, his unquestioned social status, and a fastidious appearance to his musical thought. Usually identified with church and choral music, his most successful works were large-scale oratorios. Nowhere in his music is there any sign that he wished to communicate with common, ordinary people, or that he wished to exploit native American themes and forms. Since democratic and nationalistic aspirations were always important to Ives, it is not surprising that professor and student clashed, however amicably they resolved their differences.[10]

Ives was inclined to underplay his Yale training and the influence of

Parker, but students of Ives should not. George Ives had schooled his son
on Bach and instructed him by means of the same texts in harmony and
counterpoint that Parker used, but such training was casual and limited by
the father's lack of sophistication. Ives learned original and even eccentric
musical experiments as a boy, but he could not have known much about
the history of western music or the basic techniques of composition. Parker
was not especially original, but he was an experienced craftsman. He may
well have been one reason that Ives so often created his original experiments within traditional forms—the sonata, the symphony, the quartet—
forms that George Ives had not apparently emphasized. Of Ives' organ
experiments with hymn tunes, Parker said: "The hymn tune is the lowest
form of musical life." When Ives attempted fugues with themes in four
different keys, Parker refused to take him seriously and suggested that
"hogging all the keys at one meal" was a musical gluttony that Ives should
avoid. Ives read Parker's attitude as hostility toward originality, but the
tension between Parker's training and Georges Ives' originality was fruitful. It made Ives a successful musical pioneer. The discipline and the
originality had to go together or the result would have been another
George Ives—an unknown musician of merely local repute.

Ives felt tension over the influence of his father and the influence of
Parker throughout his career. He remembered the occasion when he took
Parker a song, "At Parting," which contained several unresolved disso-
nances. Parker told Ives, "There's no excuse for that." When informed of
Parker's remark, George responded: "Tell Parker that every dissonance
doesn't have to resolve, if it doesn't happen to feel like it, any more than
every horse should have to have its tail bobbed just because it's the prevailing fashion." An even more significant example of the tension was Ives'
First Symphony, composed after George Ives' death and directly under
Parker's supervision. In the original version, Ives remembered, the first
movement was supposed to be in D minor, yet the first subject "went
through six or eight different keys, so Parker made me write another." Ives
disliked the result, and told Parker he preferred the first version. Parker
allowed him to return to his earlier draft, but he smiled and said, "But you
must promise to end in D Minor." The conflict that arose at Yale persisted
throughout Ives' life with enormous creative consequences. The earthy,
democratic, experimental originality of George Ives fruitfully combated
the European cultivation of Horatio Parker.[11]

The final important memory that the mature Ives retained from his
youth was that of his choice of career. In one sense, George Ives had not
provided an especially attractive model for his son. He had no status in the
world of music; he was not taken seriously by others, and he never earned

much money. As a result, Charles himself felt ashamed at the thought of being identified publicly as a musician; real, masculine Americans played baseball, not the piano. He also deeply feared the effect that a wife and children might have on a musical intelligence; and surely the financial demands of a family would force a composer into producing kitsch, if only to keep bread on the table. Ives therefore abandoned any thought of making music his profession, and turned to life insurance to pay the bills.

In 1898 the life insurance business had not yet been subjected to the merciless exposures of the muckrakers. On the surface the insurance business seemed to offer a genuine public service. Ives believed devoutly in the ideals of the industry, and he particularly enjoyed work in the actuarial department. He joined the Mutual Life Insurance Company in New York, and soon worked out the connections between the life-insurance business and his developing political and social ideas. The actuarial department in an insurance company is the place that makes predictions about how long a given person will live, by considering factors like age, health, and occupation. To Ives such activity was not merely a business proposition; it was, instead, a way of discovering in the data some kind of divine average man, because the experience of many men was taken into account, and the tools of science uncovered what all men seemed to have in common. The computations of the actuaries, to Ives and to other progressives, were scientific versions of Walt Whitman's lists of democratic events and objects. Once enumerated, they could be of assistance in realizing Transcendental ideals. Indeed, Ives' agency even quoted Emerson in its ads.[12]

For a decade Ives' position in the industry was precarious. He received his first job probably as the result of two family connections. The position was not an especially good one, and he soon transferred to the Charles H. Raymond Agency, the chief Mutual office in New York City. There, he met another clerk, Julian S. Myrick, and began a close friendship that lasted for the rest of their lives. The Raymond agency was involved in the rather unsavory activities exposed by the Armstrong investigations, and it was abolished; a number of good insurance men left the Mutual organization entirely. In spite of having suffered a slight heart attack in 1906, Ives planned to join Myrick in opening their own agency within the Washington Life Insurance Company, whose president had formerly been with Mutual. Chiefly because of the recession that soon developed, the agency failed, but on 1 January 1909, the agency reaffiliated with a newly cleansed Mutual for what soon became one of the success stories in American life insurance. Economic conditions brightened, and the agency eventually became preeminent in the field. Ives' health after World War I prevented him from staying in insurance, but Myrick became one of the most honored and respected men in the business.[13]

Historians who see progressives as people of science bent on organizing and streamlining a modern industrial corporation feel quite at home with Myrick, the efficient friend of Herbert Hoover and a supporter of modernization in all business affairs. But Ives, a more genuine progressive than Myrick, was more complicated. On the one hand, as Shepard Clough has pointed out, Ives and Myrick introduced classes to teach salesmen not only what to sell, but also how to sell it. "To Charles E. Ives should go much of the credit for setting an example in the Mutual Life of training agents and also for furthering the ideas of 'programming'—of integrating a person's life insurance with his financial and family needs." Ives' pamphlet, "The Amount to Carry—Measuring the Prospect," of 1912, was an early classic in the industry. On the other hand, as Morton Keller has pointed out, life insurance had a self-image that was religious to the point of unctuousness. Life insurance was associated, at least rhetorically, with the family, the church, and the state. Several leaders compared it to "missionary activity," and George W. Perkins told a correspondent: "Our profession requires the same zeal, the same enthusiasm, and the same earnest purpose that must be born in a man if he succeeds as a minister of the Gospel." Darwin Kingsley of New York Life also thought insurance agents were much like preachers, and that the deferred dividend policy was similar to the religious system of rewards and punishments. Life insurance was a gospel: "What a picture of belief—I had almost said of faith—is presented by the securities that center in the vaults of the . . . companies doing an international business."[14] Ives was much attracted to both science and religion, but any references to "science" or "efficiency" in his work and writings are subordinate to the religious, Transcendental faith that he had evolved. "Science" was a means of reaching a "religious" goal.

Ives' pamphlet, so influential in the industry that it was reprinted frequently, demonstrates how intertwined his business and his religious ideas were. "There is an innate quality in human nature which gives man the power to sense the deeper causes, or at least to be conscious that there are organic and primal laws . . . underlying all progress," it begins. "Especially is this so in the social, economic, and other essential relations between men." Superstition, he thinks, "is giving way to science," and "the influence of science will continue to help mankind realize more fully, the greater moral and spiritual values." In becoming more scientific, life insurance was aiding in the Darwinian movement toward greater progress and democracy; it was "doing its part in the progress of the greater life values." Ives idolized the "majority," the "fundamental" truths, "democracy," and the "moral," and these words operate with a kind of ethical compulsion in the pamphlet: The "great majority of men today" know if only subconsciously "that a life insurance policy is one of the definite ways of society for toughening its moral muscles, for equalizing its misfortunes," and for "supplying a fundamental instinctive want." Because of this, any normal adult "knows that to carry life insurance is a duty. . . ."[15] Progressivism did indeed value science and efficiency, but it always did so for religious and democratic reasons. Ives' business, music, and philosophy were all of a piece—a beautiful example of an integrated, progressive aesthetic intelligence.

II

The mature Ives transformed his memories into progressive ideas about human nature and politics as well as into progressive music. He was especially insistent about the inherent wisdom and goodness of human nature. In his essays he wrote of "the great transcendental doctrine of 'innate goodness' in human nature," and translated the doctrine into what he called the "great primal truths," namely, that "there is more good than evil, that God is on the side of the majority," that God "is not enthusiastic about the minority," that he "has made men greater than man, that he has made the universal mind and the over-soul greater and a part of the individual mind and soul, and that he has made the Divine a part of all." Like many progressives Ives assumed that people were basically good, that they functioned harmoniously in a group, and that God blessed the political and social results of their deliberations. Ives consciously carried the idea into his writings on music. Even though few people seemed able to understand or appreciate his music, and a great many seemed to like music he detested, he persisted in his faith that, at some distant day, every man would be a composer. The common man, "while digging his potatoes will breathe his own epics, his own symphonies (operas, if he likes it)," and would sit in the evenings "in his backyard and shirt sleeves smoking his pipe and watching his brave children in *their* fun of building *their* themes for their sonatas of *their* life."[16]

Given his ingrained habit of thinking in natural and universal laws, Ives needed only to make a short step to assert that "the law of averages has a divine source," and that this assumption applied directly to politics. Like many other progressives, Ives wanted to level the fortunes of the rich and give some of their money to the poor, presumably with the help of an income tax. He explicitly favored the limitation of the personal property a man could acquire, by means determined by a majority vote. He insisted that this leveling of wealth would "not be a millennium but a practical and possible application of uncommon common sense." He assumed that no important differences separated capital and labor, and that God benevolently guided society toward perfection. He favored voter participation as much as possible in important legislative decisions. Most progressives thought that initiative, referendum, and recall would help accomplish this, but Ives wanted more. "The initiative and referendum," he argued, is at best "remedial or corrective rather than constructive." What he wanted was some form of town-meeting-of-the-whole, which would regularly express the will of the majority on legislative issues of substance. Given such a procedure, what might look like an incorrect decision would really be correct, because according to the laws of nature, the majority was right by definition. "It must be assumed, in the final analysis and consideration of all social phenomena," he insisted, "that the Majority, right or wrong, are always right."[17]

Music provided the most important examples of Ives' innovative nostalgia. Many critics of the fine arts have believed that a work of art cannot be "about" anything. Observers tend to experience a creative work as an independent entity, ignoring the time, the place, and the life behind its creation. Ives opposed this view. "Is not all music program music?" he asked. "Is not pure music, so called, representative in its essence?"[18] For Ives the answer was always yes; his music always was about something. Ives wrote program music, and the programs he had in mind were chiefly the ideas, places, memories, and people that had impressed him in his youth, in college, or in his career. They flowed in his mind like the elements of William James' unconscious.

Ives used music to express this sense of the flow of conscious life. The apparent formlessness that distressed listeners was misleading. The music had form, but the form depended on extramusical stimuli, on ideas and memories, and on desires and yearnings—on nostalgia—rather than on formal, finished, and purely thematic material. "Everything in life is relative," George Ives had once told his son, and like so many of his memories, this memory of relativism from childhood was always with Charles Ives when he tried to compose. His rejection of conventional notions of the "absolute" and his experiments with various elements of chance composition were in this progressive faith in the natural, organic "process."[19]

This emphasis on process and program often led Ives to musical dead ends—pragmatic experiments that did not work. Ives could argue that "vagueness is at times an indication of nearness to truth," but on some occasions, at least, Ives was so close to his object that he could not see it at all, and his work was incommunicably private. Ives realized this, when he wrote that "what is unified form to the author or composer may of necessity be formless to his audience." A critic might well find in this statement a contradiction, since form is presumably universal, and its rules reducible to statements that convey the same meanings to any trained intelligence. But to be fair to Ives, he had to be taken on his own terms, and his successes had to be balanced with his own admitted or unadmitted failures.[20]

Ives' descriptions of three of his works illustrated his problems. The first, "A Yale–Princeton Game," was an obvious failure, an experiment that did not work, and it survives only in part.

> It is "Two Minutes in Sounds for Two Halfs within Bounds." It was short, only four or five pages in full orchestra. The last two pages are quite clear and fully scored—the first part [I] have only in sketch. But to try to reflect a football game in sounds would cause anybody to try many combinations etc.—for instance, picturing the old wedge play (close formation)—what is more natural than starting with all hugging together in the whole chromatic scale, and gradually pushing together down to one note at the end. The suspense and excitement of spectators—strings going up and down, off and on open-string tremolos. Cheers ("Brek e Koax" etc.)—running plays (trumpets going all over, dodging, etc. etc.) —natural and fun to do and listen to—hard to play. But doing things like this (half horsing) would suggest and get one used to technical processes that could be developed in something more serious later, and quite naturally.[21]

This description exemplifies a progressive mind at work and play. The activity was healthy and strenuous enough for Theodore Roosevelt, and as full of "nature," the "natural," and of "process" as John Dewey could have wished. The form clearly expressed the content, even when the content was unclear. The composer felt nostalgia for the atmosphere of his college days; the nostalgia created a program; and the attempt to realize the program led to an innovative form.

The same process succeeded in one of Ives' most famous and innovative works, the second section called "Putnam's Camp, Redding, Connecticut," from the longer piece, known both as "Three Places in New England" and "Orchestral Set #1." Musically, the piece used material taken from several earlier compositions, especially the "March and Overture, 1776" and some preliminary work on a stillborn opera about Benedict Arnold. Biographically, however, the piece began when Ives was a boy, on an occasion when his father invited a neighboring band and the Danbury band to play in support of their respective baseball teams. The parade that was scheduled put each band at opposite ends of the town, and assigned them pieces to play that were in different meters and keys. They approached each other, playing full force; naturally, the dissonances grew acute, and each musician had to play louder and louder so that he could follow his own score. A few of the players had difficulty, but most managed. The bands survived as musical units, gradually marched past each other, and the sounds faded out. As Ives' friend, Henry Cowell, has pointed out, this experience was most probably the germinal point of Ives' concept of polyharmony; from it, he developed "the idea of combining groups of players" in the orchestra "to create simultaneous masses of sound that move in different rhythms, meters, and keys." Thus, Ives' "polytonality may be polyharmonic, each harmonic unit being treated like a single contrapuntal voice," as, for example, when "the bands played two separate tunes, each with its own harmonic setting," and it "may also be polyrhythmic."[22]

In addition to these memories of times, sounds, and places, Ives added themes of patriotism and of an experience he himself apparently had while dreaming on a picnic. In Redding, not far from Danbury, was a small park that was a Revolutionary War memorial; it had been the winter quarters for General Israel Putnam in 1778-9. One 4 July, "a child went there on a picnic, held under the auspices of the First Church and the Village Cornet Band." The child, presumably young Charles Ives, wandered from the group into the woods, apparently hoping to find some of the soldiers. The sounds from the camp grew fainter; the child rested, and then he had a vision of the Goddess of Liberty, who pleaded with him never to forget the heritage of freedom that men like Putnam's soldiers left the country. The boy then dreamed about the soldiers marching, playing popular tunes with fife and drum; the sounds intermingled with those of the picnic, he woke from his reverie, and returned to the group.[23]

All these elements combined in the "Putnam's Camp" movement. The two bands approached each other, each playing louder and louder, trying to follow their own melodic lines over the approaching competition. The tunes that recurred sometimes came from the revolutionary period, as the Continentals routed the Redcoats, but they also included elements of other times and memories, such as "Marching Through Georgia" and some John P. Sousa marches. The narrative was not linear and chronological, but wandered off softly at times, and became quiet, because Ives believed that was the way the memory worked. Random thoughts and impressions invaded the stream of consciousness, even though Civil War tunes hardly belonged in a piece about the Revolution. But then, Ives was conveying to his listeners an experience, a program about how dreams happened, how patriotism and sense of place and childhood affected the artist, and how the attempt to recreate this nostalgia and its values led to new music.[24]

Such an attempt to evoke memory created problems for conductors and players. Nicholas Slonimsky, the Russian-born specialist in new music who had worked with Serge Koussevitzky in Boston, and who had founded the Boston Chamber Players, was giving the première of *Three Pieces* in 1931. He decided that he had to develop a new method of conducting to convey the different beats simultaneously, one with each hand. Ives had actually worked out a way for the second movement to be conducted with a single beat in 4/4 time, but Slonimsky decided to be truer still to the spirit of the score and maintain the separate rhythms. He recalled: "So four bars of my left hand equaled three bars of my right hand." Despite the complexity, "the orchestra could follow me. There was no difficulty." He simply "told one-half of the orchestra to follow my left hand and the other half to follow my right hand." One critic remarked that "my conducting was evangelical, because my right hand knew not what my left hand was doing." He believed "that this was the first time when a really stereophonic or rather bilateral type of music was played and conducted." Ives was grateful, but such willingness to adapt to a new music probably cost Slonimsky an important career in conducting, and he soon turned to writing musical reference books.[25]

Even as a mature composer, however, Ives could not always control his method, and nothing demonstrated its pitfalls quite like his failure to complete his "Universe Symphony." In the fall of 1915, he conceived the idea of a piece of music that could encourage listening in a manner analogous to looking at nature simultaneously in two different ways: "(1) with the eyes toward the sky or tops of the trees, taking in the earth or foreground subjectively—that is, not focussing the eye on it"; and "(2) then looking at the earth and land, and seeing the sky and the top of the foreground subjectively." The music would have two distinct parts, each played simultaneously: The lower parts, those played by basses, cellos, tubas, trombones, bassoons, and so on, would be "working out something representing the earth, and listening to that primarily," and the upper parts, those played by strings, high woodwinds, piano, bells, and so on would be "reflecting the skies and the Heavens." The piece would be played twice, "first when the listener focusses his ears on the lower or earth music, and the next time on the upper or Heaven music." It would fall into three parts, representing the past, the present and the future: 1. "Formation of the waters and mountains"; 2. "Earth, evolution in nature, and humanity"; 3. "Heaven, the rise of all to the spiritual." It would, in short, be a programmatic symphony that synthesized Ives' memories of religion and

Transcendentalism with his progressive ideals and hope for the future.

The piece did not work; it is hard to conceive of how it could have worked. By 1932 when Ives was recalling its conception, he still seemed to believe he could finish it, but he never did. His intentions and their musical consequences, however, were most instructive. Ives wrote: "The earth part is represented by lines starting at different points and at different intervals," producing "a kind of uneven and overlapping counterpoint sometimes reaching nine or ten different lines representing the ledges, rocks, woods, and land formations." He saw "lines of trees and forest, meadows, roads, river, etc.—and undulating lines of mountains in the distance." With this counterpoint a few instruments "playing the melodic lines are put into a group playing masses of chords built around" various sets of "intervals in each line. This is to represent the body of the earth, from whence the rocks, trees and mountains rise." The musical consequences continued for many more lines; Ives had clearly given the piece much thought. Its underlying plan, he wrote, "was a presentation and contemplation in tones, rather than in music (as such), of the mysterious creation of the earth and firmament, the evolution of all life in nature, in humanity, to the Divine." The program was clear enough; the role of memory and ideas was clear—but it was not enough to make viable music. The creative impulse failed; the experiment did not work; the innovation was stillborn.[26]

Ives' innovations included most of those now identified with twentieth-century "modern" music. Unknown in Europe, and unaffected by it during his creative period, he nevertheless managed to incorporate not only the spatially divided orchestra, polytonality, and polyrhythms, but also atonality, tone row, and aleatoric music. He did not have available the tape recorders and sound synthesizers that became important to later musicians, but he did pioneer in virtually every area of music in which pioneering was technologically possible in his day.

Yet from the first Ives received criticism and contempt for his work. He recalled vividly the time when an accomplished violinist, Franz Milcke, visited the Iveses; Ives hoped to get him to play two of his violin sonatas so that he could profit from the experience in finishing his next one. After a little confident talk, Milcke sailed into the first piece apparently at sight and came immediately to a spluttering halt. The strange rhythms and note combinations confused and irritated him. "This cannot be played. It is awful. It is not music, it makes no sense," he said. Ives played parts of his piece to show the man how, but subsequent efforts were no better. Milcke came out of the little music room of the Ives house with his hands over his ears and said: "When you get awfully indigestible food in your stomach that distresses you, you can get rid of it, but I cannot get those horrible sounds out of my ears [by a dose of oil]."[27]

The experience was unusual only in the sense that Milcke paid any attention at all to Ives' music. Most people did not, and even Ives' close personal friends more often than not expressed displeasure and bewilderment when they heard his work. Their comments, he recalled later, "had something of the effect on me of a kind of periodic deterrent, something approaching a result of a sedative." He had always felt peculiar and diffident about composing serious music, and the incomprehension of his friends produced something like a spiritual wound. It "would get me thinking that there must be something wrong with me, for, with the exception of Mrs. Ives, no one seemed to like anything that I happened to be working on when these incidents came up."[28]

Ives stopped composing about the time America entered World War I; his health and creative juices seemed to evaporate in his feverish support for the war, President Wilson, and the vision of world government and world peace that enthralled so many progressives. He had had a history of heart trouble going back to 1906. He now suffered permanent cardiac damage and diabetes, and then he developed cataracts over both eyes, which were inoperable because of the heart condition. Yet, he did have the health and energy to publish some of his music and ideas, and to attempt to distribute them wherever he thought he might find a suitable response. A few brave friends, performers, and conductors broke the ice, and there were scattered performances of compositions or parts of compositions during the 1920s and 1930s. Henry Bellamann, Nicolas Slonimsky, and John Kirkpatrick were among the most prominent men who attempted to help Ives find an audience. Even they found themselves stretched, at times, to achieve Ives' intentions. Slonimsky's inventiveness with his music soon became something of a legend. As Henry Cowell recalled: "When Nicolas Slonimsky conducted *Washington's Birthday* he gave seven beats with the baton (in itself not a thing every conductor finds simple), three with his left hand, and led two beats by nodding his head. This created both great amusement and great admiration and is a tale still often told."[29]

Two great breakthroughs were especially memorable. During the 1930s John Kirkpatrick struggled to make sense out of the *Concord Sonata,* finally playing it to an enthusiastic crowd in New York's Town Hall on 20 January 1939; a month later, during a concert devoted wholly to Ives, Kirkpatrick repeated his achievement. A 1948 recording presented his version to the world, and it astonished everyone by selling remarkably well. Equally significant was Lou Harrison's performance of the *Third Symphony* in 1947. It won Ives a Pulitzer Prize, forty years after the work had been composed.

Ives was gratified; he disliked the hubbub, however, and remarked: "Prizes are for boys. I'm grown up." He had a point. The years that followed saw the recording of virtually all his important works, and Ives took his place in the textbooks as, all things considered, probably the most important American composer.

Europeans discovered him as well. When Arnold Schönberg died in 1951, his widow mailed Ives a sheet of paper that she had found. On it he had written:

> There is a great Man living in this country—a composer.
> He has solved the problem how to preserve one's self and to learn.
> He responds to negligence by contempt.
> He is not forced to accept praise or blame.
> His name is Ives.[30]

III

Just as the consensus of music historians has been that Charles Ives was America's greatest composer of the Progressive Era, so the consensus of historians of architecture has been that Frank Lloyd Wright was America's greatest architect. A comparison of the two men provides a constructive method of determining what the concept of progressivism means when applied to the arts. They each had a dominant parent whose values were instrumental in providing creative opportunities, as well as an important teacher familiar with European academic standards. They shared a passionate devotion to the literature of the American Renaissance, as well as an inveterate sense of morality that led directly to assumptions about the programmatic nature of artistic creation. Most of all Wright's work was like that of Ives in not being genuinely modernist. It was, instead, innovative and nostalgic. It originated new art forms and theories, yet did so by looking backward. Likewise, it borrowed freely from disciplines far afield from architecture, not only from the literature and social thought that Ives also used, but also from interior decoration and from book design.

The two men differed chiefly in their orientation toward Europe and Asia. Ives had little more than an undergraduate knowledge of modern European culture, plus an intense love of composers like Beethoven. Wright was deeply influenced by several British thinkers and creators. As a child he experienced a fashionable European theory of education and intensely studied Japanese prints and buildings. Wright's interest in European social thought and educational methods, while generally foreign to Ives' provincial Yankee world, provided essential links between progressive art and the rest of the progressive world. The same educators influenced John Dewey, the same social thinkers influenced Jane Addams, and the same reformers influenced George Herron. Nor was the influence all one way. Wright's work influenced the most creative architects in Europe and ranked with Dewey's philosophy as the most important American contribution of the period to the cultural history of the world.

In contrast to Ives, Wright always remembered his father with a certain hostility. "The youth hardly had known himself as his father's son," he said in his autobiography, and the portrait he painted was certainly not an endearing one. His most vivid memory of his father was in church. His father was playing Bach in the gloom, while young Frank, "crying bitterly as he did so," pumped away with all his strength on a huge bellows, supplying the organ with the air it needed. Always in his mind was the fear of his father's disapproval if he stopped pumping. Unfortunately, the muscles of a boy of seven were not strong, and he was often tempted into dreaming by the music. The dreams and the weakening muscles would conspire, the pumping would cease, and then his father would call out in irritation at the fading sounds that he so wanted to create. Frank grew up "afraid of his father," and never really able to understand him or to sympathize with his problems. Only late in life did he become more appreciative, telling his son-in-law that he had come to value greatly his father's musical contributions to his education. In some mysterious way, his father's music helped him to conceptualize his later architecture: "Father taught him to see a symphony as an edifice—of sound!"[31]

The boy's feelings toward his father were genuine, but the portrait was unfair. William Russell Carry Wright was a minister's son who himself became a minister, but a man who could never match his own ambitions. He had tried teaching, law, and politics, always with some success but never enough to satisfy himself. During the early 1860s, his life changed when his first wife died, leaving him with three children to raise. He remarried a woman much younger than himself, and Frank was the first of three children and the only son from this second marriage. During his childhood his father made a good name for himself both for his preaching ability and his musical achievements, but the goodwill of the community never resulted in either the monetary or the social rewards that William Wright and his wife craved. The family moved restlessly from Wisconsin to Iowa to Rhode Island to Massachusetts, as William Wright grew increasingly gloomy and depressed and his wife grew discontented and hostile. She and her family were basically Unitarian in sympathy and their influence combined with this lack of success to encourage a move back to Wisconsin; in 1878 Wright became the pastor of a Unitarian church in the small Wisconsin town of Wyoming. The family pattern continued: a good reception from the community but no financial rewards and not enough cultural outlet for any kind of spiritual satisfaction. Mrs. Wright finally informed him: "I hate the very ground you walk on," and "I don't care what becomes of you." She refused him the marriage bed and even decent civility, and he reluctantly sued for divorce. She did not contest any of his allegations, but she did demand the house and custody of the three children. The divorce was granted in 1885 and Frank apparently never saw his father again. Only the music and some painfully distorted memories seemed to survive.[32]

Instead, he was the "son-of-his-mother." He had no place in his memory for the ambitious shrew who appears in the records of the divorce court. Anna Lloyd Jones had been born in Wales into a clan of people enthusiastically religious and mutually devoted to each other. They migrated chiefly to the little Wisconsin town of Spring Green in the 1840s. Her father, Richard Jones, was "an impassioned, unpopular Unitarian," who according to his grandson "preached as Isaiah preached," and was a "Holy Warrior" capable of making God in his own image. The family crest was an old Druidic symbol that meant "Truth Against the World," and compromising with an uncooperative world was never much of a family virtue. Anna's brother Jenkin soon became a leading Chicago Unitarian preacher. A friend of Jane Addams, he played an important role in local political and religious reform movements. Thus, despite its immigrant and Unitarian roots, the family atmosphere was similar to that of other pioneering Protestant Americans. It dominated the mind of the young boy. One of his childhood duties was to decorate the pulpit of the family chapel where Uncle Jenkin preached so often and so movingly. In the process family devotion, religion, and architecture fused in his mind in ways Wright never outgrew. He later wrote: "This family chapel was the simple, shingled wooden temple in which the valley-clan worshipped images it had lovingly created. In turn the images reacted upon the family in their own image. Those sunny religious meetings were, in reality, gatherings of the clan."[33]

The full weight of both clan and religion most affected the young through their schooling. Education was his mother's "passion" from an early age; "all this family was imbued with the idea of education as salvation." As an expectant mother she had decided that she would bear a boy, and that when he grew up, he would be an architect. She not only kept her mind obsessively on architecture, but she furnished his room with ten wood engravings of old English cathedrals, taken from a contemporary magazine. In 1876 she made a momentous discovery: she went to the Centennial celebrations in Philadelphia, and there discovered the "Gifts" of Friedrich Froebel in the Exposition Building. Her son had reached kindergarten age, and here was the perfect means of training him to be an architect. He remembered the "gifts" until he died:

> The strips of colored paper, glazed and "matt," remarkably soft brilliant colors. Now came the geometric byplay of those charming checkered color combinations! The structural figures to be made with peas and small straight sticks; slender constructions, the joinings accented by the little green-pea globes. The smooth shapely maple blocks with which to build, the sense of which never afterward leaves the fingers; form becoming *feeling*. The box had a mast to set up on it, on which to hang the maple cubes and spheres and triangles, revolving them to discover subordinate forms.
> And the exciting cardboard shapes with pure scarlet face—such scarlet! Smooth triangular shapes, whiteback and edges, cut into rhomboids with which to make designs on the flat table top. What shapes they made naturally if only one would let them!

Anna Wright went to Boston as soon as she could to obtain whatever training was available in the new methods and passed it along to her children.[34]

Unitarianism in America has had many connections to New England Transcendentalism and Wright's family often read the work of Emerson and his circle. The Froebel method, although developed from quite different sources in Europe, fit easily into the Transcendentalist worldview. The system of education was pantheistic, as it attempted to develop the child's powers of reason so that he could perceive natural law and the harmony and order of God. The kindergarten guide told parents that God's works reflected the logic of his spirit, and that human education ought to imitate the logic of nature. Just as the family of Richard Jones had stressed the unity of all things, so Froebel stressed "the Divine Principle of Unity" as a moral discipline. Children, for example, were to experience only complete geometrical patterns in the hope that their pleasing effect would inculcate knowledge of divine order. Designs that were capricious or arbitrary were against nature and not permissible. Such training explicitly joined religion and aesthetics, and enabled Wright to conceptualize his structures as being expressions of divine unity, and to see himself as something of a messiah leading the masses to knowledge of God. A cube, the manual said, should be taken out of its box whole, "in order to inculcate alike the sense of order and the idea of completeness." Because life does not permit isolation, no single part of the cube should ever "be left apart from or without relation to the whole. The child will thus become accustomed to treating all things in life as bearing a certain relation to one another."[35]

Thus in microcosm first appeared Wright's concern for the organic aesthetic that became so obsessive later in his writings. Just as the blocks should not be divided, nor conceptualized in any way except as entities, so should buildings not be conceptualized as separate porches, windows, or fireplaces. No single part of any building meant anything unless it related to the entire structure. At this stage, the key scientific parallel was not so much the plant as the process of crystallization. Froebel had himself been a student of crystallography, and the geometry of crystallography led directly to Wright's own earliest conceptions of architectural pattern. Wright insisted that all things in nature had a tendency to crystallize, and that Froebellian kindergarten training made him think of terms of appropriate geometrical scale, with all the pieces in either nature or house interdependent. Thoughts of crystallization led directly to the T-square and the set-square, and one of the characteristics of Wright's later preliminary drawings was their being matted with exploratory lines. "This principle of design was natural, inevitable for me. It is based on the straight line technique of T-square and triangle. It was inherent in the Froebel system of kindergarten training given to me by my mother." Emerson had already instructed Americans in how natural objects corresponded to spiritual ideas, and Wright absorbed this habit of mind and adapted it to geometry: "In outline the square was significant of integrity; the circle—infinity; the triangle—aspiration; all with which to 'design' significant new forms. In the third dimension, the smooth maple blocks became the cube, the sphere and the tetrahedron; all mine to 'play' with." For Wright his training, even at so early an age, had permanent results. It awoke the mind of the child "to rhythmic structure in Nature—giving the child a sense of innate cause-and-effect otherwise far beyond child comprehension. I soon became susceptible to constructive pattern *evolving in everything I saw.* I learned to 'see' this way and when I did, I did not care to draw casual incidentals of Nature. I wanted to *design.*" Wright himself thus established the connection between Froebel and Emerson and the organic Prairie House of Wright's maturity, with its overhanging roofs, podia, and projecting cubic forms, and the cruciform plans which have such a striking resemblance to standard Froebel grid designs.[36]

These interrelationships between his mother's clan, religion, education, and architecture not only gave Wright essential mental training, they also gave him material assistance in beginning his professional career. Uncle Jenkin Lloyd Jones, for example, was pastor of All Soul's Church in Chicago during the 1880s, and he played an important role in Wright's decision to leave Wisconsin for Chicago. The Unitarian preacher had been instrumental in luring Joseph Lyman Silsbee to Chicago to be the architect for his church and Wright soon found himself doing preliminary draughting work on the plans. He made further contributions to Unitarian architectural design when he acted as delineator for Silsbee's plan for the Unity Chapel that Lloyd Jones wanted built in Helena, Wisconsin; Wright apparently was also in charge of either the design or the construction of the interior of the building. Shortly thereafter, he published a sketch in the June, 1887, *Inland Architect and News Record*, for still another one of Silsbee's Unity Chapels, this one located at Sioux City, Iowa.[37]

Wright and his family were even more closely involved in some educational experiments being directed by his aunts Jane and Ellen Lloyd Jones at Hillside, near Spring Green, Wisconsin. Although John Dewey was still a decade away from his Chicago Laboratory School, the sisters had already planned a school along what became known as progressive lines. They were interested in developing a child's ability to grow and to express himself, and they innovated methodologically with use of field trips to study botany, farm animals to study biology, and gardens to study the production of food that the school children would soon consume. Neither woman cared much for ideology or jargon, and their basic assumptions seem to have come from sources as diverse as Plato and Emerson, but the school nevertheless developed quickly into an institution emphasizing cooperation rather than the traditional regulation. It was also enthusiastically Unitarian, and Jenkin Lloyd Jones visited regularly. Wright quickly executed a house and school combination. In 1901, when he was in his full maturity, he added another structure known as the Hillside Home School, which proved to be one of the most effective structures of the early Prairie style.[38]

Religion was an inseparable part of Wright's private life while he lived in Chicago. Four other ministers' sons worked in Silsbee's office; one of them, Cecil Corwin, was Wright's closest friend. When not at the office, Wright naturally gravitated toward Uncle Jenkin's church. It seemed never to be closed and provided young people with a gay social life. It had evening classes and concerts, meetings to discuss the poetry of Robert Browning, a neighborhood center, a kindergarten, and a circulating library that contained important books on architecture. Wright not only met prominent local citizens like Jane Addams, but he also entered the upper-middle class of Unitarians who were the core of the congregation. As the minister's nephew, he was welcome everywhere, and he soon met Catherine Lee "Kitty" Tobin, the daughter of a prosperous local businessman.

Neither family thought the relationship should end in marriage, but Frank and Kitty persevered, and Jenkin Lloyd Jones married them in 1889 when Wright was twenty-one and his bride eighteen. With his early architectural success, the backing of the influential Lloyd Jones clan, and a wife more comfortable in the upper middle class than himself, Wright seemed to have a world of infinite possibilities before him. For many years he proved quite capable of taking every advantage of his fortunate circumstances.[39]

For all the difficulties which he experienced studying music, Charles Ives went through Yale in what seemed to be a sober and conventional way. He adjusted to its institutional demands and within a reasonable period after graduation he was a promising young insurance executive. Wright's education was fragmentary and without evidence of such adjustment. Family transiency made his elementary education disorganized. He apparently attended Madison High School in Wisconsin but did not stay long enough to graduate. He then entered the University of Wisconsin as a special student in the scientific course. He briefly studied civil engineering, but apparently remained at the university only for parts of two academic years and received grades in only two courses. He left for Chicago early in 1887, without a college degree and with academic achievements that would normally have qualified him only for acceptance elsewhere as a college freshman. Once in Chicago he worked briefly for Silsbee and another firm. He heard from one of his friends that Louis Sullivan, of the firm of Adler and Sullivan, was looking for someone to finish the drawings for the interior of the auditorium which he was designing. Wright already felt that he had learned what he could in his year with Silsbee and jumped at the chance. When they met, Sullivan was characteristically brusque and imperious, but Wright impressed him. Sullivan accepted him with a minimum of formality, and Wright joined the firm that would be his substitute for Yale College.[40]

At the time Wright went to work for Sullivan, he was on the verge of his most productive phase as a designer of commercial architecture in Chicago. Chicago was becoming the preeminent city in the world for the development of what came to be called the skyscraper, and Sullivan played a major role in its development. The skyscraper, as its name implied, was a tall building, but it could not have developed without a coalescence of population density, business needs, and technological innovation. The skyscraper obviously needed cities in which to develop, with enough available capital and people to make such elaborate structures feasible. If such structures were to be usable, they needed adequate lighting, space, air, and strength. The great Chicago fire had underscored the need for fireproofing, as it cleaned out many of the old structures, and threatened to do the job again if their successors were not adequately insulated from the flames. Obviously too, the elevator was an essential component of any building over five stories, and between 1850 and 1870, the necessary technology produced a workable model. Finally, architects had to master new techniques with iron and steel. They had to learn to use frames for buildings that took the weight entirely off the walls and allowed the exteriors of buildings to be simply curtains or envelopes. This also led to more satisfactory lighting, air, and better use of available square footage for offices or display areas than older buildings had. With William Le Baron Jenney's Home Insurance Building (1884–85), Chicago commercial architecture took the decisive step which made Sullivan's career possible.[41]

Jenney was more of an engineer than an architect. The architect who made a decisive impression on Sullivan, and through him, on Wright, was Henry Hobson Richardson. Although known chiefly for his Romanesque designs in the area around Boston, Richardson left one indisputable masterpiece in Chicago, the spectacular Marshall Field Wholesale Store (1885–87). Richardson was intolerant of the fussiness of much of the architecture of his time, and the plain, quiet grandeur of the Field store also fit in with Marshall Field's sober business attitude. Sullivan was impressed with the building, at least in part because of the man who had designed it. He was fond of discussing buildings in terms of the men who had created them, using a moralistic and humanistic aesthetic that had much in common with the ideas of writers such as Emerson, Ruskin, and Whitman. In one of his kindergarten chats, Sullivan told his pupil about Richardson: "Here is a *man* for you to look at . . . a real man, a manly man; a virile force—broad, vigorous and with a whelm of energy." There the building stood, "a monument to trade, to the organized commercial spirit, to the power and progress of the age, to the strength and resource of individuality and force of character." Spiritually it stood "as the index of a mind, large enough, courageous enough to cope with these things, master them, absorb them and give them forth again, impressed with the stamp of a large and forceful personality." Artistically, the building stood "as the oration of one who knows well how to choose his words, who has something to say and says it—and says it as the outpouring of a copious, direct, large and simple mind." Thus, some of the values of Sullivan's aesthetic were revealed: a good, imaginative, and sensible man should discover the values and needs of his time and produce structures that solved the building problems in such a way as to represent himself and his spirit. Both the architect's intention and his personal qualities were important in the judgment of his work.[42]

Despite all the praise that Wright showered on Sullivan later in his life, he probably did not learn much of strictly architectural significance from him. In spite of his own great devotion to Sullivan and his theories, their mutual friend Claude Bragdon felt that Wright probably learned more from Sullivan's partner, Dankmar Adler. Grant Manson, one of the leading Wright scholars, found the idea that Wright was a disciple of Sullivan "nothing more than an amiable fiction." The true impact involved personality and inspiration. Just as the H. H. Richardson who designed the Marshall Field store had inspired Sullivan, so the Sullivan behind the Auditorium or the Carson, Pirie Scott store inspired Wright. "His function is not to teach, but to inspire," he wrote. "Instead of a 'form follows function' scientist, I shall give you a great lyric poet." Wright felt that the buildings erected by Adler and Sullivan did not make form and function one, and that with the exception of their use of terra cotta, they did not adequately take into account the nature of the materials in determining organic form. "But the buildings if considered on their own in time and place went so far beyond contemporaries in point of enlightened countenance as to prophesy a new integrity so far in advance of the work of the period as to arrest the sentient passer-by with prescience of a new world." Before Louis Sullivan high buildings had lacked unity. "They were built up in layers. All were fighting height instead of gracefully and honestly accepting it." In recognizing that form and function were one, Sullivan meant that "a *building can only be functional when integral with environment and so formed in the nature of materials according to purpose and method* as to be a living entity true each in all to all." If we believe this, "we will gradually learn to express and expand the thought of the great lyric poet—that was Louis H. Sullivan."[43]

In another one of his kindergarten chats, Sullivan once warned against "the fugacious nature of words, their peculiar tendency to transformation in meaning while they retain the same outward form." With both Sullivan and Wright, words changed meaning with alarming frequency and great fuzziness. Both read deeply in the literature of the American Renaissance and much of Wright's architectural theory verges on the incomprehensible unless it is understood in terms of the literary and religious ideas that dominated New England during the mid-nineteenth century. When his theory is examined against this background, his buildings not only take their rightful place with the mature creations of Charles Ives as progressive creative achievements, they also provide a key tie between progressive artistic creation and the parallel current that flowed into John Dewey's pragmatism. The moralism that related successful building to its site and function was essentially the same moralism that related a successful idea to the solving of a democratic social problem. The enemy was Beaux-Arts classicism as well as Platonic idealism; the goal was an open-ended environment in which the demands of nature presented problems and the democratic architect or philosopher solved them organically.[44]

Functional theories were as old as art. In architecture they were the theories "which make strict adaptation of form to purpose the basic guiding principle of design and the principal yardstick by which to measure the excellence or the beauty of architecture." Ornament was permissible, but it was to have function and not merely delight the eye. Writers have often equated the concepts of functionalism and organicism, although not always appropriately, since a building was not an organism. Either term tends to imply that architects should seize new problems and invent new forms to solve them efficiently. Elaboration of the concept might be complex, and might include the needs of occupants, the expression of structure, social functions, and the intended symbolism. Applied by a moralist like Wright, functionalism meant that "architecture should reflect and contribute to the moral or ethical ideals of man. A building should be true, not dishonest. Forms must be what they seem to be." Therefore, a building "should be a true expression of its purpose and of its age. Materials and structural systems should be used with integrity and be honestly expressed. The society of forms should achieve its goals through harmonious cooperation."[45]

The humanistic tradition of the Renaissance in Europe had emphasized the impression that a building ought to make on an observer. The builder had an ideal that his building was to embody, and his spatial values were those of order and proportion. They had nothing really to do with the site of the structure and its natural environment. The surrounding landscape and the intended use of the building were essentially irrelevant and definitely subordinate as considerations to its visual effect. Pillars need not support anything; fronts could be false; and windows need not open. As far as American cultural history is concerned, this attitude found its first serious challenge in England during the early nineteenth century in the romantic criticism of Samuel Taylor Coleridge, William Wordsworth, and Thomas Carlyle. In an 1818 lecture on Shakespeare, Coleridge presented the new attitude. He stated that a form was mechanic "when on any given material we impress a predetermined form, not necessarily arising out of the properties of the material," as for example "when to a mass of wet clay, we give whatever shape we wish it to retain when hardened." An organic form, by contrast, "is innate; it shapes, as it develops, itself from within, and the fullness of its development is one and the same with the perfection of its outward form." Form, in other words, should not be applied from the outside; it ought to grow from within, according to its own organic nature.[46]

These English romantic premises of taste soon took root in America. Emerson, the American thinker most aware of European developments of this kind, used them in his rebellion against classicism in religion and philosophy. For Emerson, the Transcendentalist loved beauty as an aspect of the Deity. He tended always to identify God and Nature as a means of weaning people away from churches and other institutions and permitting them to perceive God directly through their own faculties. For the Transcendentalist Protestant, freeing art, whether in literature or in architecture, was closely related to freeing the conscience. Classical architectural theories were as inorganic as classical theologies—all European standards should give way to more organic American ideas, and each individual should retain the right of personal judgment. Emerson's language was generally imprecise and inconsistent, and as one of his critics has pointed out, by the word "Nature," he seems to mean "variously man, not-man, not-society, out-of-doors, essence, universe, and God," but his definition of beauty was reasonably clear: "We ascribe beauty to that which is simple; which has no superfluous parts; which exactly answers its end; which stands related to all things; which is the mean of many extremes." In architecture this meant that the architect needed to be aware of the proportion, function, and place of the building in the universal scheme of things if he wished to create anything beautiful. He should also be democratic in his appeal and avoid any unnecessary distinction between the fine and applied arts. The architect who built this way was an instrument of the Divine: "To create is the proof of a divine presence. Whoever creates is God. . . ."[47]

Emerson's friends likewise commented on art in ways that lingered in architecture well into the twentieth century. Wright several times mentioned Henry Thoreau, and in several places Thoreau attacked buildings that suggested neither their origin nor their purpose, as well as ornamentation that had no genuine relationship to the building or organism. He advocated instead a beauty that grew from within and thus produced truth and nobility unconsciously. But for both Emerson and Thoreau, architecture was a minor and infrequent consideration. For their friend Horatio Greenough, it was far more important, and a number of critics have suggested that Greenough was the source of his friends' remarks on architecture. Like Emerson, Thoreau, and Wright, Greenough was an evangelical Protestant aesthetician trying to formulate American principles in rebellion against European standards. He, too, was hostile to institutions, academies, dreary teaching, and anything else that did not strike him as autochthonous and democratic. He wrote privately to Emerson that his theory of structure included "a scientific arrangement of spaces and forms to functions and to site"; an emphasis on "features proportioned to their *gradated* importance in function"; ornament and color "to be decided and arranged and varied by strictly organic laws, having a distinct reason for each decision"; and the entire "banishment of all make-shift and make-believe." He wanted "to strike a blow for this style now because the aesthetical world abhors a vacuum, and ours is just sucking in hostile elements. I mean the excremental corruptions of foreign and hostile systems."[48]

Wright, thus, had available an American tradition of criticism that he could apply to architecture. The comments might have a "fugacious nature" as Louis Sullivan had noted, but they were based largely on literature and philosophy, and they were embedded in specific times, places, and arguments that might not be directly relevant to Chicago architecture in the 1890s, but the words were there, and Wright knew them and admitted their influence. He, too, was a Protestant preaching on art, teaching people to look at buildings and their place in nature, and through them to perceive Divine unity. He too would see in the intimate interrelationship of form and function a message from God about how Americans should cope with their very special environment.

IV

If Wright shared his devotion to the New England heritage with Charles Ives, he shared an equally important European heritage with many other progressives. Ives rarely showed much interest in Europe aside from its music, but the work of John Ruskin, William Morris, and the Arts and Crafts Movement affected many progressives. Wright's aunts Nell and Jane Lloyd Jones had given him Ruskin's *Seven Lamps of Architecture,* along with more conventional children's adventure stories. As a child, he also read *Fors Clavigera, Modern Painters,* and *The Stones of Venice.* He knew at least some of the work of William Morris, Owen Jones' *Grammar of Ornament,* and most of the important volumes of the French architectural critic Eugène Emmanuel Viollet-le-Duc. In the early years of the twentieth century, Wright was active in the Arts and Crafts Movement in the Chicago area, at least in part because of his friendship with Jane Addams and the centrality of Hull-House to such activity in America. Some Englishmen were influential in person as well as through their writings. Walter Crane, a well-known British artist active in the movement, spent five weeks in Chicago in 1891–92, and lectured at the Art Institute immediately after Wright's uncle, Jenkin Lloyd Jones. Charles Robert Ashbee came to America and met Wright late in 1900. At the time Wright was already on the membership list of the Chicago Arts and Crafts Society and the two men quickly formed a life-long friendship. Its major fruit was the introduction Ashbee wrote to one of Wright's European publications, *Frank Lloyd Wright: Ausgeführte Bauten* (Berlin, 1911). Whatever the native American qualities of Wright's Prairie Style, they clearly were influenced by this important and seemingly paradoxical European tradition: important because of its great impact on modern building and decoration; and paradoxical because, contrary to casual assumptions about modernity in art, as a tradition it helped usher in the future while devotedly studying the past.[49]

Wright had become aware of British aesthetic attitudes at the very time these attitudes were undergoing probably their greatest change in a century. Appropriately enough, the issue that most clearly divided British aesthetic thinkers was the role of nature in art. Both major groups assumed that the artist was important, a "religious" figure who played the role of preacher to a society that was rapidly losing its conventional religious faith. They differed over whether art should represent the natural world, whether that natural world symbolized anything that might be called divine, and whether the role and perceptions of the artist ought to have serious social consequences. The major thinkers of the mid-nineteenth century, like John Ruskin, disliked the unnatural clutter that seemed to vitiate most Victorian creative achievements. Ruskin wished to enable his audience to perceive the natural world directly and to stop worrying about whether, for example, a painted scene reminded a viewer of a pleasant holiday or told an uplifting story. Ruskin firmly believed that nature was tied closely to God, that understanding nature was a moral act, and that the emotional well-being of the individual and the progress of society depended on improving this aesthetic perception. As he grew older and less religious, Ruskin came to demand that the world be made fit for artists and to feel that if society did not yet want art, then critics should educate society until it did. Only then, in a moral, beautiful, and socialist society, would life be genuinely rewarding. Opposing this view, and all but unnoticed at first in England and America, was the view that overwhelmed the whole Ruskinian attitude toward the arts and became a central theme of early twentieth-century modernism. Identified at first with France and a figure like Théophile Gautier, it spread to England with writers like Walter Pater and Oscar Wilde, and first touched the American sensibility through expatriates like James A. McNeill Whistler. In this view nature was inferior to artifice, and morality as commonly understood was the way the middle class kept themselves and others from appreciating true beauty and worth. Artists were inherently more valuable than nonartists, and they ought to ignore society as much as possible and cultivate their precious sensibilities. Throughout his life, Wright remained loyal to the position of Ruskin, and scornful of much of what he understood about the principles of modernism. Even when his own work had an important effect on leading modernists, he remained essentially a Victorian figure. Like Charles Ives, he remained tied to assumptions about nature, morality, and the improvement of society that seemed outmoded to later generations.[50]

A person who believed that art was worship could hardly avoid seeing buildings as sermons in stone. In the early nineteenth century, Britain had experienced a great revival of the Gothic forms of the Middle Ages, stimulated by an act of Parliament that had authorized the building of churches throughout the country to help stem the tide of the atheism sweeping Europe in the aftermath of the French Revolution. Gothic revivalist leaders believed that architecture was inherently an ethical art, whose practitioners should be concerned with the expression of truth and personal morality. Some British minds feared that the revival of Gothic forms would lead to the truths of Rome instead of Canterbury, since Roman Catholicism had been the religion of those who had created the Gothic style. The threat was real, symbolized vividly by the conversion to Rome of the leading Gothic advocate, Augustus Welby Northmore Pugin. It soon seemed less threatening because of the formation of the Anglo-Catholic, or Oxford, Movement in 1833, devoted as it was to a nationalistic and ritualistic middle way between evangelical Protestantism and Roman Catholicism. Ruskin adjusted himself to these problems by arguing that what was best in medieval Catholicism was its growing spirit of Protestantism. He seemed genuinely to believe that all Gothic creative acts were praiseworthy because they had been executed by happy masons sure of divine sanction for their work and their place in an organic society. Such architecture was good, beautiful, and true, as well as an instrument of social progress. Ruskin was thus the perfect bridge between the best Victorian aesthetic thought and Protestant American sensibilities that were fearful of anything sensuous or immoral. Not only was Ruskin loudly and even bigotedly Protestant, but he believed that nature was emblematic of the divine in ways that were very similar to those of Emerson and the Transcendentalists. Through Ruskin and the Gothic, Americans could become conscious of art and remain convinced that morality and religion would be improved rather than harmed by their interest.[51]

Ruskin could be an eclectic and paradoxical writer, his views changed markedly as he grew older, and he was not especially well-organized or consistent even throughout one book. But with this said, three of his specific positions seem relevant to any discussion of Wright. Beauty, Ruskin thought, "is either the record of conscience, written in things external," or the "symbolizing of Divine attributes in matter," or the "felicity of living things," or the "perfect fulfillment of their duties and functions." In every case it "is something Divine; either the approving voice of God, the glorious symbol of Him, the evidence of His kind presence, or the obedience to His will by Him induced and supported." The artist, Ruskin wrote, had "the responsibility of a preacher," and he had "to kindle in the general mind that regard which such an office must demand." Wright believed that Ruskin was correct about the serious nature of the role of the artist. Finally, and more directly architectural, Ruskin at times could sound like Horatio Greenough when he touched on the connection between form and function and the organic analogy. "That building will generally be the noblest, which to an intelligent eye discovers the great secrets of its structure, as an animal form does, although from a careless observer they may be concealed."[52]

Ruskin did not write as extensively on architecture as on other subjects, and even when he did he often ignored topics that were central to later critics. He paid no attention to the overall plan of a building or to its masses or proportions. He disliked largeness of scale. He could occasionally sound like a functionalist in his principles, but he did not really believe that function yielded beauty. He tended to be most functionalist when he was criticizing contemporary excesses, and he seemed to write best when concerned with tangential matters such as ornament, color, or sculptural detail. He was more concerned with the happiness and enjoyment of the artist than with strictly formal aesthetic matters. He did share certain critical idiosyncrasies with Wright. Both were preoccupied with architectural surfaces and the use of rich building materials. Both tended to visualize a building as a series of planes, leading to Wright's emphasis on horizontality and his effective use of cantilevers. Both were charmed by color and beautiful surfaces, by patterns of light and shade, and by ornament. Wright did not share Ruskin's belief that ornament was the principal part of architecture, but he may well have internalized Ruskin's assumption that an architect's treatment of detail was one index to his moral worth.[53]

The next important link between British thought and Wright's work occurred when William Morris discovered Ruskin's ideas while studying at Oxford and spent his life putting into practice what he felt to be Ruskin's key principles. Like so many American progressives, as a young man, Morris was interested in religion, and for a while he intended to take Holy Orders. Instead, under Ruskin's influence, he became part of the

Pre-Raphaelite circle, taking up painting, furniture, and wallpaper design. Early in life he had also studied English Gothic architecture and he thus found thinking in Ruskin's categories easy. He decided that the art of previous generations in England had been rigid, cold, and inhuman as well as derivative, and that it insulted the poor in every line. He came to detest mass production and the machine, and to wish instead to recapture the virtues of the medieval craftsman who had lovingly handcrafted designs with deep personal and social meaning. Morris was obsessively energetic, and his interests soon spread over a broad area of creativity. He wrote poems and stories, carved wood, modeled clay, illuminated manuscripts, and even embroidered. His interests led to the founding of a small company to manufacture the items necessary to furnish a house in such a way as to revive the medieval spirit of design. Morris, Marshall, Faulkner and Company advertised themselves as "Fine Art Workmen in Painting, Carving, Furniture, and the Metals," and made themselves a name in stained glass as well. In 1875 the firm reorganized as Morris and Company and during the next decade, specialized in the production of wallpapers and textiles. Morris also was active in a well-publicized association formed to protect the English Gothic heritage against restorers bent on "modernizing" many of the most memorable examples of English architecture.[54]

Morris thus absorbed Ruskin's attitudes about art and architecture and effectively expanded them to the whole range of design. Not an architect himself, what he did could have an immense effect on the way someone like Frank Lloyd Wright thought about his work. Morris conceived of a home or larger building as an artistic whole, and he cast a religious meaning and value over every aspect of interior design, from the wallpaper to the stained glass windows. Furthermore, in elaborating his designs, Morris proved to love nature in a way that appealed to one of the basic tendencies of American aesthetic thought. Morris loved plants, and he patterned all of his designs on some natural form. He thought that by imitating nature a designer could protect himself against the danger of imitating an earlier and now inorganic form. The result was soon evident in wallpaper designs called Daisy, Fruit, and Pomegranate, and in the chintzes called Tulip and Willow. Morris felt that if the designer's hand moved in accordance with the natural forms he designed, it would itself always remain free and natural. The worker as well as the design would thus be in accord with the divine scheme of things.

By the early 1880s, Morris had developed his ideas to the point where he became active in politics. He studied the work of Karl Marx and became a publicist for a number of socialist organizations. He wanted to make the world and the work of every citizen beautiful and rewarding. Socialism seemed to offer the only hope. What Morris could not face was the unbridgeable gap between his private feelings and desires about art and the requirements of the machine, public taste, and poverty. The carefully designed pieces of handicraft in which he exulted were by definition expensive and rare. Morris could not face the thought that the masses might not share his own refined tastes, or that the machine might be capable of turning out beautiful designs cheaply enough so that the masses might benefit from them. He could repeat his golden rule, "Have nothing in your house which you do not know to be useful or believe to be beautiful" as often as he pleased, but it was meaningless to anyone who was not wealthy, educated, and eager to be creative. Wright understood Morris' mistake immediately, and always avoided it, even while he retained his admiration for Morris and acknowledged his influence.[55]

By the late 1880s, the major themes and organizations that influenced Wright were converging. In 1888 the term "Arts and Crafts" came into use with the founding of the Arts and Crafts Exhibition Society by rebellious young Englishmen tired of the concentration of the Royal Academy on the fine arts and its neglect of the applied and decorative arts. Just as John Dewey had attacked philosophy and education for being too detached from life, these workers wanted to reunite artists, art, and the people. They wanted their ideas and their creations to permeate society. Through books and the visits of figures like Walter Crane and Charles Ashbee, the Arts and Crafts Movement came to Chicago in the 1890s, and there, as in England, it was closely associated with the growing settlement movement. Just as Toynbee Hall had played an important role in the British Arts and Crafts Movement, so its offspring, Hull-House, was central to the American movement. The Chicago Arts and Crafts Society was founded at Hull-House in 1897, crystallizing a movement that was already well underway. Both Jane Addams and Ellen Starr were deeply committed to it. Frank Lloyd Wright not only delivered one of the most influential addresses of his career there, but he soon made a practice of designing his own interiors and furnishings along the lines established by William Morris. Progressive American art had absorbed this important European movement, and the Prairie School of architecture was its direct descendant.[56]

V

Oak Park, Illinois, was an appropriate place for Wright to settle, with his head full of the Unitarianism of the Lloyd Joneses and the aesthetic moralism of Emerson and Ruskin. Even so, Wright's mother could not face the idea of her young son going away alone. She followed him to Chicago and sought the advice of Augusta Chapin, the Universalist pastor of Oak Park and an old friend. Miss Chapin invited the Wrights to stay with her while they made more permanent arrangements. The nickname of the town at the time was "Saint's Rest," and Wright discovered there a rapidly expanding community of upper-middle class professional people, mainly white, Anglo-Saxon Protestants. The town was legally dry, full of churches, and provincially proud of its righteousness. Thus, Wright found himself beginning his career in the same kind of New England-style environment that had produced Jane Addams, John Dewey, Charles Ives, and so many other progressives. A generation or two later, Oak Park produced prototypical figures for the 1920s: Bruce Barton, the advocate of success, advertising, and Christianity, and Ernest Hemingway, the rebel against these values whose books gave the post-war world a few of its more notorious models. The town was small enough so that everyone was acquainted with one another. Wright's sister Maginel sang with Grace Hemingway in the choir of the local Congregational church; its pastor was Bruce Barton's father.[57]

In 1889 Wright built a home and quickly settled in with his wife and a rapidly growing family. Wright's autobiography, as well as the memoirs of his sister and his son all paint much the same picture of the creative chaos that ensued. The house was full of spontaneity and laughter and at first the family was close. Wright appeared to be recreating the extended family of the Lloyd Joneses, and his wife was a social and professional asset to his growing career. Like Morris and Ruskin, Wright wanted his children "to grow up in beautiful surroundings. I intended them all to be infected by a love for the beautiful." Unfortunately, money was never plentiful, and no one could cheaply satisfy the tastes of a William Morris. Wright was blasé about the whole problem: "So long as we had the luxuries, the necessities could pretty well take care of themselves so far as we were concerned." He found money for tickets to the symphony and tasteful clothes for the children, even if the grocer went unpaid. "This love for beautiful things—rugs, books, prints, or anything made by art or craft or building—especially building—kept the butcher, the baker, and the landlord always waiting. Sometimes waiting an incredibly long time." That same taste proved integral to the Prairie house and some of the finest examples of art ever conceived.[58]

Wright's early work showed little evidence of the Prairie homes he designed a decade or so later. He tended to overemphasize his debt to the romantic character of Louis Sullivan, but in fact, late in the 1880s and early in the 1890s, the most notable quality in much of his work was its academicism. One of the most fruitful designs along the East Coast, for example, was what Vincent J. Scully has called the Shingle Style, which developed out of the traditional Queen Anne of Norman Shaw and reached a brief maturity in a few of H. H. Richardson's houses. Enhancing the vernacular building styles common in America since the early nineteenth century, the Shingle Style houses emphasized space, piazzas, lightly scaled woodwork, and rough shingles. Above all, the best buildings in the style interwove interior and exterior spaces in ways that Wright quickly adapted to the different landscape and building materials available in the Midwest. Wright apparently first encountered such designs when working for Silsbee in 1887; they also appeared in the periodical *Building,* which had just published the work of Bruce Price in Tuxedo Park, New York. Silsbee had·already introduced the Shingle Style to Chicago a few years earlier, and at least for a time the style seemed to influence Wright far more than Sullivan did. The Price houses made such an impact on Wright that he seems to have designed his own home in Oak Park directly upon Price's precedents. Despite his devotion to Ruskin and Morris, and despite their popularity with many Americans, the Gothic Revival had little influence on building practice. Americans adapted the words of the revival to the best classical and academic building traditions then available in America.[59]

Influences from more exotic areas also affected Wright's developing style. Stylistic devices from Asia and Latin America all soon made their mark, however much he would deny these influences later. Perhaps the first of these came from India. The "bungalow" was originally from the Bengali: word *bāngla,* meaning a low house with galleries or porches around it. The English in India adopted the term in the late seventeenth century, and the word generally came to mean a primitive, temporary shelter that was also simple and utilitarian. The style finally came to America about 1880, where it became associated with the Stick Style and then the Shingle Style that grew out of it. Not until shortly after Wright began practicing, in the middle 1890s, did the word come to mean what it generally means today: a house with simple horizontal lines, wide projecting roofs, numerous windows, a porch or two, plain woodwork, and without an attic, second story, or basement. At about the same time, Wright became aware of other influences acquainting him with the bungalow. The French and Spanish settlers in the deep South had developed a similar form and Louis Sullivan, using Wright as his chief designer, had already built structures of this kind in Ocean Springs, Mississippi. The style made a distinct impression in Chicago at the 1893 Exposition when the state of Louisiana chose to ignore the prevailing classicism of the other buildings and to build a structure based on the raised-cottage style customary in a Creole plantation residence. Within a decade after the fair, a large number of similar buildings were constructed along Chicago's North Shore. Wright himself featured homes of remarkably similar qualities in two articles in *The Ladies' Home Journal* published in 1901 and made a significant contribution to the widespread vogue for bungalows that occurred in pre-war America when he built the W. A. Glasner house (1905) in Glencoe, Illinois, and the Isabel Roberts house (1908) in River Forest, Illinois.[60]

The same 1893 Exposition that informed the Middle West about bungalows has come down to students of Sullivan and Wright largely in germ imagery, thanks to the pathological description in Sullivan's autobiography. Masterminded by Daniel H. Burnham, the White City was more Roman than American. To its viewers, Sullivan wrote, "it was a veritable Apocalypse, a message inspired from on high. Upon it their imagination shaped new ideals." They returned to their homes, "each one of them carrying in the soul the shadow of the white cloud, each of them permeated by the most subtle and slow-acting of poisons." After a brief period of "incubation in the architectural profession and in the population at large," the country showed "unmistakable signs of the nature of the contagion. There came a violent outbreak of the Classic and the Renaissance in the East, which slowly spread westward, contaminating all that it touched. . . ." Sullivan's analysis was not only an exaggeration that overlooked a number of other important features of the classical revival, it also ignored the contributions that style made to himself and to Wright. Further, it passed over the other structures at the fair that received less publicity, but which later affected Prairie School architecture.[61]

In addition to the Louisiana bungalow, the building erected by the Turkish government could easily have influenced subsequent Prairie Style architecture. It used a flat, overhanging roof that contrasted with the surrounding classical structures. The building as a whole had simple rectilinear massing and was geometrical in its detail and its placement of the stairs, doors, and windows. This kind of geometric structuring later became something of a hallmark in the Prairie style.[62]

Even more significant to Wright's work was the Japanese influence. Japanese art began to influence Europe in the late 1850s and soon affected the fine arts in America and the work of Americans living abroad: the painting of James A. McN. Whistler, the music of Charles T. Griffes, and the architecture of Wright all showed a Japanese influence. Within architecture, Wright was not an isolated case. A number of publications, led by Edward Morse's *Japanese Homes and Their Surroundings* (1885), were spreading the influence along the East Coast, and the same 1876 Centennial Exhibition that had acquainted Wright's mother with Froebel blocks also featured the first noteworthy display of Japanese art ever shown in America. Wright was not clear about where he first encountered Japanese art, but certainly he was professionally aware of it by the time he was working for Silsbee, because Silsbee was already an enthusiast of Oriental art. The Fair of 1893, in other words, had a foundation upon which to build and the Hō-ō-den Palace, the Nippon Tea House, and the exhibition of Japanese art in the Fine Arts Palace could only have made the greater impression because of their being surrounded by so much classicism. Sullivan also showed the impact of Japanese art. It was compatible with his organic rhetoric of horizontal lines, banked windows, low and widespread roofs, and thin timbers embedded in plaster surfaces, all of which appeared in his few post-1893 houses.[63]

Japanese art had clearly influenced Wright, and after the fair, he began his life-long habit of collecting Japanese prints. He always denied that the Japanese architecture at the fair affected him directly, but if so the denial was true only in the narrow sense. His early houses tended to resemble those of Silsbee, Sullivan, and Richardson, and little in their design gives much hint of the forms he later made so famous. But by the time Wright built the B. Harley Bradley/Warren Hickox pair of houses in Kankakee, Illinois, and the Ward W. Willitts house in Highland Park, all designed about 1900–2, the Japanese influence seems unmistakable. While remodeling his own Oak Park house in 1895, Wright altered the original Queen Anne design of 1889 to include a typically Japanese device of dividing the rooms of structures by rectangular panels instead of permanent, fixed walls; the Willitts house made even more effective use of this concept. The Bradley/Hickox houses displayed specific Japanese influences. As Clay Lancaster has written: "They are Japanese in their intimacy with nature, inconspicuous entrances, and asymmetrical massing, as well as in the plastered terrace walls with wood copings, slender timber bands in stucco on the vertical planes of the houses, and reduplication of deep obtuse gables, which are thin and thrust outward at the peak as in some Shinto shrines, a type known in Japan as *kirizuma*. " The houses also display some evidence of Wright's Froebel block training in the windows, and the furniture is pure Arts and Crafts—designed by Wright himself in good William Morris style as part of the total architectural concept and then executed by one of the leading Arts and Crafts furniture manufacturers in America, L. and J. G. Stickley. Wright made the Japanese influence a permanent part of his style by visiting Japan for the first time in 1905 and assimilating his experience to the aesthetic of Horatio Greenough: "I saw the native home in Japan as a supreme study in elimination—not only of dirt but the elimination of the insignificant. So the Japanese house naturally fascinated me and I would spend hours taking it all to pieces and putting it together again. I saw nothing meaningless in the Japanese home." In this fashion did home training, Arts and Crafts social art, and Japanese influence all come together as a way of demonstrating the concept of functional form.[64]

In the same year as the Exposition, William Herman Winslow commissioned the first important private home that Wright built. The building normally appears in architectural history because of its primacy in Wright's chronological development, and because its roof gives a strong indication that the Prairie Style is developing. But in cultural history, the Winslow house and its place in Wright's career were important as well for the way in which they brought together in one relationship the whole international, interdisciplinary nature of Wright's mind. Just as Japanese prints and buildings could combine effectively with Transcendentalist rhetoric and Arts and Crafts furniture to help create a Prairie house, so the Winslow relationship demonstrated the cultural ambience within which such houses could happen. In this case the focus of attention should not be on the house so much as on a book that was published at the same time.

In addition to Wright, two individuals were important to the Winslow story. Winslow himself was a prosperous businessman whose early career was chiefly with the Hecla Iron Works. Winslow was fascinated by the ornamental side of his business and in how iron and bronze could be shaped decoratively; during his career he did such work for both Sullivan and Wright. A sometime inventor fascinated by mechanical devices, he also liked to build cabinets and print books in his spare time. These interests led him to friendship not only with Wright but with William C. Gannett, a Unitarian minister who had been a friend of Wright's family for many years and a man whom Wright much admired. Gannett was the author of a book called *The House Beautiful,* which urged its readers to make their homes simple and honest and to throw out all the gaudiness of Victorian materialism. In true William Morris fashion, Gannett said that the house decorated simply and beautifully would influence its inhabitants to have sound values and to love each other. The prose tended to be purple and the ideas sentimental, but they appealed to Wright, newly and happily married and a believer in the possibility of the environment shaping society.[65]

Such a book illustrated key influences perhaps more obviously and clearly than a building, however well described. Good printing techniques in Britain had gone into a precipitous decline after 1830 and feeble, tasteless designs had been the rule throughout the period of the Gothic revival in architecture. William Morris, omniverous in his interests as ever, turned to book design toward the end of his remarkably productive career and in

1890 founded the Kelmscott Press with the cooperation of the great printing technician, Emery Walker. Arts and Crafts was already deeply indebted to the Middle Ages for many of its artistic principles, and so
naturally Morris returned to the dark, compact hand-written books of the
period for his precedents. By the middle of the 1890s the influence had
spread to America and next to Boston was strongest in Chicago. Publishing
firms like Stone & Kimball and Way & Williams, and printers like the
Lakeside Press experimented with the new ideas and techniques, although
not always in imitation of Morris. *The House Beautiful,* the result of the
relationships between Wright, Gannett, and Winslow, has been described
by the best historian of the subject as "a quintessential Arts and Crafts
book, advocating a return to nature, simplicity and beauty."

> The book is lavishly decorated in red and black by Wright. All text pages are
> within borders with Morrisian margins and there are several double-page
> spreads of solid ornament with no text, similar to the "carpet pages" of Celtic
> manuscripts. The double-spread title page is in red old-style caps (as in the text)
> within a black border with a repeating frieze of figures and leaf ornament,
> recalling in its rectilinearity the work of Ricketts and the Glasgow School.[66]

Thus once again did the Lloyd Jones heritage, the sentiments of New
England Unitarianism, and the Arts and Crafts ideas of William Morris
combine with business interests in Wright's career. The Gannett book, like
the Bradley/Hickox houses, symbolized this vital aspect of progressive
artistic creation. It was international, interdisciplinary, and embedded in
the business affairs of Chicago at the turn of the twentieth century. Winslow reportedly had to survive considerable kidding from his more philistine friends because of the design of his new house, but that in no way
indicated that either Winslow or Wright were detached from their culture
or in some way opposed to it. Like Theodore Roosevelt or Woodrow
Wilson, Wright wanted to reform society and reinvigorate its morality; he
preferred the artistic to the political path for accomplishing this

VI

Ives made occasional attempts to attract a larger public but was unsuccessful for two generations. Wright, on the other hand, succeeded. He
made speeches, published short articles, found his work the subject of
several analyses in America, and by 1910 found himself featured in European publications as one of the leading men in his field. Both his words

and his works made him successful at home and abroad until the vicissi-
tudes of his private life put him into artistic eclipse.

Two of his publications are worth particular attention. On 6 March
1901, Jane Addams invited Wright to speak at Hull-House to the opening
meeting of the Arts and Crafts Society. "The Art and Craft of the Machine" was one of the more important speeches of his career. In it Wright
established his debt to William Morris and the Ruskin tradition and then
described how he had gone beyond both men. He acknowledged the
hostility Morris felt toward the machine because of its inherent hostility
to the artistic impulse. According to Wright, Morris "miscalculated the
machine," but it did not matter. Morris, Wright said, had great faith in art
and its possibilities for renewing and improving democratic life, and as a
result, "all artists love and honor William Morris." In his time Morris did
his best for art "and will live in history as the great socialist, together with
Ruskin, the great moralist: a significant fact worth thinking about, that the
two great reformers of modern times professed the artist." They rightly
detested the machine because of its ties to greed and its hostility to crafts,
but society, according to Wright, had now progressed to the point where
the machine was able to undo the damage it had caused.

The machine had "dealt Art in the grand old sense a death-blow," and
Wright did not deny it. The art that had depended on the handicraft ideal
was indeed dead. The poet as architect, a figure as dear to Wright as to
Morris, who "summed up in his person the sculpture that carved his
facades, painting which illuminated his walls and windows, music which
set his bells to pealing and breathed into his organs"—that great figure
began to die somewhere in the fifteenth century with the invention of the
book: "Architecture is dethroned" and "the book is about to kill the
edifice." With the book, true invention began to die, and it became all too
easy to establish what was classical and correct; architecture became imita-
tive and sterile. Poets converted to print and left architecture without their
inspiration. What no one noticed was that the machine developed without
them, and soon made possible a whole new attitude toward the arts. "The
Machine is Intellect mastering the drudgery of earth that the plastic art
may live; that the margin of leisure and strength by which man's life upon
the earth can be made beautiful, may immeasurably widen; its function
ultimately to emancipate human expression!"

Wright's words and logic were never as clear as they could have been,
but the essence of his argument was the same as that of his fellow visitor
to Hull-House, John Dewey. Just as Dewey attacked the sterility of old,
formalized education and dualistic philosophy, so Wright attacked the
formal principles of art and their detachment from the life of the commu-

nity. Dewey would hardly have joined Ruskin in his devotion to the practices of the Middle Ages, but the burden of their arguments within the terms of Wright's speech was the same. Art was a vital means of personal expression for everyman, and a democratic society needed to have all of its citizens able to express themselves creatively. The idea expressed by the solitary Gothic carver through his love of ornament, had become in 1901 the idea that a democracy needed art, that all citizens needed to express themselves, and that the machine could eliminate the drudgery of life and permit this democratic expression. We should deny Greek art, Wright said, "because we insist now upon a basis of Democracy." We must distinguish between the old art, which had no machine, and the new art, which insists on the machine. The machine is "the tool which frees human labor, lengthens and broadens the life of the simplest man," and becomes by so doing "the basis of the Democracy upon which we insist."

Morris always had pleaded for simplicity, and rightly so. But the machine that he so hated was also the great simplifier. Morris was right in many of his feelings, but his protest was more properly against needless complication and artificiality rather than against the machine as Wright now understood it. Simplicity was a quality of the organic. Wright deftly intertwined his British and American roots: "A thing to be simple needs only to be true to itself in organic sense." Thus, the machine that helped people create and live according to their functions was an aid to the organic life as well as to democracy. Years before the same tradition brought Dewey to his work on art as expression, Wright as a practicing artist had come to a similar position, expressed in the terms of the creator rather than the thinker. Misused, the machine could turn out countless cheap and fussy reproductions, perverting the taste of democracy. Properly understood, the machine could be an adjunct to the simple and the organic: it taught that "the beauty of wood lies first in its qualities as wood," and so virtually any carving or reshaping by man or by machine intrinsically distorted it. Such wood-carving was "apt to be a forcing of the material, an insult to its finer possibilities as a material having in itself intrinsically artistic properties": its markings, its texture, and its color. "Rightly appreciated, is not this the very process of elimination for which William Morris pleaded?"[67]

A decade later Wright was in Europe preparing for the press an important review of his work that would introduce him as an influence on European modernism. In the introduction he explicitly stated how the words—as "fugacious" as ever—of the Gothic revival spirit could be adapted to the new architecture. "I suggest that a revival, not of the Gothic style but of the Gothic spirit, is needed in the Art and Architecture of the modern life of the world." In no sense did Wright want to fasten exhausted forms upon modern life, nor did he wish to teach subservience to tradition. But what modern students of architecture had to realize is that "the Gothic spirit," properly understood, possesses "the Organic-character of form and treatment" better perhaps "than in any other period." The infinite variety of Gothic forms was more "literally organic," and "the Spirit in which they were conceived and wrought was usually one of integrity of means to ends. In this Spirit America—other nations no less—will find forms best suited to her opportunities, aims and her life." The Spirit of America, especially the grass-roots America of the Middle West, was one where "breadth of view, independent thought; and a tendency to take commonsense into the realm of Art, as in life" were characteristic qualities. "It is alone in an atmosphere of this nature that the Gothic spirit can be revived in building. In this atmosphere, among clients of this type, I have lived and worked." In this fashion the four generations, from Pugin to Wright, took root in the singularly improbable soil of the American Middle West, and their fruits were carried back again across the sea to Berlin. The architecture of autocracy and Roman Catholicism had become a part of the spirit of democracy: "Organic architecture is thus the only form of art-expression to be considered for the human faith that is Democracy." The Gothic was the progenitor of organic form.[68]

The words would have made little difference without the buildings. Between about 1900 and 1910, Wright went through what Grant Manson has rightly termed his First Golden Age, establishing in the process both a style and a school of American architecture.

Wright tended to animate his landscapes and his houses, so organic did he find the world. He also animated what he hated, and much of the Prairie Style developed in opposition to what Wright hated. He hated the old houses because they *"lied* about everything." They "had no sense of Unity at all nor any such sense of space as should belong to a free man among a free people in a free country." Each house was stuck on a hill inorganically, a box "cut full of holes to let in light and air," or else "a clumsy gabled chunk of roofed masonry" that any minimally aware technician could have put up without a thought about what he was doing.

The buildings standing around there on the Chicago prairies were all tall and all tight. Chimneys were lean and taller still—sooty fingers threatening the sky. And beside them, sticking up almost as high, were the dormers. Dormers were elaborate devices—cunning little buildings complete in themselves—stuck on to the main roofslopes to let the help poke their heads out of the attic for air. Invariably the damp, sticky clay of the prairie was dug out for a basement under the whole house and the rubble stone-walls of this dank basement always stuck above the ground a foot or so—and blinked through half-windows.

Everywhere decoration was overdone. Walls were "be-corniced or fancy-bracketed up at the top into the tall, purposely, profusely complicated roof." The roof "was ridged and tipped, swanked and gabled to madness before they would allow it to be either watershed or shelter." The exterior of the house "was bedeviled," "mixed to puzzle-pieces with corner-boards, panel-boards, window-frames, corner-blocks, plinth-blocks, rosettes, fantails, and jiggerwork in general."[69]

Wright's style was a direct response to these structures. He got rid of the attic and the dormer. He lowered the ceilings and removed the basement. He loved fireplaces and chimneys, but insisted on only having one central one, to unite the family around the comforting warmth. He wanted one or two broad generous roofs, gently sloping or even flat. He was himself slightly over 5'8" and so he found a building organic when it was scaled to about the height proper for a man his size. He admitted that had he been taller, perhaps his ceilings would have been higher. He placed the walls at ground level on a cement or stone-water table, but then stopped them at the second story to let the windows appear in a horizontal band, just below the broad, comforting eaves. He made light and air parts of the design, and introduced casement windows that did not need to be closed when it rained, and that seemed to merge happily with the outdoors. Inside he attacked the walls, hating the effect of boxes beside boxes, and retained a sense of privacy only in the utility and sleeping areas. Most of the lower floor became a large room, with screens to divide areas for dining, reading, entertaining, and playing. He attempted to design appropriate furniture and suggest suitable hangings for the walls and found it hard to relinquish the organic, living result of his buildings to a mere purchaser whose taste he rarely shared and who would always be tempted to alter the living design.[70]

The resulting Prairie Style thus came organically from the architect's personal prejudices, but it was not uninfluenced by its predecessors. The Japanese had taught Wright about the massing of a structure, the value of low, long hipped or gable roofs, and the clearing away of inside, permanent walls. They also confirmed his prejudices about the organic relation of structure to site and the need for functional form. The Shingle Style architects along the East Coast had picked up some of this Japanese influence and combined it with more massive styles derived from the Colonial and the Queen Anne. Wright's roofs, his horizontal bands of windows, his use of porches and the continuity of his lines, edges, and surfaces—so often reminiscent of Froebel block patterns—all were in sharp contrast to the insistent verticality of earlier Midwestern styles. The emphasis on the horizontal gave Wright's house a geometrical air with clear and precise detailing and few ornaments. Instead of old-time jiggerwork, he insisted on natural materials, especially using wood and stone, although he was also occasionally willing to experiment with exotic materials like stained glass. The textures and the colors of nature, and of the patterns caused by the moving sun, so important to an Emersonian intelligence, became the organic replacement for the inorganic ornament of tradition. He also experimented with brick, plaster, and concrete, but in most of his houses, he felt no need or desire to pioneer anything new in engineering.[71]

The Prairie Style was only the outward and visible sign of a Prairie School of which Wright was the preeminent preacher.[72] A great many other architects and designers were involved, in one way or another, with Wright and the Prairie Style. Walter Burley Griffin and his future wife, Marion Mahony, were among the most important; George Grant Elmslie and William Gray Purcell ran one of the most influential partnerships; Barry Byrne and Hugh M. G. Garden had distinguished careers after exposure to Prairie ideas; the list could be longer. Nor were the products of the Prairie School necessarily limited to houses of a distinctive kind. The Griffins became world famous city planners deeply involved with the artistic life of Australia; Barry Byrne was one of the most successful architects of churches of his generation; the repertoire of Prairie School designs included banks, libraries, courthouses, and club houses. The direct influence of the school and the style was great; the indirect, incalculable.[73]

Wright himself was most creative between 1905 and 1910, but the Prairie School as a whole enjoyed its greatest vogue between about 1901 and 1915, after Wright had left Chicago, first for Europe, then for his new home, Taliesin, in rural Wisconsin. For a number of complex reasons, it then went into decline. Wright branched out when he worked at all, preferring Wisconsin, California, and Japan to Chicago, and his influence was more of a memory than anything else until his astonishing reappearance in the 1930s. Sullivan was in eclipse, chiefly for temperamental reasons. The Griffins became involved with Australia. Perhaps less noticeable but more significant, American society and taste changed. The taste of the Ivy League and of New York society became fashionable throughout the country, and it preferred the established forms of the revivalist architects working in the tradition of McKim, Mead, and White. Women became better educated and came to play a far greater role in the choice of their husbands' architects, and the universities taught women to respect Europe and to prefer its forms and standards. Chicago lost its vitality in fiction and poetry as well as in architecture, and the most innovative artists settled in New York and California for the next three generations. New York scorned Chicago as raw and ill-bred and ignored her architecture. Only in the performance of music did Chicago retain status of the first rank, and while

Wright himself claimed to owe much to music, its effect seemed to be different on the architectural taste of everyone else.[74]

But Wright himself proved more durable than anyone could have predicted. He became embroiled in a series of marital scandals; for twenty years his commissions waned, and his touch became correspondingly uncertain. He seemed to have exhausted himself, much like Charles Ives. But Wright recovered his ability, managed finally to find marital stability, and metamorphosed into an aging preacher of progressivism in the arts. Critics like Lewis Mumford insisted on the greatness of his work, and scholars like Henry-Russell Hitchcock and Sigfried Giedion agreed. Repeating the principles of his mother, of Emerson and Greenough, and of Ruskin and Morris, he published book after book of repetitious exhortation, urging the world to convert to organic thinking and functional form. Theodore Roosevelt and Woodrow Wilson may have died with political progressivism, but progressivism in all its Protestant glory did not disappear in the arts until Wright's death, in an incredibly productive old age, on 9 April 1959; he was ninety-one, and had just accepted a new commission.

VII

The Progressive Era in the arts was thus different in certain key respects from the climates of creativity that preceded and succeeded it. Before progressivism dominated the country, the biological paradigm of Charles Darwin and Herbert Spencer taught artists to study nature with the minute care of science, collecting information about the slow but inevitable progress of the race. The emphasis was on the present and on those scientific laws that taught people how to encourage an organically rewarding future. Realism and naturalism were terms that described painters like Winslow Homer and Thomas Eakins as well as writers like William Dean Howells and Stephen Crane. By the 1890s, the Darwinian climate was exhausted by pessimism, determinism, and despair. Alcohol and disease claimed several talents; others proved unable to complete new work. The sense of inevitable progress through adaptation to nature disappeared.

The progressives were younger and Darwin was no longer news. Realism and naturalism inspired only a few, chiefly those involved directly with journalism. Instead, progressives rebelled against their fathers by rediscovering their grandfathers. The seamy practicalities of modern industrialism did not have half the appeal of Transcendentalist Nature and the opportunity to reexperience the sublime. By making a fantasyland out of their childhoods, by paying homage to Emerson and Lincoln, the progressives could escape their immediate artistic heritage and still retain a sense of belonging. The least innovative were those most attracted to realism—the writers of progressive novels urging reform, the poets yearning for a lost hero, and the painters laboriously experimenting with the shades of gray in the New York air. The most innovative were those like Ives and Wright, who seemed largely untouched by realism and who could joyfully pillage the heritage of their families and use it to forge new styles. They remained believers in progress, demanded social reform, and stressed environmental determinism over biological determinism.

Yet, to call them modernists is wrong. The essence of modernism in most of the arts was a life of art for its own sake and a body of work that took forms, colors, and designs as being inherently of value and independent of moral meaning. Modernists did not pillage their youth and the family past for stories to tell the purchasers of their paintings or listeners to their music. They chose any subject or any past that stimulated them: the Greece of *Ulysses* or "The Waste Land," the Provence of the early Ezra Pound, or the eighteenth-century Italy of Pergolesi for Igor Stravinsky. The typical modernist emphasized himself, his own consciousness, and his own needs, not society and its needs. He was often undemocratic in politics and elitist in taste, and he had no interest in compelling others to change their ways. A modernist took no responsibility for anything but his art; a progressive could not accept that posture for a minute. A Darwinian artist sought to depict society; a progressive wanted to reform it; a modernist preferred to ignore it.

Chapter 6

It Is Sin to Be Sick:
The Muckrakers and the
Pure Food and Drug Act

NEXT TO political activity, journalism was the most visible progressive profession. Progressive journalists have been congratulated for inclining Presidents toward reform and attacked for not advocating radical social change. Scholars have frequently recognized the evils that progressive writers exposed and wondered at the way many of those evils seemed to persist long after reform laws were enacted to end them. The subsequent careers of a number of writers, as they joined the service of the very economic institutions they once condemned, have brought charges of bad faith, inconsistency, naïveté, and economic self-interest. Some later students of the period seem to have implied that a desire for inflated magazine sales, a need to achieve personal celebrity, and even a desire to deflect real reform all played their parts in the careers of progressive journalists.[1]

Much of this analysis fails to consider the progressive journalists in the context in which they thought, wrote, and acted. Like anyone else in any other period of history, they grew up believing in values that led them to certain questions and goals. They could not predict the direction of the economy or the evolution of government. Likewise, they did not share many of the compulsions so dear to subsequent analysts: to be radical, to take an adversary posture toward the middle class, to be the creators of a perfect literary form, to be hostile to wealth of any kind, to be hostile toward anything smacking of patriotism. They lived in the only world they had, and their economic, political, literary, social, and sexual values inevitably reflected what they learned as children, and what their friends and colleagues assumed was true. As with the novelists and painters, their fate was to possess one system of values while social conditions were rapidly making other systems of values more appropriate. Understanding this predicament provides insight into the progressive achievement.

Within progressivism one important group of writers were "muckrakers," a term President Roosevelt used in a mood of hostility when he felt that they had gone too far in their criticism of American life and its politicians. About forty journalists have been subsequently identified as muckrakers; about twenty of these were consistently active. As a rule they were attached, at one time or another, to several large-circulation magazines, although working for newspapers and writing fiction also attracted many of them. Any understanding of the progressive world, its laws, and its successes and failures, requires a thorough knowledge of muckrakers, their ambivalent relationship to several eminent progressive politicians, and their central role in the writing of progressive legislation. The key to any such understanding of the muckrakers, as with other progressives, lies in the pervasive Protestantism that was so much a part of their childhoods.

The most famous of the muckraking editors, and the man who commissioned several of the most famous exposés of the period, for example, was S. S. McClure, the founder of *McClure's Magazine* and the inspirer of Lincoln Steffens, Ida Tarbell, Ray Stannard Baker, and others. Writing about his childhood, McClure could remember only three books in his home in Ulster, Ireland: *Pilgrim's Progress,* Foxe's *Book of Martyrs,* and the Bible. As the loyal son of a Scots-Irish Protestant and a French Huguenot, he never could forget them, and it was thus peculiarly appropriate that the muckraker, in Bunyan's book, became the chief symbol of the period. Theodore Roosevelt knew the work as well, and meant his usage of the term to be critical; he was clearly fighting men from within his own moral frame of reference with words and images they could readily understand. Indeed, when a group of distinguished muckrakers left *McClure's* to form *The American Magazine,* they began almost immediately to run a column called "In the Interpreter's House," thus implicitly retaking possession of the book for the cause of righteousness. No ideas or policies coming from such a background could avoid being tinctured with all its values.[2]

The muckrakers generally acquired their values in the Middle West. Most were born between 1857 and 1878, with 1868 about the median. They grew up in small towns and were vicarious if not always actual farmers. Chiefly of English, Scots-Irish, or German backgrounds, muckrakers were a racially homogeneous group. About 72 percent were native Americans. They came from families that had long been settled in America, sometimes with roots in colonial New England, and from families whose fathers were well-educated and sometimes professional men. They took great pride in their pioneer backgrounds. Their religious values were comparable to those of other progressives. Many were nominally Presbyterians and Methodists, but few were serious churchgoers.[3]

They thus found themselves equipped intellectually with religious categories when they had to confront economic, social, and political problems. This is where much of the modern problem of analysis lies. Because of the muckrakers' backgrounds, they tended to regard exposure of an evil as an end in itself. Once a sin was recognized, surely decent citizens would repent and a good world could develop. Likewise, public agreement was important: if the public agreed with the preacher/journalist, then it was converted, and to the descendants of the Puritans, conversion was a self-justifying event. Good works would presumably follow, but few worried nearly as much about good works as about conversion. Here, too, was the birthplace of so many of the red herrings of progressivism: the drive for prohibition, for control of prostitution, for the regulation of human behavior. Here, too, was a faith in mechanism for its own sake, as if the initiative, referendum, and recall were self-fulfilling, good things. After all, the good puritan was more interested in being a saint than in performing good works. If the heart of the body politic were pure, then its actual achievements were often of minor interest.

Any number of progressive laws offer good examples of how muckrakers worked and how they pushed for specific legislation. Numerous studies of local reform, of railway regulation, of tariff reduction movements, of financial reorganizations, and of antitrust activities are already available, as are more specialized studies of state scandals, such as the investigations into life insurance in New York. But one piece of legislation conspicuously stands out as an archetype of how moral indignation could lead to progressive legislation. The background of the enactment of the Pure Food and Drug Act of 1906 shows muckraking in all its variety and detail: it includes the best piece of literary progressivism, Upton Sinclair's *The Jungle*; it includes muckrakers' feuds with their progressive president Theodore

Roosevelt; it shows how public opinion could be aroused, shaped, and used to force the enactment of a law, thus demonstrating how much a part of their culture the muckrakers were, and why they should not be separated from the body of average voters; and it shows, unfortunately, why progressive laws so often accomplished so little: honesty became a substitute for effective medicine, and inspected meat all too often became a substitute for pure, wholesome meat.

II

Any attempt to establish a "typical" muckraker or "typical" progressive encounters problems with Upton Sinclair. Even though most students of the Progressive Era acknowledge that Sinclair had great impact on American public opinion, legislation, and on the image of America in the minds of foreign readers, they often refuse to see him as a genuine progressive. He was a socialist, a food faddist, and believer in various forms of mental telepathy. Unfriendly journalists made his private life into an object of public scorn, expecially on matters of sex. He was not from the Midwest, but from the South. His father was not an old, abolitionist yankee, but rather, a frank sympathizer with the Confederacy whose family was deeply involved in the Confederate navy, and who cursed Republicans until the day he died. Furthermore, Sinclair never lived for long in any small town: he was born in Baltimore, spent much of his early adulthood in New York City, and took occasional retreats to isolated cabins, farms, and a utopian colony. Born in 1878, he was ten to twenty years younger than his more important colleagues. Even odder, the family religion was a genteel, class-conscious Episcopalianism, on the surface innocent of any of the evangelicalism so obvious in other nurseries of progressivism. But with Upton Sinclair, the surface is most misleading; because of the peculiarities of his upbringing and his remarkable progress toward radicalism, he proves to be the exception that establishes the rule.

Sinclair's father was one of the few male Sinclairs who avoided the navy. He remained proud of his family heritage, but hardly contributed to it by making a precarious living as a salesman of whiskey and hats. He retained his Southern pride, courtesy, good manners, regional loyalty, and a generosity often bordering on irresponsibility. He was that pathetic figure, a professional Southerner, living on past glory and unwilling or unable to come to terms with modern industrial life and its northern values. Where a typical progressive could idolize his father and even confuse him at times with God, Sinclair never could. His father was a weak, irresponsible alcoholic who shamed Sinclair and his mother with pathetic bouts of debauchery. Because his father was such a hopeless model, Sinclair rejected everything he stood for. The South became a Slavocracy that taught its youth to be lazy, immoral, and exploitative; it deserved to lose in 1865. Sinclair, in other words, came to adopt abolitionist values backing away from their enemies.

Sinclair's mother was another story. Sinclair worshipped her and internalized her values. Priscilla Harden was the daughter of a prosperous railroad official who had also been a Methodist deacon. Conscious of status, she miscalculated badly when she married Sinclair. She never forgave her husband for his unsuccessful career, lack of money, and love of alcohol. Indeed, she came to hate any kind of "stimulant" and refused even tea or coffee. Instead, she developed a habit of retreating to the homes of her father or his close relatives, all of whom were wealthy, lived well, and shared a staunch Methodism. Upton Sinclair thus found himself living a schizophrenic life, shuttled between wealth and poverty. At home he was poor and lived in rooming houses so ill-equipped he rarely had a bed of his own. With relatives he was well fed and coddled by aunts and servants. Some of his prejudices began to form at an early age. He hated business and capitalism, because he associated them with his father's shabby status and alcoholism. He had a great ignorance and fear of sexual life, because his mother overprotected him and saw that he never encountered the "real" world of men and instincts until he was beyond puberty.[4]

The Episcopal Church has always had its high church, or sacramental, wing, and its low church or evangelical, wing. Sinclair somehow managed to straddle them. Every Sunday he went dutifully with his mother to the church in their neighborhood with the most status, and yet like her, he seemed to be at heart a Methodist untouched by the gentility, tolerance of alcohol, and intellectual sophistication that generally marked members of the high-church group. The stern, humorless Methodism of his maternal grandfather ruled Sinclair and his mother, and evil to the boy became definable by men drinking and women smoking. Sinclair remembered later that he was conscious of the devil and a devoted reader of the *Christian Herald* while still young. More to the point, he was intensely bookish, and lived most of his psychological life in literature, especially the Bible and *Pilgrim's Progress.* He tried to model his own character, so lacking in male models, on Jesus and Christian. Only as a college student in New York did he finally find one that seemed appropriate. The Reverend Mr. William W. Moir was a young assistant at Sinclair's church; he filled the role of surro-

gate father for Sinclair and as many as fifty other young men who came
to him regularly for advice. Moir came from a wealthy background, but
had abandoned any striving for money and success when he converted to
the social gospel. Unfortunately, he was also obsessed by sexual inhibitions, and he felt called upon to preach constantly to his young men about
the need for chastity, the horrors of self-indulgence, and the diseases that
could result. He thus reinforced many of Sinclair's attitudes and in effect
institutionalized Mrs. Sinclair's fears, teachings, and prejudices. Even to
Sinclair himself as an adult, this period in his life struck him as confused,
intense, and abnormal: "It was my idea at this time that the human race
was to be saved by poetry. . . . Any psychiatrist would have diagnosed me
as an advanced case of delusion of grandeur, messianic complex, paranoia,
narcissism, and so to the end of his list." Phrased more calmly, Sinclair had
acute psychological problems that he had sought to resolve first through
religion, then through creative writing; reform came later.[5]

The alcohol that symbolized evil to Sinclair became an obsessive terror
that inspired all or part of a number of his books. "Whiskey in its multiple
forms: mint juleps, toddies, hot scotches, egg-nogs, punch—was the most
conspicuous single fact in my childhood." His father's drinking created
innumerable social and financial problems, forced him to change careers
and go from selling whiskey to selling hats, and led the family from
Baltimore to New York. It produced many of the inhibitions in Sinclair's
mother, her abnormal dependence on her son, and his own acute dependence on her. It seemed tied to the contrast between wealthy Hardens and
impoverished Sinclairs. As Sinclair grew older, it seemed as well to be a
link between the forces of capitalism and the forces of political corruption,
and ultimately, prohibition became an important part of his progressivism,
as it did for many other progressives, from Jane Addams to William Allen
White. "It made an indelible impression upon my childish soul, and is the
reason why I am a prohibitionist, to the dismay of my 'libertarian'
friends."[6]

Sexual license, the other great symbol of religious evil, likewise tied the
problems of the home to the political issue of society. His mother, Sinclair
later recalled, taught him "to avoid the subject of sex in every possible
way; the teaching being done, for the most part, in Victorian fashion, by
deft avoidance and anxious evasion." He certainly had a strong sexual
drive, and was frequently tormented by it, yet he always accepted his
mother's standards. He thus found himself, more than once, on the verge
of sexual union with a woman, only to draw back and begin feverishly to
lecture his partner on the moral dangers of their actions. He fell in love
with a family friend who was also enamored with literature and convinced

of Sinclair's "genius." They embarked on a marriage that must have seemed doomed from the start. The couple resolved that physical union was a desecration of the purity of their love, and that marriage was merely a social convenience that would prevent people from gossiping about the time they spent together. Sinclair hints strongly that the marriage was not consummated for several months, and only then because a doctor told him that celibacy could make his wife psychologically disturbed. But Sinclair never reconciled himself to normal sexuality, and forever after sexual temptation remained "like some bird of prey that circled in the sky just above him—its shadows filling him with a continual fear, the swish of its wings making him cringe."[7]

With these inhibitions Sinclair discovered prostitution with horror. Apparently, his mother and his wife were not the only kind of women, and "the truth, finally made clear, shocked me deeply, and played a great part in the making of my political revolt." He turned to politics when he discovered a reformer named William Travers Jerome, who was attempting to stop prostitution, liquor abuses, and a whole network of political corruption in New York City. Sinclair was still not a socialist, but his campaigning for Jerome marked his entrance into the moralistic politics that attracted so many progressives. The problems of his personality had found a political as well as a literary outlet. He wrote: "So it was that, in my young soul, love for my father and love for my mother were transmuted into a political rage, and I sallied forth at the age of twenty, a young reformer armed for battle."[8]

For all his awareness about problems with alcohol, sexual depravity, and political corruption, Sinclair was a remarkably innocent and rather conservative citizen of William McKinley's America. After graduating from City College, he attended Columbia University graduate courses for several disorganized years, never worrying much about satisfying degree requirements. Since age sixteen, he had been supporting himself by writing potboiler novels for children, for magazines like *Argosy* and the *Army and Navy Weekly,* and he continued to write frantically both for mass audiences and for more sophisticated ones as well. In neither literature nor politics did any trace appear of his radical future. Late in life he claimed to be thankful that he had not come across anything modernist in literature until he "was to some extent mature, with a good hard shell of puritanism to protect me against the black magic of the modern Babylon." Since 1900 depraved literature had flooded the land, and the young writers adopted it as their own: "They had no shells of puritanism, but try fancy liquors and drugs, and play with the esoteric forms of heterosexuality and homosexuality, and commit suicide in the most elegant continental style."

Those, like Sinclair, who preferred "to remain alive are set down as old fogies. I must be one of the oldest." He was equally benighted in politics, if more briefly. He regarded William Jennings Bryan and populism as "vulgar, noisy, and beneath my cultured contempt," and categorized himself as "intellectually a perfect little snob and tory."[9]

In the fall of 1902, Sinclair's life suddenly changed. He met Leonard D. Abbott on a visit to the offices of the *Literary Digest,* and then John Spargo, the editor of *The Comrade.* Pleasant, and committed socialists, they quickly attracted Sinclair to their cause. Through them Sinclair met Gaylord Wilshire, the wealthy Los Angeles entrepreneur and editor of the socialist *Wilshire's Magazine,* and George D. Herron, by this time a notorious ex-minister, whose divorce and remarriage had made him one of the best-known and most viciously attacked socialists in the country. Herron's charm of character and his suffering appealed to Sinclair, who had severe marital problems of his own. When William Moir died suddenly, Herron replaced him as the chief father-figure in Sinclair's life. Herron explained the new religion of socialism to him and suggested that Sinclair's work had not been properly appreciated in America because capitalism had made the people blind to art and beauty. When Sinclair told him how poor he was and how desperate to write his next book, Herron offered to advance him enough money to survive. Sinclair not only remained true to Herron's ideas, he also remained grateful to Herron for this material aid.[10]

The book Sinclair wrote while living on Herron's money was *Manassas,* projected to be the first volume of a Civil War trilogy. In it he followed a crooked path toward the progressivism that came more easily to those born within a conventionally northern, evangelical heritage. Sinclair abandoned the personal world of his earlier novels, discovered a heritage worthy of his new socialist ideas, and in the process, renounced forever the mint-julep Southernism of his inebriated father. The plot took a Southern slave plantation heir to Boston, where he was exposed to abolitionism and literate Negroes; it then returned him to the South, where he became convinced of the iniquity of slavery, the corruption of Southern whites, and his own estrangement from his people. Naturally, given Sinclair's own past, the most horrible of the curses brought by slavery were sexual license, especially between white men and black women, and excessive drinking. The plantation, he found, "was simply a house of shame," and every slave woman on the place was a harlot bidding for favor. Half-naked girls were everywhere, most of them pregnant by the time they reached fifteen. The boys grew up "steeped in vice to their very eyes." The hero became associated with a Quaker abolitionist, met Frederick Douglass and John Brown, joined the Union army, and by the end of the book, had gone through the battle at Manassas with a socialist German freedom fighter at his side. The writing of the book in effect joined abolitionism and socialism for Sinclair, and enabled him to write his most successful fiction to date.[11]

Indeed, while most progressives tended to find their historical roots in abolitionism, the Union, and the North in the Civil War, Sinclair characteristically carried the tendency to extremes. He decided that chattel slavery and wage slavery had a historical parallel, and that, therefore, abolitionism was another kind of progressivism and socialism. He worked out a list of the individuals of earlier days who had prefigured the reformers and reactionaries of his own time. Daniel Webster was "the conservative reformer" like Grover Cleveland; John C. Calhoun was "the unwilling prophet" like Marcus A. Hanna; Henry Clay was the "great compromiser" like Theodore Roosevelt; Charles Sumner was the "statesman of radicalism" like William Jennings Bryan; Wendell Phillips was the "orator of the revolt" like George D. Herron. The list continued until the final sublime parallel, all too prophetic of some of Sinclair's later insights: Abraham Lincoln was "the untried hope," like William Randolph Hearst![12]

Once initiated into socialism, Sinclair scarcely paused for breath. He began to read omnivorously the most significant radical writers: Europeans like Karl Marx, Karl Kautsky, and Peter Kroptkin and Americans like Edmund Kelly, Jack London, and Edward Bellamy. He became fascinated by *The Appeal to Reason,* a populist weekly published in Girard, Kansas. Socialist ideology gave him a doctrinal substitute for his old religious views, and he swallowed it whole. Fred D. Warren, editor of the *Appeal,* was impressed by *Manassas,* and offered Sinclair another subsidy to do for wage slavery what he had already done for chattel slavery. The offer appealed to Sinclair, who was desperate for the money and eager to tie literature to socialism. A recent strike in the Chicago stockyards had drawn the attention of many socialists to Packingtown, so Sinclair decided to go there and study meat-packing the way he had studied slavery. He also formally joined the Socialist Party. Even so, he never abandoned his old religious vision; over a decade later, in a book almost unremittingly hostile to organized religion, he could still write: "I count myself among the followers of Jesus of Nazareth. His example has meant more to me than that of any other man, and all the experiences of my revolutionary life have brought me nearer to him."[13]

The book that resulted from this commission remains one of the best examples in the history of literature of a limited but often influential aesthetic. Sinclair was too young fully to have articulated his views on literature, but when he finally did so in 1925, he gave, in effect, the rationale for his art in *The Jungle.* He clearly stated his dislike for any art that insisted that it should be judged solely for its own sake, or that was esoteric, traditional, dilettantish, or amoral. All art, he insisted in italics, was propaganda: *"It is universally and inescapably propaganda; sometimes unconsciously, but often deliberately, propaganda."* Artists should always be in tune with the spirit of their own times, and their art should be *"a representation of life, modified by the personality of the artist, for the purpose of modifying other personalities, inciting them to changes of feeling, belief and action."* He fully intended his book on Packingtown to stand with other great examples of this kind of art, from *Pilgrim's Progress* to *Resurrection, Germinal,* and *L'Assomoir.* [14]

In gathering material for this propagandistic art, Sinclair soon learned of other progressive currents that helped to shape his own perceptions and prepared public opinion for his own writing. Chicago political corruption was already legendary. The traction industry had been particularly obnoxious, and the machinations of Charles T. Yerkes were even then in the process of exciting Theodore Dreiser to the writing of *The Titan.* Other food investigators, like Charles E. Russell and Algie M. Simons, had fingered some of the unlovely products of the meat industry, with little obvious public impact. Sinclair was already familiar with much of this story; he knew of Russell's work in particular, and congratulated him for one of his articles in January 1905. He heard much more during the few weeks he spent exploring the industry and interviewing everyone available about working conditions, sanitation, and government inspection methods. He roomed at Mary McDowell's University Settlement House, where he met a constant flow of informed visitors. He visited Jane Addams at Hull-House and discovered what he could from her infinite supply of data on urban problems. Perhaps most significant for his insights and his self-confidence, he also met another man on a similar mission: Adolphe Smith, an investigator from the respected and influential British medical journal, the *Lancet.* Smith not only confirmed much of what Sinclair discovered, but he assured Sinclair that the unsanitary conditions were all unnecessary, and that in Europe under state regulation, conditions were different. Sinclair wrote frantically for about three months and serial publication of *The Jungle* in the *Appeal to Reason* began even before he was finished writing. [15]

Sinclair's announced intentions in writing the book differed in some detail from the result, and enormously from the impact on most readers. He had wanted to "set forth the breaking of human hearts by a system which exploits the labor of men and women for profits." He planned a book with an implicit socialist message that would be fundamentally "identical with" *Uncle Tom's Cabin* in what it tried to do. It would depict an exploitative system "slaughtering women and children." The action would turn on a Chicago stockyards strike, with workers, employers' associations, grafting politicians, and foreign laborers all entangled in the scramble for survival and profit. It would include the white-slave traffic and impure food; the impact these make on the hero would drive him to socialism.[16] Sinclair, in short, did not seek federal legislation. Now converted to the new religion of socialism, Sinclair planned to write a socialist tract in the form of a novel. His goals were the conversion of the reader through empathy with the workers of Packingtown, votes for socialist candidates, and ultimately, a socialist America. Pure food and drug laws received one unstressed sentence.

The serialization attracted some attention among the devoted readers of the *Appeal,* but none among a wider audience. George Brett, an influential editor at Macmillan, had offered a $500 advance for book republication, but when the full text came in, he demanded cuts that amounted to serious censorship. Sinclair refused to have his work disemboweled like one of J. Ogden Armour's cattle, and after several refusals from other publishers, he persuaded Walter Hines Page, a liberal, Southern, progressive editor at Doubleday, Page Company to print the work. Page himself investigated the problem to be sure that libel suits would be unlikely, and that Sinclair had reasons for his indictment. When the investigation confirmed what Sinclair himself had seen, Page went ahead. The book version appeared on 16 February 1906 and was an immediate sensation. The public, however, ignored the book's chief message; its stomach was so upset it had no time for socialism. Twenty-five thousand copies were sold in the first forty-five days of publication.[17]

To Sinclair's dismay and disgust, the public read his book as an exposé intended to arouse support for pure food and drug legislation. Such legislation was again pending in Congress, and it seemed on the verge of success during 1906. In effect, they took a book of well over 300 pages and concentrated on perhaps 10 or 20 pages meant by the author to be true and effective, but hardly central.

The book contained only about three references to drugs, but on the issue of pure food, Sinclair could become lyrically nauseating. The passages are few, but brutally specific and evocative. There are suggestions that people occasionally fell into vats, and that all but their bones went forth to the public as high-grade, edible lard. He hints strongly that Packingtown children were ill because they drank milk that had been watered and then doctored with formaldehyde, and because peas they ate had been colored an appetizing green with copper salts, and jams with aniline dyes. Meanwhile, the adults, used to pure European meat processing, were eating large quantities of smoked sausage, and thus slowly poisoning themselves with the chemicals designed to hide the spoilage.

MINISTERS OF REFORM

The book's instant fame gave it a life of its own. Once he published it, Sinclair could not control it. It became a source of poetry, jokes, and indignation. It affected pending legislation, foreign relations, and, of course, the nation's eating habits. It showed how moral indignation about alcohol, sexual relations, and their Christian meaning could be translated into progressive ideas and, ultimately, into progressive legislation—even without the cooperation of the author. To Sinclair the brouhaha was distressing. "I aimed at the public's heart, and by accident I hit it in the stomach," was his famous lament. Jack London was more in the spirit of the actual writing of the book: "Here it is at last!" he wrote enthusiastically. "The *Uncle Tom's Cabin* of wage slavery! Comrade Sinclair's book *The Jungle*! And what *Uncle Tom's Cabin* did for black slaves, *The Jungle* has a large chance to do for white slaves of today." Alas, it was not to be. The *New York Evening Post* summed up the more popular response unforgettably:[18]

> Mary had a little lamb,
> And when she saw it sicken,
> She shipped it off to Packingtown,
> And now it's labelled chicken.

III

For an examination of the development of that key piece of progressive legislation, the Pure Food and Drug Act, the story of how Sinclair came to write about Packingtown was only part of the narrative. Working almost simultaneously with him was a disparate group of editors, writers, government officials, and doctors who were trying to mobilize public opinion on a closely related subject: the quality, efficacy, composition, and purity of American drugs. These people were increasingly convinced that Americans' health, not to mention their pocketbooks, suffered because of the lax legislation that passed for government regulation. The crusaders for pure drugs were not people with any great grievance against American life, nor were they neurotics looking for an issue to exploit for chiefly psychological reasons. They also were not conservatives trying to throttle reform, nor were they opportunists trying to make reputations or increase magazine circulation. They were, instead, men of moralistic character and values who uncovered a serious problem in American life, analyzed that problem, helped to shape public opinion, worked for legislative change, and saw a law enacted that went at least part of the way toward effective reform. Like

Sinclair, they, too, provide an example of how moral indignation led to progressive legislation.

The problem of how to regulate drugs was as old as the country. Part of the problem, of course, was the primitive state of chemistry; part was the lack of scientific medical theory or training; part was the anarchy that accompanied the growth of a new country with few legal precedents for regulation and a jealous tradition of separating individual, local, and federal rights. Benjamin Franklin's mother-in-law believed she had developed an ointment that would help relieve itching; Cotton Mather thought that urine and dung had therapeutic value; Benjamin Rush, one of the best-known and most respected doctors of the revolutionary period, was convinced that all fevers and diseases were the product of excited blood vessels, and he thought that purging and bleeding his patients would calm the vessels. Modern medicine did not exist, and well into the age of Andrew Jackson, most Americans believed, as one writer has put it, in their "inalienable rights to life, liberty, and quackery."[19]

In the absence of scientific knowledge and any official public guidance, quackery indeed flourished. The patent and copyright laws gave legal protection to the names, bottle shapes, and labeling materials on medicines, while the ingredients and formula remained closely guarded secrets. People bought the medicines at their own risk. The local newspapers were an excellent place to advertise the various drugs, and personal endorsements were a popular sales tool. The mails became more efficient, so advertising circulars as well as drugs could go throughout the country. Promoters became especially adept at making lists of vague symptoms that their products could alleviate, as well as diseases for which they were specifically effective. Weakness, headaches, impotence, irregularity of the bowels, night-sweats, and depression bothered most people at one time or another, and yet they were difficult for even the most reputable physician to diagnose and cure. The result was the proliferation of useless, even harmful, drugs, and a great many people who did not recover, or who did recover for reasons unrelated to the drugs. The elder Oliver Wendell Holmes, one of the most eminent medical teachers in the country, observed, "If the whole materia medica, *as now used,* could be sunk to the bottom of the sea, it would be all the better for mankind—and all the worse for the fishes."[20]

Quite aside from the uselessness of most drugs from a purely medical standpoint, other issues began to arouse professional, legislative, and public concern during the later part of the nineteenth century. Many medicines, for example, made preposterous curative claims. One of the most flamboyant of the medicine men, Henry T. Helmbold, was extreme in style but not in substance when he began to advertise his African plant, buchu, as a cure-all. The medicine had been known in Africa mostly as a cosmetic that, applied to the skin, made it smell like peppermint; it also was thought to be a mild diuretic when taken internally. Helmbold's extract included buchu, cubebs, licorice, caramel, molasses, and a little extra peppermint. He also became known as a purveyor of exceptionally bad advertising taste, with his striking ads showing his ornate "Temple of Pharmacy" on Broadway in New York, as grotesque a building as any architectural abortion of the Gilded Age. But it is not Helmbold's success in vending his nostrum to important figures from Boss Tweed to John Jacob Astor that is worth remembering; it is, rather, the claim the medicine made to cure almost all known diseases. One ad claimed that buchu could cure "General Debility, Mental and Physical Depression, Imbecility, Determination of Blood to the Head, Confused Ideas, Hysteria, General Irritability, Restlessness and Sleeplessness at Night, Absence of Muscular Efficiency, Loss of Appetite, Dyspepsia, Emaciation, Low Spirits, Disorganization or Paralysis of the Organs of Generation, Palpitation of the Heart, And, in fact, all the concomitants of a Nervous and Debilited state of the system."[21]

Drug manufacturers soon followed the general course marked out by other industries during the Gilded Age. They began as small, independent concerns, but in time, as some medicines failed and others succeeded, the businesses expanded and began to resemble other ventures. In 1881, the newly formed Proprietary Association of America gave a unified voice to the new industry, and served as an effective lobby in local, state, and national legislatures. In time the organization came to symbolize the drug trust, which well deserved membership in the ranks of those industrial giants then exciting the attention of reformers. Like the others this group soon controlled prices, restricted competition, and centralized methods and channels of distribution. Even had their products been honestly labeled and medicinally effective, the drug trust still deserved regulation for many of the same reasons as the steel trust—even more considering its dishonest advertising and the ineffectuality of its products.[22]

Governmental bodies made some efforts to regulate food purity and drug effectiveness. Even before the Civil War, the federal government had attempted to keep artificially colored tea out of the country, and after the war, several states enacted early regulatory measures. Between 1879 and 1898, six states enacted a law on the subject, and several states enacted more than one. By 1906, when the campaign for federal regulation reached its height, most states had at least some legal guidance for the production and content of food, drinks, and drugs, however ineffectual these laws might be. Thus, the muckrakers were not working in a vacuum. They had discovered a salient issue, and they concentrated on informing and shaping public opinion.[23]

A similar combination of interest and ineffectuality dominated the medical profession. The first attempts to organize an American Medical Association began during 1847 and 1848, spurred by the understandable desires of reputable professionals to distinguish themselves from such once respectable havens for charlatans as homeopathy and hydro-therapy. For decades the organization devoted its attentions to narrow concerns strictly within the profession. Medical opinion about most of the issues of the years before 1900 was either silent or amorphous. The profession offered little protest against unhealthy industrial conditions, misleading advertising, or the adulteration of food. Likewise, it was inefficient as a lobby, and had little influence on either state or local legislation throughout the nineteenth century, although it did attempt to obtain better statistics on medical and public-health matters, and it pushed for a greater role in the federal regulation of public health. At the turn of the twentieth century, however, the profession, and the AMA as a group, reorganized itself significantly, and a number of reform movements developed, especially in the area of medical-school administration. Physicians felt they needed some kind of lobbying power to assert professional medical influence upon relevant legislation, and perhaps even elect doctors occasionally to public office. In 1899 the Committee on the National Legislation was formed as a first step toward establishing a permanent lobby, and various watchdog committees began to review all relevant health legislation. After 1904 a Committee on Medical Legislation met regularly, and after the enactment of the Pure Food and Drug Act, it worked to have state laws enacted to conform to the federal standard.[24]

Thus, at the time when the muckrakers went to work, public opinion was ill-informed about the medicines people used, doctors were either ignorant or apathetic, and legislatures dealt with other issues that seemed more pressing. Certainly, editors and journalists were at first neither better nor worse on this issue than any of these other groups, and many of them had good reasons for refusing to examine the problem closely. Not only did leading medical and religious journals freely accept the most preposterous advertisements, but also the drug trust had devised a peculiarly effective means of maintaining a good press. Mr. F. J. Cheney, the proprietor of Hall's Catarrh Cure, had devised the so-called "red clause" to muzzle press criticism. The "red clause" was generally a clause printed in red in advertising contracts stating that the contract was void if any hostile regulatory legislation were enacted in the state in which the ad appeared.

The assumption was that newspapers and magazines would write editorials or print articles opposing such laws and otherwise lobby with state legislators in order to save their lucrative contracts. In practice the newspapers sometimes needed nudging from the drug trust, but when pushed, they frequently responded, and regulatory legislation was thereupon drowned by what seemed to be a flood of hostile opinion from throughout the state.[25]

Even before the 1890s, however, a few sensitive publishers became aware of the risks involved to their readers and their own consciences. Orange Judd's *American Agriculturist* had been exposing nostrums and refusing their patronage since 1859, and after 1877 Wilmer Atkinson's *Farm Journal* did the same. The *Popular Science Monthly* ran Dr. S. Weir Mitchell's *The Autobiography of A Quack* in 1867, a book so vividly successful in exposing the racket that it reappeared later in another journal as a contribution to muckraking. The first real breakthrough that led, however erratically, to legislation did not come until the 1890s, when Cyrus H. K. Curtis, publisher of the *Ladies Home Journal,* overrode the objections of his treasurer and banned all advertisements for proprietary medicines. He found a supporter in his son-in-law Edward Bok, who was willing at regular intervals to include critical articles in the journal he edited for the family. The high alcoholic content of many of the medicines under examination gave Bok an effective means of reaching middle-class America, especially members of or sympathizers with the Women's Christian Temperance Union, many of whom apparently supported the journal. But excitement did not peak until 1904, when Bok and his journal were sued by the owners of Doctor Pierce's Favorite Prescription over some inaccurate chemical analyses that Bok had printed unknowingly from outdated sources. Thoroughly aroused, Bok printed a retraction, prepared for the trial, and hired a young journalist named Mark Sullivan to examine the problem and perhaps also to find a dusty old bottle of the drug that would substantiate Bok's data.[26]

Sullivan could not find such a bottle. Bok lost the suit, and had to pay $16,000, but once started, the research continued. S. S. McClure had recommended Sullivan to Bok, and Sullivan immediately began to pursue the problem with the thoroughness one would expect from someone from *McClure's.* He advertised for experts in making, selling, and promoting drugs, pretended to be interested in hiring such people, and then interviewed them in detail about what they had done for competitors. He discovered many of the more squalid tricks of the game. Two that received great attention were the widespread use of personal letters full of an individual's problems, letters that were then sold wholesale to other nostrum makers despite assurances of privacy to the writers, and Lydia Pinkham's grave. Lydia Pinkham's comforting face on countless labels and advertisements assured users of her medicine that she would happily answer letters from suffering women. Sullivan and a friend found her tombstone near Lynn, Massachusetts, and a picture of the gravestone was soon in the *Journal.* William Allen White, editor of the *Emporia Gazette,* provided Sullivan with contracts containing the "red clause," and they, too, soon appeared in print. Sullivan even managed to obtain the minutes of a meeting of the Proprietary Association of America, with fascinating details about lobbying and the fear of regulatory legislation.[27]

Bok approved of the conclusions of the resulting piece, but thought it too long and technical for his magazine. He knew that Norman Hapgood at *Collier's* was deeply disturbed both by impure drugs and the alliance of the nostrum makers and the press, and thought he might take it. Almost simultaneously with Sullivan, Hapgood himself had been at work on the issue, and *Collier's* also had stopped accepting advertisements for patent medicines. Hapgood printed Sullivan's article in November 1905. Meanwhile, he found himself more and more concerned about the problem. He decided that previous work was inadequate, and that he needed a new reporter who could spend more time on the matter. Hapgood, too, had long admired McClure and his stable of investigative reporters, and he wanted one of his own to do this job. He made the perfect choice in Samuel Hopkins Adams.[28]

Adams was in the typical muckraking group, as well as being closely connected to *McClure's* and its traditions. He had attended Hamilton College, and had long been interested in science and a possible career in medicine. He spent several years as a reporter after graduation, chiefly for the *New York Sun,* before joining *McClure's* to become the house expert on public health. In addition to articles that he contributed on subjects such as modern surgery, tuberculosis, hookworm, and typhoid, he was also the advertising manager. He was especially aware of the many connections between depressed social conditions and disease, and of how Europeans managed these problems better than Americans. Some of his views did not reach print until after the enactment of the Pure Food and Drug Law, but their appearance only confirmed his belief in the connections among social conditions, science, and public health. He was also an indignant moralist. "Sam Adams we considered a born muckraker," his roommate at the time, reporter Will Irwin remembered. "A character inherited from a line of insurgent theologians gave him firm conviction in his beliefs," and he apparently "gloried in combat." This son of a minister found plenty of combat when he engaged the drug trust for *Collier's.* [29]

Given his own background and interests and the few pieces that had already been published about impure or useless drugs, Adams was interested in the subject even before Hapgood gave him a job. Ray Baker had already suggested to Adams that he investigate nostrums for *McClure's,* but the unpredictable S. S. McClure was only lukewarm about the project, so when Hapgood appeared, Adams felt free to shift his allegiances to *Collier's.* He proved to be a superb investigative reporter. He interviewed widely, and found a few knowledgeable men willing to give out professional secrets. He purchased many remedies and had them tested by professional chemists. He looked up the writers of prominent testimonials to find out how genuine they were. He researched the red clause and its effect. He even found himself investigated by private detectives and threatened with blackmail on occasion. He was especially irritated by the laxity of the AMA and medical journals, which seemed to offer no protest about false advertising and unhealthy preparations; the use in many papers of advertising material printed as if it were locally generated editorial matter; the ineffectuality of many laws; and the apparent unwillingness of government bodies to use or publicize even the sound data that they did possess.[30]

The first three articles in the Adams series, printed on 7 October, 28 October, and 18 November 1905, were the most influential on public opinion, and did the most to push legislators to their duty. The first opened with this unforgettable paragraph:

> Gullible America will spend this year some seventy-five millions of dollars in the purchase of patent medicines. In consideration of this sum it will swallow huge quantities of alcohol, an appalling amount of opiates and narcotics, a wide assortment of varied drugs ranging from powerful and dangerous heart depressants to insidious liver stimulants; and, far in excess of all other ingredients, undiluted fraud. For fraud exploited by the skillfulest of advertising bunco men, is the basis of the trade. Should the newspapers, the magazines and the medical journals refuse their pages to this class of advertisements, the patent medicine business in five years would be as scandalously historic as the South Sea Bubble, and the nation would be richer not only in lives and money, but in drunkards and drug fiends saved.

He presented a methodical overview, classifying most of the nostrums to be either "harmless frauds or deleterious drugs." Laxatives could well be effective, but taken too often without medical advice, they could cause physical damage and dependence. Acetanilid could indeed relieve the symptoms associated with headaches, but was "prone to remove the cause of the symptoms permanently by putting a complete stop to the heart action." Cocaine and opium did indeed stop pain, but concealed in soothing syrups and catarrh powders in unknown quantities could also produce drug addicts. Alcohol certainly had valid uses, but placed in medicines given regularly to unsuspecting women and children, could easily lead to alcoholism and related complications. As for medicines that were essentially worthless, the problems were different only in degree: "In the case of such diseases as naturally tend to cure themselves, no greater harm is done than the parting of a fool and his money. With rheumatism, sciatica and that ilk, it means added pangs; with consumption, Bright's disease and other serious disorders, perhaps needless death."[31]

In his second article, Adams concentrated on "Peruna and the Bracers," or all the nostrums that had significant amounts of alcohol in them. Peruna was perhaps the foremost drug of this kind, and possibly of any kind, in the country. People of all ages took it for whatever "ailed" them, and it was reputed to be especially popular with ministers and temperance workers whose consciences did not allow them to take alcohol in any nonmedicinal way. In dry areas and on Indian reservations, it was often the chief alcoholic beverage, and the term "Peruna drunk" was quite common for the resulting antics and hangover. Adams found the beverage without medicinal value. "Anyone wishing to make Peruna for home consumption may do so by mixing half a pint of cologne spirits, 190 proof, with a pint and a half of water, adding thereto a little cubebs for flavor and a little burned sugar for color." It cost 15 to 18 cents to make and sold for about one dollar, thus yielding handsome profits. Peruna was advertised throughout the country as a cure for catarrh: a modest claim since, in practice, catarrh was "whatever ails you." The key book that told people why they should use Peruna found catarrh the basis of most medical ills: "Pneumonia is catarrh of the lungs; so is consumption. Dyspepsia is catarrh of the stomach. Enteritis is catarrh of the intestines. Appendicitis—surgeons, please note before operating—is catarrh of the appendix. Bright's disease is catarrh of the kidneys. Heart disease is catarrh of the heart." And so on. Upon analysis Peruna was about 28 percent alcohol combined chiefly with water. That made it roughly half as strong as normal whiskey, and roughly three times as strong as the typical claret or champagne of the day. Its primary competitors were Hosteter's Stomach Bitters, which contained a solid 44.3 percent alcohol, and Paine's Celery Compound, which was a mere 21 percent alcohol. It was no wonder that people who thought they drank no alcohol were receiving diagnoses as being hopeless alcoholics.[32]

In his third article, Adams concentrated on Liquozone, a reputed germicide and bactericide. Upon chemical analysis Liquozone turned out to be nine-tenths of 1 percent sulphuric acid, three-tenths of 1 percent sulphurous acid, and 98.8 percent water. It was thus medically worthless, and yet it prided itself on its professional image and its endorsements by presumably reputable medical men. Its advertisements claimed to cure the following:

Asthma	Gallstones
Abscess—Anemia	Goiter—Gout
Bronchitis	Hay Fever—Influenza
Blood Poison	La Grippe
Bowel Troubles	Leucorrhea
Coughs—Colds	Malaria—Neuralgia
Consumption	Piles—Quinsy
Contagious Diseases	Rheumatism
Cancer—Catarrh	Scrofula
Dysentery—Diarrhea	Skin Diseases
Dyspepsia—Dandruff	Tuberculosis
Eczema—Erysipelas	Tumors—Ulcers
Fevers	Throat Troubles

—All diseases that begin with fever—all inflammations—all catarrh—all contagious diseases—all the results of impure or poisoned blood. In nervous diseases Liquozone acts as a vitalizer, accomplishing what no drugs can do.

Adams had the Lederle laboratories run extensive tests, which showed that Liquozone had no curative powers and seemed instead to lower resistance to disease and death.[33]

IV

The muckrakers were inherently intellectuals. They preferred to research and write rather than to apply their ideas. They rarely ran for public office, and when they did, they tended to offend people because they seemed too detached from daily concerns. A muckraker might well devote his time and energy to the "democracy"; an actual citizen was far more likely to devote his or her time to child rearing or to keeping food, pure or otherwise, on the family table. The muckrakers proved to be good at arousing, informing, and directing public opinion, but they often had problems in directing legislators to the enactment of good laws.

During the year before the enactment of the Pure Food and Drug Act, two discrete influences tended to flow together, joining the many weaker ones that had been building for years. On the one hand, the muckrakers joined to form what amounted to an indignant and aroused moral lobby, pressuring every politican from President Theodore Roosevelt to the local alderman. On the other a dedicated group of government officials, led by Harvey W. Wiley, the highest ranking scientific bureaucrat in Washington, finally perfected their own case for reform, supported it with adroit publicity and hard data, and provided the political savvy essential to see the legislation through and then administer it.

Between 1879 and 1906, 190 measures connected to pure food or drugs were introduced into the Congress, but only 8 ever became law. Most bills died after their introduction. Several patterns emerge from the study of these bills. It was easier, for example, to enact special laws concerning one product, such as oleomargarine, lard, or glucose, than it was to enact a general law about food. It was easier to regulate foreign-made products than those made in America. It was easier to control goods sold in the District of Columbia than goods involved in interstate commerce. Too many old ideas and new special interests were involved. Southern states' rights Democrats were often worried about expanded federal powers; many legislators did not regard the issue as especially pressing or serious; others did not want great power in the hands of the Secretary of Agriculture or an appointed committee. Some legislators, especially a few visible figures in the Senate, were clearly the mouthpieces of economic groups, such as whiskey rectifiers from New York, beef packers from Illinois, or codfish shippers from Massachusetts. Before the muckrakers did their work, public opinion was not aroused to the dangers involved; accurate chemical information was insufficient; and no one, simple as it sounds, had really paid enough attention to the matter because of the press of other concerns that, at the time, seemed more important.[34]

The office of Harvey W. Wiley, head of the Bureau of Chemistry in the Department of Agriculture, was the center of government efforts to regulate food and drugs. Like other muckrakers, he came from a Protestant background. His parents had been strict members of the Disciples of Christ Church, and Wiley's father had been a lay preacher for over fifty years. The Wiley home had been heavy with a sense of morality, and of duty to home, church, community, and God. His family applied their values to political life. They read *Uncle Tom's Cabin* eagerly, and became a contact point on the underground railroad. As Wiley looked back in 1930, he emphasized this religiosity. He recalled the strictness of the Sabbath, when he "was not allowed even to whistle" much less "to indulge in any harmless frivolity in the way of play." Fishing "was a heinous sin. Sunday-school and church, memorizing Bible verses, listening to long sermons and being as quiet as possible all the day was the kind of Sunday discipline to which I was subjected." The cardinal principle of child rearing at the time was "obedience to parents," and Wiley could not "remember ever having knowingly disobeyed either my father or mother," with but a single exception. "One may say what he believes about the Presbyterian theology, I think that everybody will agree that the method of bringing up children by the old Presbyterian faith is a most excellent character-forming basis."[35]

Wiley received most of his education in obscure schools and colleges in Indiana, with a brief period at Harvard. He read and briefly taught the classics, but he was attracted to medical studies, especially those emphasizing chemistry. After a period of ill health, due chiefly to overwork, he became a professor of chemistry at Purdue University in 1874, where he also taught physics, physiology, and political economy. He became both a reader of Darwin and an "evangelist of chemistry," displacing the energies once put into orthodoxy onto a new and more contemporary field. He was particularly enthusiastic about conveying new chemical knowledge to physicians and farmers, as if he were trying to convert them to the new faith. He then studied in Berlin, traveled in Europe, and returned to begin studies in sugar, manufactured glucose, and the related subject of adulteration. By 1883 he had become a leading expert in sugar chemistry, and so unorthodox in his religious beliefs that he was an embarrassment to his college. Eager to leave, he accepted a job as a chemist with the U.S. Department of Agriculture.[36]

An inconspicuous figure in Gilded Age Washington, Wiley slowly acquainted himself with a whole range of products that might require regulation. By the late 1880s, he was reporting on dairy products, spices, condiments, alcoholic beverages, lard, and baking powders, and during the early 1890s, he went once again into sugar and its substitutes, then into tea, coffee, and cocoa, as well as canned vegetables. Eager to publicize his findings, he managed to have Alexander J. Wedderburn, an experienced journalist and an early advocate of pure food, appointed as a special agent to prepare a popular compilation of the findings of the Bureau of Chemistry. Wiley also became convinced that food adulteration was getting worse. The need for comprehensive federal legislation, standards, and regulation became clearer with every day.

Conventional categories of liberal and conservative, reformer or supporter of the status quo, fail with Wiley as with many other progressives. Wiley was basically a Republican, for reasons that had something to do with the Civil War and his family, and also because the Republicans had a national policy that called for governmental intervention to foster intelligent economic expansion. Under Republican Administrations Wiley found his work expanding and his superiors encouraging; under Democratic ones retrenchment was the word. Wiley hoped for effective regulation and reform, yet he was also a staunch supporter of William McKinley in 1896, much like other leaders in the ranks of the muckrakers. His nemesis, insofar as one can be singled out for special attention, was the seemingly immortal James Wilson, McKinley's Secretary of Agriculture, and Wiley's superior. For sixteen years, surely one of the longest tenures of any cabinet member, Wilson pushed departmental expansion, tried to increase services to agriculture—and showed a total incapacity to understand Wiley's position or take a progressive stand on matters of public health. Wilson, Wiley wrote later, "had the greatest capacity of any person I ever knew to take the wrong side of public questions, especially those relating to health through diet."[37]

The situation that Wiley and his growing staff faced was not a hopeless situation of wicked deception, which is the way it sometimes appeared in contemporary accounts. The changing technology of food distribution meant a far longer time period between the berry and the jam. Most honest manufacturers and distributors had few precedents to guide them; competition was often fierce; and the pressure to control costs often irresistible. Likewise, there was vast ignorance of nutrition and of the chemical problems involved with certain preservatives. Alcohol was often used as a solvent and preservative. Many citizens felt that medical care was risky and expensive and that home remedies were part of the right to life, liberty and the pursuit of happiness. Refrigeration, which ultimately became one key solution, was still in its infancy. The use of borax, salicylic acid, and formaldehyde to retard spoilage was common, and few people seemed to die from it. No one had given much thought to the possible long-term, deleterious effects of continued small doses of such chemicals until Wiley and a few men like him came along. Even then, public disbelief was strong; other problems seemed more pressing; and little could happen without public education, a receptive political climate, and a catalyst like the publication of *The Jungle*.[38]

Nevertheless, considerable deception did exist. Manufacturers sold flavored glucose as fruit jelly, they added chicory to coffee, and no one, including Upton Sinclair, knew exactly what went into many processed meat products. For Wiley, these additions and mislabelings were clearly frauds by moral or scientific standards, and he thought they deserved exposure and prohibition. In 1901 Wiley found the platform he needed and the prestige he deserved when the Bureau of Chemistry was established, and he could shift his campaign into high gear. His emphasis was not on the evils of the drug trust, or on the life of the workers; it was on "the ethics of pure food." His upbringing and moral principles had brought him to his own version of progressive reform: "What we want is that the farmer may get an honest market and the innocent consumer may get what he thinks he is buying."[39]

Wiley's concern with the ethics of pure food led him to a kind of chemical fundamentalism, and to some of the most famous scientific investigations of the Progressive Era. In his annual report as Chief Chemist in 1899, and in his subsequent testimony before Congress, Wiley consistently stressed chemical *honesty.* He said repeatedly that under certain unusual circumstances, such as when food had to travel great distances, last for an unusual period of time, or endure extreme climatic conditions, certain preservatives were essential. But he insisted that in America, most of these conditions existed only rarely, and that the law should be framed with this in mind. More important was honest labeling. All chemicals tended to interfere with digestion and related activities of the body. Healthy people could certainly tolerate a small amount of this interference, although the larger the quantity and the longer the period of ingestion, the greater the danger to health. But weak or diseased stomachs might well have trouble coping with certain unnatural chemicals, and serious injury to health might well result from an ignorant, unthinking use of some chemical a person never even knew he was ingesting. The only forthright way of dealing with this problem was full, clear, and explicit labeling of all contents. In that way, a person conscious of his health or fearful that certain products might be dangerous to him could avoid a particular food. In that way as well, the entire investigation of pure food and drugs suddenly became a recognizably progressive issue: it entered the realm of ethics and honesty, distracting attention from the essential problem of which chemicals did the most harm, which had unfortunate but masked effects, and which were unnecessary. It also diverted attention from basic issues like the size and power of the trust involved.[40]

Wiley combined this chemical fundamentalism with his new position as head of the Bureau of Chemistry to begin his legendary experiments on his "Poison Squad," experiments that ultimately produced the strongest scientific evidence available of the need for a pure food and drug act. Twelve healthy male volunteers from the Department of Agriculture volunteered to take all kinds of drugs and chemicals involved in the public issues under debate. They were isolated in a boarding house, their food carefully measured, their excreta preserved and analyzed, and their general health monitored. Somehow, the thought of these twelve men risking their bodies for the public caught the nation's fancy. Women became concerned and protective of all those healthy young males in danger, and women's clubs began to pay attention to the issue. Combined especially with the material Edward Bok was printing in the *Ladies' Home Journal,* the work of the Poison Squad demonstrated to mothers what unknown chemicals could do to the health of their own children. Wiley was also effective when speaking to such groups: "His large head capping the pedestal of broad shoulders and immense chest, his salient nose shaped like the bow of an ice-breaker, and his piercing eyes, compelled attention." He was soon widely known, affectionately or otherwise, as "Old Borax."[41]

The results of Wiley's research appeared in an enormous volume of five parts and 1,499 pages. It examined in detail boric acid and borax; salicylic acid and salicylates; sulphurous acid and sulphates; benzoic acid and benzoates; and formaldehyde. Two other parts were completed on sulphate of copper and saltpeter, but were withheld from publication because of the opposition of the Secretary of Agriculture, who remained unfriendly to Wiley and his work. The results, in brief, were that boric acid and borax, whether given in small doses over a long period or in large doses in a brief period, "create disturbances of appetite, of digestion, and of health"; that salicylic acid and salicylates in general do not have "favorable effects upon the metabolic processes," but rather, have "deleterious or harmful effects"; that "the administration of sulphurous acid in the food, either in the form of sulphurous acid gas in solution or in the form of sulphites, is objectionable and produces serious disturbances of the metabolic functions and injury to health and digestion," and may cause serious kidney damage; that benzoic acid in any form "is highly objectionable and produces a very serious disturbance of the metabolic functions," with injury to digestion and health; and that formaldehyde tended to "derange metabolism, disturb the normal functions, and produce irritation and undue stimulation of secretory activities." With these results, and his combative, effective personality, Wiley entered the fray as a close advisor to the politicians who had become aroused over this key progressive issue.[42]

In the Fifty-ninth Congress, no fewer than eleven bills were introduced on the issue in the early days of the session. In the House William Peters Hepburn of Iowa and James R. Mann of Illinois led the fight, while in the Senate Porter J. McCumber of North Dakota and Weldon Heyburn of Idaho persistently pressed for action. Things were stalled, as they usually had been in previous years, when in the middle of February 1906, *The Jungle* landed on the dinner tables of the nation. The resulting outcry was so great that Senator Albert Beveridge of Indiana, then moving toward a progressivism greater than he had previously shown, seized the issue. He told the president about the book and then introduced in Congress a meat-inspection bill of his own as a rider to a pending agricultural appropriation bill. Beveridge's rider, and the essential parts of the bills supported by Heyburn and Hepburn, all survived to become law in May and June of 1906.[43]

The sources that survive are more meager than one would expect for such a major piece of legislation, but they nevertheless reveal two fascinating aspects of progressive lawmaking in action. On the one hand, Harvey Wiley seems to have been ubiquitous in Washington. He helped Senator Heyburn write the bill, which, in altered form, became the key law. He was in constant contact, in person and in writing, with members of Congress. He supplied enormous quantities of data. He spoke to public meetings and helped to arouse the women's clubs and the AMA. He was even the chief contact for lobbies from the packers, rectifiers, and others who felt they would be damaged by possible legislation. They understood that if they could persuade Wiley to compromise, he could probably persuade Congress. If ever a man were a vehicle for the transmission of an aroused public opinion, it was Harvey Wiley between late 1903 and the middle of 1906.[44]

Such influences were essential, because on the other hand, Theodore Roosevelt was preoccupied with other matters. The President was concerned with the issues that were in the headlines: the Russo-Japanese War and the possible menace of Japan, the Algeciras Conference, the tariff, and above all, railroad regulation.[45] The issue of pure food and drugs seemed tinged with fanaticism and journalistic excess, and the President was unclear about his commitment to the issue. Thus, following him as he slowly perceived the seriousness of the problem not only provides an excellent case study of progressive legislative change, it shows how public opinion could galvanize a president to moral indignation and result in progressive legislation.

Roosevelt was the consummate politician, but his motives, as a rule, were honest and decent. As the articles of the muckrakers had their impact, members of Congress and finally the President sensed the increased public concern and came to recognize the stakes. Just as Wiley worked closely with Congressional leaders like Heyburn, so, too, did Adams. Indeed, it was apparently Adams' influence on Heyburn, and Heyburn's on Roosevelt, that led Roosevelt to commit himself on the issue. The President also was already in close personal contact with Lincoln Steffens and Ray Baker. Soon this network of influences had their effect. Despite his preference for railroad regulation and foreign affairs, Roosevelt found the space to include the following commitment in his annual message to Congress on 5 December 1905: "I recommend that a law be enacted to regulate interstate commerce in misbranded and adulterated food, drinks, and drugs. Such law would protect legitimate manufacture and commerce, and would tend to secure the health and welfare of the consuming public. Traffic in foodstuffs which have been debased or adulterated so as to injure health or to deceive purchasers should be forbidden." Having adopted such a position,

Roosevelt found it attracted attention, focused public opinion, and seemed a sensible part of his own reform package. He had nothing to do with the food and drug reform movement at its conception, but he was willing to adopt it when the times were ripe.[46]

Roosevelt continued to dwell on railroad problems until late January 1906, when Attorney General William H. Moody informed him that legal counsel for the beefpackers, then under an antitrust indictment in Chicago, had attempted to bribe local reporters into giving the public news that was slanted favorably toward the packers. The government could pursue no legal remedy, but Moody thought that some distorted news had been printed that could have shaped local opinion more favorably toward the packers than they deserved. Roosevelt was predictably upset at this violation of ethics, and the next day ordered all relevant documents published so that public opinion could have more accurate information.[47]

The Jungle appeared in the middle of February. Secretary Garfield called Roosevelt's attention to the book and shortly thereafter Senator Beveridge did as well. Garfield thought the book "too pessimistic," but was generally sympathetic; Beveridge immediately spotted a popular issue. Roosevelt's reaction was complex. He disliked the "ridiculous socialistic rant" in the book, and tried to pressure F. N. Doubleday, its publisher, to make Sinclair cut much of the blatant propaganda. Doubleday replied that they already had succeeded in excising 30,000 words from the original manuscript, but that the rest of the "unfortunate sermonizing" had to remain because Sinclair would budge no farther.[48]

Whatever his skepticisms about socialism or his distaste for reformers like Sinclair, Roosevelt was clearly aroused in the same way that the general public was aroused. Filthy meat was disgusting, its packers unethical, and the consequences possibly lethal. He wrote to Sinclair suggesting a meeting with Secretary Garfield. He instructed Secretary of Agriculture James Wilson to begin an investigation that would go well beyond what had already been done in Chicago. "The experiences that Moody has had in dealing with these beef trust people convince me that there is very little that they will stop at. You know the wholesale newspaper bribery which they have undoubtedly indulged in." No ordinary or perfunctory investigation would do. "I would like a first-class man to be appointed to meet Sinclair, as he suggests; and then go to work in the industry, as he suggests." The choice of investigator would remain secret and would be done in direct consultation with Roosevelt. "We cannot afford to have anything perfunctory done in this matter. I wish you would take this letter and read it over with Garfield."[49]

At this point Roosevelt assumed a sincere and characteristic pose, that of the interested and aroused citizen holding off the radicals and lecturing them while trying in all honesty to get the facts so he could act sanely. He often found Sinclair petulant and irritating. Sinclair gave the impression of being a hysteric even at a distance, as he peppered the White House with his doubts, fears, and intuitions about the packers and the government. He was especially perturbed at the thought that the government investigation would examine only meat—a white, middle-class, respectable concern— and entirely avoid the condition of the workers so important to Sinclair and the socialists. Roosevelt assured him that the working conditions would also be investigated, and he did issue instructions to that end.[50]

On 10 April, to cite only the most extraordinary day, Sinclair sent one telegram and two separate letters, in an agony of distress over rumors he had heard about a possible "whitewashing" of the packing industry. He claimed to have "perfect confidence" in Roosevelt's appointees, but he clearly thought them vulnerable to easy deception. He also claimed to have confidence in Roosevelt's "sincerity and fairness"—a confidence unex- pressed in the tone of his letters. The next day his confidence further expressed itself with a personal phone call to the White House. For his part Roosevelt could be equally moralistic and censorious. In one letter, he preached, for example, about his literary peeves—Tolstoy, Zola and Gorki —and lectured Sinclair about the superiority of individual initiative to socialism. His true attitude came out more clearly in a letter of a day earlier to Ray Baker, where he talked in a way that covered not only the railroads and pure food, but his whole orientation toward reform: "I want to let in light and air, but I do not want to let in sewer gas. If a room is fetid and the windows are bolted I am perfectly contented to knock out the window, but I would not knock a hole into the drain pipe."[51]

The individuals Roosevelt appointed to make his own investigation of Packingtown were Charles P. Neill, the United States labor commissioner, and James B. Reynolds, a New York reformer. Neill was apparently the man closest to the president, and Roosevelt passed Sinclair's suspicions and allegations along to him. At the same time, Roosevelt continued to press Secretary of Agriculture Wilson to maintain secrecy about the inves- tigation, and to keep the packers and newsmen as ignorant as possible about the Neill and Reynolds mission. On the surface Sinclair was right: the packers and the press had seemed rather too well-informed for there not to have been collusion, or at least carelessness, on someone's part.[52]

Sinclair's suspicions about Roosevelt's commitment to reform had more basis than Roosevelt was willing to admit. The two progressive moralists disagreed on many levels of tone and substance, and the disagreement would soon be public. Word began to filter out to the muckrakers that

Roosevelt was about to turn on them, in spite of his having worked closely with a number of them, and in spite of the impact they had had on his and the public's perceptions of reform. Roosevelt would forever deny that this was the case: there were good and bad journalists, just as there were good and bad trusts, and journalists who were serious, moderate, and responsible were not muckrakers. But since Roosevelt never made public lists of his targets, many muckrakers could be forgiven if they felt their powerful friend had suddenly turned away. In all probability Roosevelt did not have them in mind; he had Sinclair in mind, along with Thomas Lawson of the "Frenzied Finance" series and David Graham Phillips of "The Treason of the Senate" articles. These articles not only tended to attack the very roots of American capitalism and democracy, but they came uncomfortably close to men who were personal friends and supporters of Roosevelt and of the Republican Party. If they were guilty, then by implication Roosevelt himself might be covered with muck, and that to him was inconceivable. It may be, as Roosevelt once suggested to William Howard Taft, that certain writers worked with a revolutionary spirit—although few were even as socialistic as Upton Sinclair, and none were violent—but it certainly was not true that they had done much evil and little good. They had forced both Roosevelt and America to face hard problems, and their tone was hardly more passionate than that of Roosevelt himself on many issues. They were clergymen, as it were, of different faiths, but all were preachers within an evangelical Christian moral environment.[53]

Roosevelt's imagery and message in the speech, finally given on 14 April in Washington, were replete with his Manichean vision of the forces of good fighting the forces of evil. The man with the muckrake, regrettably, helped the forces of evil because of the violence of tone and the lack of discrimination in his attacks. He set a tone, in other words, that created an immoral climate. In discussing it, Roosevelt took a tradition dear to the McClure band, and in effect, tried to turn it against them:

In Bunyan's "Pilgrim's Progress" you may recall the description of the Man with the Muck-rake, the man who could look no way but downward, with the muck-rake in his hand; who was offered a celestial crown for his muck-rake, but who would neither look up nor regard the crown he was offered, but continued to rake to himself the filth of the floor.

In "Pilgrim's Progress" the Man with the Muck-rake is set forth as an example of him whose vision is fixed on carnal instead of on spiritual things. Yet he also typifies the man who in this life consistently refuses to see aught that is lofty, and fixes his eyes with solemn intentness on that which is vile and debasing. Now, it is very necessary that we should not flinch from seeing what is vile and debasing. There is filth on the floor, and it must be scraped up with the muck-rake; and there are times and places where this service is the most needed of all the services that can be performed. But the man who never does anything else, who never thinks or speaks or writes, save his feats with the muck-rake, speedily becomes, not a help to society, not an incitement to good, but one of the most potent forces for evil.

Roosevelt then went on to detail what this interpretation of a classic meant. To anyone familiar with the past lives of most muckrakers, his indictment was ludicrous. Here were people whose lives had been devoted to religion, to writing novels, to exalting poetry and love, who had been so conservative as to have been more favorable to William McKinley than to Theodore Roosevelt, McKinley's own vice-presidential running-mate! To Roosevelt these reform journalists exaggerated; they did not discriminate; and words such as "hysterical," "sensational," and "lurid" categorized their work. As a result, public opinion became numbed, and so color-blind that it could not distinguish good from evil. It also became cynical, and nothing made Roosevelt himself verge closer on hysteria than cynicism about democracy, patriotism, or his own righteousness.[54]

Despite the injustice of this indictment, the muckrakers were never the same after Roosevelt's speech. The progressive president with wide popular following had called reform journalism extremist, and any further exposures could be dismissed as mere "muckraking." Reform journalists sensed immediately that they had received a possibly fatal blow. In fact they had peaked in their influence anyway, and could not have sustained the pace of exposure much longer. Roosevelt merely shoved them more quickly down the slope, bruising sensibilities convinced that they deserved better from the White House, let alone from a man who had been in some cases a personal friend for years. In future years the term took on reverse connotations, reform returned to favor, and *Pilgrim's Progress* no longer was vivid in the minds of most Americans. But that was small solace at the time.

Even as they received the fatal wound, the muckrakers triumphed in Congress and in the very White House that had so hurt them. Even as Roosevelt dismissed them, he began to work with redoubled energy to push the investigations to a close and see remedial legislation enacted quickly. Because of his personal affection for Senator Beveridge, Beveridge's meat inspection rider received Roosevelt's closest attention and support. Given Roosevelt's dislike for Harvey Wiley, food and drug legislation on matters other than meat might well have been allowed to die. But in practice both the public and Congress seemed to identify the various measures as being part of a single effort, and the force behind the Beveridge rider was enough to carry the entire Pure Food and Drug Law.

The bills did encounter considerable opposition. Roosevelt's steady pressure on his officials quickly produced corroborative evidence from Neill and Reynolds for most of Sinclair's allegations, as well as a basic agreement between Secretary Wilson and Senator Beveridge on the wording of the meat inspection rider. The bill required the packers to submit to continuous investigation by representatives of the Secretary of Agriculture, and to pay the costs of this inspection; they also had to put the date of canning on each tin. The only appeal was to the Secretary of Agriculture, whose decision would be final; nothing could be appealed in the courts. The packers fiercely opposed these proposals. They wanted the government to pay for its own inspection; they wanted undated cans; and they wanted the right to appeal to the courts. The right to appeal to a court of law seemed to them so basic a privilege of any American citizen or company as to be undeniable. They insisted that meat remained wholesome for years after it had been canned, and that people would see the date and refuse to buy it out of ignorance.[55]

Between them Beveridge and Roosevelt seemed to have victory in their hands by the last week in May. The report that Roosevelt received from Neill and Reynolds was a potent weapon. An official and apparently unbiased document, it could not be dismissed as socialist rant as easily as *The Jungle* could. An example of the kind of pressure Roosevelt was quite willing to apply appeared in a letter to James W. Wadsworth, the chairman of the House Committee on Agriculture and a bitter foe of the legislation. The report, he informed Wadsworth, "is hideous, and it must be remedied at once. I was at first so indignant that I resolved to send in the full report to Congress," but he decided not to for several reasons. He was particularly afraid that innocent growers of cattle would have their livelihoods threatened if the market for American meat collapsed; likewise, the whole American food-export industry might well suffer. The president emphasized that he did not want to punish the packers, but rather to see "the immediate betterment of the dreadful conditions that prevail, and moreover, the providing against a possible recurrence of these conditions." The legislation must prevent any recurrence of these wrongs. The president would withhold publication of the document unless pushed to it, but "it must be distinctly understood that I shall not hesitate to cause even this widespread damage if in no other way does it prove possible to secure a betterment in conditions that are literally intolerable." He suggested that Wadsworth confer with him and Neill immediately.[56]

The packers caved in briefly, and the Beveridge amendment easily passed the Senate, but then resistance toughened in the House. Led by Wadsworth, a wealthy New York farmer, and William Lorimer, virtually the packers' own man from Chicago, the House resisters won strong support from many politicians who feared such a great extension of federal power at the expense of state and local control. Roosevelt became increasingly furious when his threats did not have the desired effect, so he sent the Neill-Reynolds report to the House on 4 June. The packers no longer appeared to care. They had received so much bad publicity that they seemed numb at the thought of more. Even the House no longer cared, and for two weeks, Wadsworth and Lorimer held out for their severely weakened version.[57]

In his efforts to achieve a suitable compromise, Roosevelt confided to Beveridge and Wadsworth that he was willing to give in on the matter of paying for the inspection, but he remained opposed to judicial review and undated tins. He also continued his policy of making blunt threats. He informed Wadsworth that much data remained unpublished, but that he would submit more to Congress if opposition continued. On 14 June he told Wadsworth that he had "gone over your bill very carefully" and was "sorry to say the more closely I investigate your proposed substitute the worse I find it. Almost every change is one for the worse." Roosevelt was especially upset that the Wadsworth bill did not give inspectors access to the plants twenty-four hours a day. Some of the other provisions "are so bad that in my opinion if they had been deliberately designed to prevent the remedying of the evils complained of they could not have been worse." Wadsworth promptly assumed a tone of high indignation, replying in a manner that permanently alienated the president. Roosevelt was wrong, "very, very wrong" in condemning the committee's bill. It was, in fact, "as perfect a piece of legislation to carry into effect your own views on this question, as was ever prepared by a Committee of Congress."[58]

Roosevelt remained unconvinced. He wrote back immediately, clearly concentrating his attention on the matter of appeal to the courts, and more or less tacitly hinting that he would be willing to compromise on the other issues by not bringing them up. "The court provision is the one to which I most object." In regard to it, "I wish to repeat that if deliberately designed to prevent the remedying of the evils complained of, this is the exact provision which the friends of the packers and the packers themselves would have provided." Congress had no intention of taking from the packers their unchallenged, constitutional rights to redress in the courts, but they did feel that the Secretary of Agriculture was the most obvious court of last resort for any kind of sensible and efficient administration, and they quite rightly felt that the secretary would be far more knowledgeable than any sitting judge. The right to appeal to a court would mean governmental paralysis, which was presumably what the packers really wanted.[59]

At this point party leaders realized that the Republican Party was in danger of coming apart. The man who broke the deadlock was Speaker of the House Joseph G. Cannon. Cannon was not always in agreement with the president on reform issues, and coming as he did from Illinois, he would probably have preferred the Wadsworth substitute, but he put party unity before his own preferences. He conferred with Roosevelt, and then sent over Representative Henry C. Adams, a member of Wadsworth's committee who had been a supporter of pure-food legislation during a term as food commissioner in Wisconsin. Adams, Roosevelt, Reynolds, and George P. McCabe, a solicitor for the Department of Agriculture, reached a basic agreement rather quickly. Roosevelt described the result to Wadsworth: "In each case Mr. Adams stated that he personally would accept the alterations we proposed." He agreed "that the court review proposition should be excluded" and was able to accept a "dozen other changes which we think should be made." If these changes were actually adopted, the president concluded firmly, then "your amendment will become as good as the Beveridge amendment," and he would be happy to accept either one. Squabbling about the bill continued over the next few days, but a compromise had finally been reached. Wadsworth and Lorimer remained hostile, but Cannon's influence prevailed. Roosevelt succeeded in keeping the courts out of the procedures for administering the bill, while the packers succeeded in eliminating the requirement for the dating of tins. The bill was the best Roosevelt could get and certainly preferable to no bill at all.[60]

To Roosevelt the Beveridge amendment was the focus of interest. He did not care that the Beveridge measure dealt only with the packers and left out all other foods and drugs. Roosevelt was friendly to Beveridge in a way he was not with Wiley, Heyburn, Mann, and the other leaders of the movement. Roosevelt had been slow to anger on the issues, and when he did anger he was selective, seizing the issue with the greatest public interest and the support of his personal friends. But the pressures of public opinion proved great enough to push the larger measure through as well. The nostrum lobby fought as hard as the packers' lobby, and states' rights legislators were upset at the expansion of federal powers, but all for naught. Throughout the battle Harvey Wiley and Samuel Hopkins Adams conferred closely with supporters of the bill. They supplied useful data that made the case for regulation overwhelming. Roosevelt signed the final version on 30 June, the same day he signed the appropriations measure that carried the Beveridge amendment.[61]

The resulting law both limited the rights of food and drug manufacturers and expanded the power of the federal government. At the time these issues seemed paramount, and in a purely medical or legislative sense, they also remain paramount to later historians. But in terms of the bill as a document of progressivism, it also stands as a fine example of institutionalized moral indignation. It was *wrong* as well as unhealthy to include certain preservatives in food packages; it was *unethical* to claim efficacy for drugs that were really ineffectual compounds of water or alcohol; it was *immoral* to include drugs in a compound that might damage the heart or cause unpredictable addiction. The resulting law was moral in the sense that it did not require that drugs work, or that food taste good or keep for any specified period of time. It merely required that manufacturers tell the truth on their labels. They could not claim that their drugs did something they could not be proven to do; they must not claim ingredients if those ingredients were not present. They were not forbidden to include harmful ingredients, but if they did so, they must say so plainly. The United States Pharmacopoeia or National Formulary established standards for the use of certain terms, and the terms on the labels must coincide with these national standards unless there was clear indication of the way in which the given medicine was different or substandard. No one could mix or pack so as to deceive purchasers, and no use of chemicals, powers, or other ingredients could camouflage spoilage or inferiority. Filthy or putrid animals or vegetables were unacceptable. Misbranding or mislabeling was forbidden. Officials in the government would make uniform rules and regulations as the need arose; the Bureau of Chemistry would make the needed scientific analyses; and the Secretary of Agriculture would work through the appropriate United States district attorney to obtain enforcement. The overriding assumption was that honest businessmen would make, label, and sell what they wished to competent, progressive, decent citizens. As long as everyone behaved correctly, society functioned successfully.[62]

V

The food and drug legislation was typical of progressive laws in many areas. It was the product of outraged public opinion; it was primarily moral in its premises and prohibitions; it was administrative and prohibitory more than it was interventionist or socially reconstructivist; it warned and regulated vested interests, but it scarcely diminished them; it gave the appearance of radical reform without the substance; it gave the intellectuals of the country a feeling that they had influenced the politicians, while the politicians, in turn, seemed really to fear them and found them useful only as a stimulant to voter manipulation. The intellectuals were progressives in their moral indignation at the genuine wrongs they uncovered; Roosevelt and Beveridge were also progressives in taking this outrage when it aroused and informed public opinion and carrying it through to successful reform legislation.

Many progressives proclaimed their belief in the ventilation of wrongs and the beneficent impact of an awakened public opinion. The transformation of Packingtown, which had in fact had much resemblance to Sinclair's evocation, began almost as soon as the publicity became overpowering. Even before the law was signed, a confidant of the President visited the major packing houses and informed Roosevelt about current life at Libby's, Swift's, Armour's, and Morris'. His experience, he wrote the President, could be described as the "Awakening of Packingtown." "It is miraculous." Soon, the companies expected to have "still more new lavatories, toilet rooms, dressing rooms, etc. Cuspidors everywhere, and signs prohibiting spitting." In most of the places he visited, "there was indication of an almost humorous haste to clean up, repave and even to plan for future changes." A visitor could see "new toilet rooms, new dressing rooms, new towells, etc., etc." The Swift and Armour plants "were both so cleaned up that I was compelled to cheer them on their way, by expressing my pleasure at the changes." The sausage girls had left their dank quarters and were now upstairs "where they could get sun and light," and had the use of dressing rooms. "I asked for showers and lockers for the casing workers at Armour's and got a promise that they would put them in. The canning and stuffing room, chip beef and beef extract at Armour's seemed really quite good." The girls at Libby's now had attractive uniforms, which they were able to buy at half price; new toilet facilities were installed and more planned. The haste toward reform would have been amusing if it had not been so nearly tragic.[63] An enlightened public opinion and the threat of legal restriction had indeed improved American life.

Given the revivalistic way in which many progressive reforms came into existence, it was perhaps only natural that there should be something of a moral hangover after the preacher left town and the new converts had to go about the duller business of daily living. Harvey Wiley, one way or another, managed to keep the faith until put in his grave in 1930, at age eighty-six and as fundamentalist in his chemistry as his father had been in religion. He had feuded once again with Roosevelt, this time over saccharin, and had slowly been forced from power and ultimately from government service, but he wrote two books and a shelf of chemical reports that secured his reputation. The muckrakers quickly moved to other issues. Only Samuel Hopkins Adams retained much interest in pure drugs, and returned to the fray occasionally for years with follow-up articles. Roosevelt disposed of the issue he had never sought as quickly as he decently could. On 28 June, he instructed Neill to "quietly reinspect the buildings of the beef-packers at Chicago and report to me, for my personal information, what the conditions then are compared with the conditions as you found them on the occasion of your recent inspection," on the next 1 January. On 30 June he signed the new laws. Occasionally, during the next few weeks he mentioned them, and then silence descended. Soon Wiley irritated him, and he took steps to see that Wiley never again could exercise his old authority. Wiley was in fact too fundamentalist and purist in his approach to chemistry. Indeed, if his general bias of mind could be summed up in a single statement, it would be one of the last sentences he ever wrote: "It is *sin to be sick.*" But Roosevelt was also too temperamental, too intolerant of views he disliked, and too committed to the politically popular. He was grand in his fight for the meat inspection amendment; he was not up to his own standards in the record he left of its enforcement.[64]

One impact of the laws was ironical indeed. In view of the hysterical opposition of the drug manufacturers and the packers, an onlooker might assume that great harm would come to these industries with the advent of regulation. Yet nothing of the sort happened. Sales, which had been badly hurt by the adverse publicity, rebounded. After all, in the eyes of the public, the government and the law suddenly seemed to offer protection. The labels now had to be honest in describing the drugs as they were; the meat had to pass inspection, so it was presumably pure and wholesome. It is simply not true, as modern radical historians have implied and sometimes claimed, that the law was really developed by capitalists to strengthen capitalism and raise profits.[65] Such a position gives entirely too much intelligence, credit, and foresight to the capitalists. Indeed, the record of most of them on these issues usually ranged from the inarticulate to the shabby. Most of them did not deserve so much from the government they attacked so fiercely. Like most progressive laws, and indeed most laws everywhere, the food and drug laws could not anticipate what judges would say, administrators would interpret, and science would create. The laws were important, innovative, and necessary, but they were not the last word. They put too much trust in the sense of the people, and too little money and authority into the hands of Harvey Wiley and his successors. They could not anticipate the development of new drugs, such as barbiturates, or of cosmetics capable of severe side effects. The language defining food adulteration eventually proved vague and ambiguous, and the authority of the government to set standards of quality was not strong enough. No thought at all had been devoted to the problems associated with advertising—in the modern sense it was an invention of the 1920s. Clearly new laws would be needed. But a first step had been taken, and it was taken in 1906 because so many sons of the puritans found themselves putting things in their mouths that offended their morals as well as their stomachs.[66]

Chapter 7

Compromise at Armageddon

BEFORE IT OPENED, careful observers could find faint signs that the 1912 Progressive Party Convention was going to be different. The *New York Times* called attention to "one novel feature of the coming convention," the "experience meeting." Imitating the practice of Protestant sects, the Progressives planned to allow every delegate "an opportunity to speak as the spirit moves him for at least five minutes." After the roll call, "anybody in the delegations desiring to give 'testimony' regarding his political conversion will be permitted to 'confess.'"[1]

Two days later the *Chicago Tribune,* the leading newspaper supporting the new party, noted without editorial comment the behavior of the Massachusetts delegation as its train headed for Chicago. The delegates "held religious services" and "prayed for the success of the new party in the same spirit that their Pilgrim ancestors appealed for divine help 300 years ago as they neared the New England Coast in the Mayflower."

Many states held conventions to prepare for the national gathering. New York attracted the most attention, and the eyes of a number of political analysts were on Buffalo on the Saturday before the trip west. They noted with amusement that male delegates no longer smoked or wore their hats when ladies were present, and that proceedings seemed more genteel than was customary. The atmosphere seemed more spontaneous than that of other conventions. The most illuminating event came when Wallace Thayer of Buffalo precipitated the biggest demonstration of the morning while comparing the Progressive movement to the crusade for the abolition of slavery. "He told of the unpromising beginnings of that fight and how William Lloyd Garrison, in a lonely attic, and with only a hand press, started it going, and he predicted similar results from the Bull Moose start." Then an unknown delegate yelled:

"And Garrison's granddaughter is here in this room now."

Everybody began to cheer and shout that she should show herself. Finally, Eleanor Garrison, "a demure, prim-appearing lady, with a lurking twinkle in her eye and a general air of being a pretty nice sort of person, arose and bowed her thanks." Then, as a tribute to Garrison, another delegate moved that the visitors from Massachusetts in the convention hall be invited to the platform, "and they went up there, headed by Miss Garrison and Miss Alice Carpenter, who is a delegate from the Bay State. They were wildly cheered." In this fashion delegates set the mood for the Chicago convention. They seemed to be political amateurs from the middle class, tolerant of an increased role for women in politics, and given to sudden outbursts of religious enthusiasm. They found their antecedents in the puritan experience and the crusade to abolish slavery.

They also could be as silly as any other group in the history of the American national convention. While Massachusetts prayed and New York remembered the abolition of slavery, the California delegation was the first to arrive in Chicago. Capitalizing on one of Roosevelt's casual remarks, that he felt as fit for the campaign as a bull moose, delegates spoke of themselves as Bull Moosers. The Californians replaced the Teddy Bear that had been their mascot at the Republican convention with the Bull Moose, and recorded their change in song. They left the train in Chicago Saturday loudly singing over and over again:

> I like to be a Bull Moose,
> And with the Bull Moose stand,
> With antlers on my forehead
> And a big stick in my hand.

Like many another camp meeting in American religious history, this one, too, had its elements of farce.

II

As the delegates arrived, the organizers of the convention wrestled with problems that threatened the success of the convention. The same irritating issues preoccupied both Theodore Roosevelt in Oyster Bay, New York, and Senator Joseph M. Dixon's committees in Chicago.

The issue that attracted the most newspaper notice was also the one that caused the Progressives the most agony: the role of the Negro. Roosevelt himself shared many of the prejudices of the white elite of his day, and had doubts about the equality of the Negro to white people. The Negro was never a major issue in the New York politics of Roosevelt's day, and he was clearly baffled by Southern white hostility to better treatment for the Negro. He did not share this hostility, but he had to deal with it, and as president, he had made several gestures that had won him broad Negro support. Long before the Progressive Party convention, he came to the conclusion that, at least in the South, the best hope for the Negro was not to seek complete equality with whites. Such a goal would cause increased racial tension and be counterproductive to genuine Negro progress. The best course for Negroes to follow would be to support a white man's party in which the Negro played a definite but subordinate role. The white men of high character would control politics but protect the Negroes. Negroes would progress under this orderly protection and enjoy a reasonable number of public offices. This position, originally stated in a private letter to Booker T. Washington in 1904, was never satisfying to white liberals, but it was clearly articulated, honestly held, and pragmatically feasible, and Roosevelt held to it throughout his career.[2]

Roosevelt was in a difficult position. His growing social consciousness had committed him to welfare measures that were intended to help everyone. He wanted the votes of Northern Negroes and social welfare-oriented whites, and felt that he deserved these votes. Yet, with a combination of idealism and calculation, he wanted to be President of a united country backed by a party that included a substantial element of Southern white support. He tried to avoid taking any stand on the issue, knowing, like any devoted reader of Joel Chandler Harris' Uncle Remus stories, that the issue was a tarbaby: you could not touch it anywhere without becoming defiled. He greatly wanted to end the chancre of sectionalism. As he wrote to John M. Parker of Louisiana, his leading Southern white supporter, if he could even carry a single one of the former Confederate states, "I should feel as though I could die happy."[3]

Newspapers speculated on the stand Roosevelt might take. His mail contained both pleas and threats. William H. Maxwell, Jr., the editor of *The Jersey Spokesman,* spoke for his race when he wrote Roosevelt that "the colored people of this country are in dire need of a champion, you have ever been friendly to them, you have ever striven to give them a fair chance and I know that they appreciate it." The Reverend Mr. J. Gordon McPherson, the editor of the *Voice of the West,* wrote in a much less conciliatory vein when he told Roosevelt that if the Progressive Party expected to have any place in the South, then it had to recognize "the superior ability of the white man and his superior civilization, attained through countless centuries of struggle and endeavor, peculiarly fitting him to lead and direct." During the same period, the Negro "has been perfectly content to remain the ignorant savage devoid of pride of ancestry or civic ambition. The South cannot and will not under any circumstances tolerate the Negro," and it insisted that the Progressive Party emphasize its position.[4]

Roosevelt believed he had a chance to carry the South. He knew that Progressive organizations had grown out of next to nothing in every state but South Carolina, and that in Georgia, Alabama, and Arkansas, people were especially enthusiastic. He knew that B. F. Fridge, the white supremacist leader of the Mississippi Party, had written to Senator Joseph Dixon of the "old line Democrat" who told Fridge on 24 July that "if Colonel Roosevelt could make a few speeches in this state he would set the roads on fire." Roosevelt knew that he had devoted supporters in Parker and Julian Harris of Georgia, and he let himself be carried along by this enthusiasm. Even so, the policy he finally decided upon was consistent with his pronouncements over the past decade.[5]

Roosevelt had delayed decision on the Southern Negro until mid-July. Not only were the heirs of the abolitionists in Northern Republican ranks exerting strong public pressure, but events in the South were reaching the critical point. Many Negroes were entering Progressive ranks, and in several states, they made strong attempts to send delegations to Chicago. The situations in Alabama, Florida, and Mississippi were especially serious, because in those states, two slates had been named, and the convention faced the job of choosing between them. Roosevelt might leave that problem to his lieutenants in Chicago, but he had no honorable way of avoiding a public statement of his position.[6]

With exquisite symbolism, he chose to write a letter to one of his most important Southern white supporters: Julian Harris of Georgia, the son of the man who invented the Tar Baby. He praised Joel Chandler Harris as a man with a gentle soul and broad sympathies "whose work tended to bring his fellow countrymen, North and South, into ever closer relations of good will and understanding." Clearly, Roosevelt wanted his own words to continue this tradition. Roosevelt praised the traditional American values of freedom for all men regardless of skin color, but he then insisted that men of goodwill could not act "in such a way as to make believe that we are achieving these objects, and yet by our actions indefinitely postpone the time when it will become even measurably possible to achieve them." Roosevelt claimed that his overriding commitment was to the social goals of the Progressive Party, that the achievement of these goals would benefit both whites and blacks, and that neither of the two older parties had a defensible position on the issue. The Democrats segregated the Negro and refused him any role; the Republicans corrupted the Negro and only allowed him a subservient role in the Southern segment of a Northern party.

Roosevelt wrapped himself and his party in the garb of morality, the same clothes that he wore whenever he found himself on dangerous ground. He often sounds hypocritical to later generations, but Roosevelt honestly assumed that his positions were moral if only because he took them honestly. A generation raised in the religious manner of late-nineteenth-century Protestantism often accepted him at his word. "We have made the Progressive issue a moral, not a racial, issue," he insisted. Wherever "the racial issue is permitted to become dominant in our politics, it always works harm to both races, but immeasurably most harm to the weaker race." Negroes must moderate their demands and adjust to the dominant powers in their local areas. The man who had an infinite contempt for states' rights in the mouths of Democrats discovered such local rights to be a convenient refuge in time of crisis. Negroes and their white friends must learn to adapt "our actions to the actual needs and feelings of each community; not abandoning our principles, but not in one community endeavoring to realize them in ways which will simply cause disaster in that community, although they may work well in another community."

If politics is the art of the possible, then Roosevelt's solution to this problem was sensible if not morally satisfying. The Progressive Party planned to have Negroes as genuine, voting delegates at their conventions, and to have them participate in their administration in ways more valuable than the token postmasterships of the Republicans. He insisted that if people of goodwill really had the interests of the Negro race at heart, they "can best help the negro race by doing justice to these negroes who are their own neighbors." Echoing the public sentiments of Booker T. Washington, he insisted that the greatest progress had been made "along industrial and educational lines," and that Negroes should rely on each other and on the best white people and concentrate for the short run on nonpolitical goals. He was offering them equality in the North, where they had already won a responsible place, but subservience and discrimination in the South, where their situation was politically hopeless. Again and again he repeated the phrase about "the best white men in the South" as a kind of litany. If repeated often enough, it might somehow make everything come out all right. Everyone should remember "Emerson's statement that in the long run the most unpleasant truth is a safer traveling companion than the pleasantest falsehood."[7] He neglected to point out that the best white men in the South were all Democrats hostile to Negro equality, and that he as president had followed such a policy with no discernible improvement in Negro social conditions.

With characteristic Western exuberance, the *San Francisco Examiner* discussed Roosevelt's Negro problems under the headline, "Dark Moose Bothering Others in T. R. Herd," and its relevance to the question of Roosevelt's vice-presidential running mate. In Oyster Bay Roosevelt said nothing publicly. Because of the Negro issue and his concessions to the "best white" Southerners, many observers thought Roosevelt might choose a Southern Democrat to emphasize that the progressives were genuinely national and not simply a Republican rump. John M. Parker seemed a possible choice, and Parker was clearly interested. The second choice seemed to be Luke Wright of Tennessee, a Democrat and former Confederate soldier who had served under President Roosevelt as Governor General of the Philippines, Ambassador to Japan, and Secretary of War. Beyond these two men, Roosevelt's Southern "best white" support was so thin that no one else was visible.

Roosevelt apparently sensed the fruitlessness of his Southern gambit and turned to the West for a running mate. The great impetus to the Progressive Party in the first place had been moralistic Republicanism, and the nomination of a Southern Democrat could cause many Progressives to return to the Republican fold, vote for Woodrow Wilson or Eugene Debs, or go fishing on election day. Ben Lindsey, the well-known "kids' judge" of Denver, offered a compromise of sorts: his background was both Southern and Democrat, but he had been a Western progressive long enough to be reasonably qualified on that ground as well. He wanted the job, and no one could doubt his personal, long-term loyalty to Roosevelt and his principles. But Lindsey was too overweening in his egotism and too small a man physically and intellectually to have been a good candidate. The most obvious candidate was Governor Hiram Johnson of California: fanatically moralistic, long devoted to Roosevelt, a Progressive of measurable achievement, and from as far away from Roosevelt's New York as a politician could get at the time. Johnson claimed that he had unfinished business in California and did not want the job, but pressure from Roosevelt changed his mind, and he was the clear choice soon after the convention began.

In addition to the Negro and vice-presidential problems, a third issue was of deep concern in Oyster Bay. One of the major concerns of President Roosevelt's years in office had been whether to break trusts into smaller corporations, leave them alone, or regulate them through commissions. As president Roosevelt had wanted to regulate only those trusts that acted irresponsibly. Nonetheless, he had obtained the reputation of a "trust-buster" for only a moderate amount of initiative. President William Howard Taft, his chosen successor and a man Roosevelt now condemned as a reactionary fraud, had actually presided over more meaningful trust-regulating activity than Roosevelt had. On this issue he deserved more of a reputation as a progressive than did Roosevelt. Indeed, certain of the Taft administration's initiatives, particularly action against the U.S. Steel Company, seemed to implicate not only Roosevelt but also his chief financier George W. Perkins, in unethical activities. Like most moralists Roosevelt could not tolerate having his moral purity questioned. At this point for him, the trust issue was personal as well as political. It also divided his supporters more seriously than did the Negro.

On one side of the argument were Roosevelt's chief financial supporters, George Perkins and Frank A. Munsey. Munsey was traveling in Europe, but Perkins was a constant presence at Oyster Bay, the "Dough Moose" to whom Roosevelt had strong personal bonds in addition to his financial ones. Perkins' enemies made much of his importance on Wall Street but, in fact, he had at least as much right to be at the center of the movement as they did, and in some ways more; only his extraordinary wealth and success made him seem out of place. His father, also named George, had wanted to be a Presbyterian missionary, and only poverty had forced him into business. He was active in the YMCA as well as his church, and carried this zeal into social work as superintendent of the Chicago Reform School and Warden of Joliet Penitentiary. Eventually, he went into the insurance industry, paving the way for his son's career. The younger George inherited what his biographer has called the tendency to be "too much the Presbyterian evangelist," and found that such zeal worked well in a business career. J. P. Morgan discovered him, and like Morgan he displayed a remarkable ability to reorganize chaotic businesses to be both more efficient and more responsible. Perkins was in fact an honest man, transparently sincere in the few ideas that he had, a close friend of progressive reformer Senator Albert J. Beveridge, and personally attached to Roosevelt.[8]

Any number of issues could bring a man into the Progressive camp, and Perkins' presence there was a sign of possible Roosevelt support in the business community. Business had learned to live with Roosevelt as President, although businessmen were deeply concerned that antitrust and corruption investigations might get out of hand and harm the business climate. Roosevelt all but forfeited much of this support, however, when he continued to attack the judicial system, which was something of a sacred cow along Wall Street. The idea that voters could recall a judicial decision or a judge horrified many businessmen, fearful of a resurgence of populism. Perkins, however, had little interest in constitutional or judicial issues. For him Roosevelt and the trust were the key issues. He was devoted to Roosevelt and believed that large businesses could be regulated into social responsibility. For the well-being of the country they ought not be harassed and broken up into inefficient smaller units. Munsey agreed. Radical enough himself to be willing to support a national minimum wage law, he insisted that Roosevelt had been a good President, had grown more conservative with age, and that people who opposed him were mistaking zeal for radicalism. He pictured a future President Roosevelt restoring business confidence, regulating prices and competition, and eliminating unnecessary waste. In fact neither Roosevelt nor his advisors ever proved capable of determining the difference between a good trust and a bad trust, and it seems unlikely that in office their policies would have differed much from those followed by Woodrow Wilson. Perhaps Simeon Strunsky made the most perceptive comment on the whole issue. He liked to improvise political skits to entertain his friends. Roosevelt's daughter could never forget his "Through the Out-looking Glass with Theodore Roosevelt," in which Tweedle Dee was the good trust and Tweedle Dum the bad trust. "The only way to tell them apart was to poke them." The bad trust said, "Tee-hee, Tee-hee," but the good trust said, "T.R., T.R."[9]

The more radical of Roosevelt's followers never understood the roles that Perkins and Munsey played in the campaign. Along with William Flinn of Pittsburgh and Dan R. Hanna of Ohio, they supplied the money that financed the party. Their funds were not replaceable, and Progressives were naïve to assume that these men were somehow tainted by their proximity to J. P. Morgan or by their ties to industry. Roosevelt had proved capable in the past of resisting the pressures of the wealthy when he wanted to, and he remained capable of it. The fact was that he agreed with Perkins, and had said so in public and in private for years. Henry L. Stoddard tells a little story that puts Munsey's influence into perspective. Munsey, he reports, was a militant diet faddist and not above telling former Presidents that they were eating incorrectly. On one occasion,

Roosevelt sat down happily to a meal of cold roast beef, potatoes and salt, and Munsey went to work to try to keep Roosevelt from eating this unhealthy meal. Roosevelt listened to his friend politely enough, hardly attempted to refute him, and kept on eating without missing a mouthful. He was not the sort of man to be influenced when he did not want to be.[10]

Oddly enough, the loudest opponents of Perkins in the inner circle were the wealthy and aristocratic Pinchot brothers. Gifford Pinchot had been Roosevelt's chief forester and a member of the "tennis cabinet." His younger brother Amos had attended Yale and received a law degree before meeting Roosevelt. He dissipated his energies in a scattershot series of reform activities, but when his brother had a famous feud with Richard Ballinger and lost his job with the Taft administration, Amos became something of a liaison man for his brother. He was friendly with a number of fellow members of the National Progressive Republican League, and absorbed ideas from Senator Robert M. La Follette and Louis D. Brandeis. Amos was devoted to both La Follette and Roosevelt, and for a time tried valiantly to bring the two men together. He failed, and when the La Follette campaign seemed to falter, he quickly joined the Roosevelt forces. Despite this shift of allegiance, he retained his hostility to the trusts and the tariff, as well as his preference for direct democracy. Both brothers tended to be unstable. Gifford hated Perkins, and Amos hated most of what he thought Perkins stood for. On one occasion Gifford and Perkins fought so bitterly that Roosevelt insisted that they be placed in separate rooms to keep the peace, while he ran back and forth working out an essential compromise.[11]

Amos Pinchot had a suspicion of the House of Morgan and of U.S. Steel that bordered on paranoia. He was convinced that they had become the dominant powers in American politics and that the distinction Roosevelt tried to make between good trusts and bad trusts was really Morgan's doing. He was suspicious of anything emanating from Wall Street, and he suspected that Taft's antitrust prosecutions and Wilson's antitrust platform had driven Wall Street to a desperate search for a leader who would protect them. Unable to control the Republican Party, they used Perkins to funnel money to a third party. Modern students know that J. P. Morgan, Jr., and his associates were actually putting strong pressure on Perkins to stop his political activities, but even if confronted with such evidence, Pinchot would never have believed it. If Morgan had not done it, then Judge Elbert Gary of U.S. Steel had, and Pinchot termed Munsey and Perkins "Judge Gary's two lieutenants," men who were the "accoucheurs of the third party child." As far as Pinchot was concerned, Roosevelt was responsive to pressure and willing to compromise, and only Gifford's constant pressure kept the former president at all progressive.[12]

While each side of this argument attempted to influence the candidate, representatives of their views were trying to shape the platform. Neither Perkins nor the Pinchots were especially adept at words, however much they might argue among themselves, but Roosevelt had influential platform writers at his command from each side. Herbert Croly, whose book, *The Promise of American Life,* Roosevelt embraced as a good statement of his general position, represented the Perkins position. Croly's father had been advocating a positive attitude toward wealth, large corporations, and government regulation of the economy for forty years, and had mentioned Roosevelt as a possibly ideal presidential candidate as early as 1884. Under his father's influence and that of Auguste Comte, Croly developed a great dislike for Jefferson's influence on American life; he preferred a Hamiltonian emphasis on business, centralization, and nationalism, all ideas that Roosevelt had long shared. When Roosevelt congratulated Croly on his book in July 1910, Croly became associated with the progressive spirit in the country. He himself did not share the moralistic background that permeated the minds of most leaders of what was then called Insurgency, and regularly opposed their demands for more popular democracy and the breaking up of large corporations. He was, nevertheless, loyal to Roosevelt personally, and to the new Progressive Party when it finally formed. Throughout July 1912, he worked on the proposed convention platform with Learned Hand and George Rublee, and he was personally responsible for answering the Woodrow Wilson–Louis Brandeis position that was already popular. Wilson and Brandeis were convincing Americans that Jeffersonian ideas about small business, the "little man," and decentralization would be more effective in controlling irresponsible wealth than Roosevelt's Hamiltonian approach.[13]

The most effective opponent of the Perkins–Croly position on the trust within the Roosevelt camp was Charles McCarthy, a Wisconsin Progressive of long experience. On 14 July 1912, Roosevelt had written McCarthy to ask if he and Professor John Commons could come to New York immediately to help complete the drafting of the platform position on "Trust, Labor, and Industrial betterment questions." McCarthy left shortly to do what he could, meanwhile sending Dean William Draper Lewis of the University of Pennsylvania Law School, Roosevelt's chief formulator of legislative positions, a long analysis of related La Follette material. In 1911 McCarthy had helped draft the La Follette antitrust bill, although he differed in certain ways from other La Follette advisors such as Brandeis and Commons. Essentially, he demanded that the "Sherman Act must be strengthened" to include "real punishment rather than a mock dissolution of the guilty trust," and that Congress create "a powerful trade commission" that would have "full power to protect the people from the aggressions of all such powerful combinations engaged in unfair production and competition." Roosevelt apparently said he agreed with the essentials of McCarthy's position, and McCarthy had every reason to believe that he would have the controlling word on those planks of the Progressive platform with which he was chiefly concerned. In all probability, Roosevelt was not clear in his own mind as to which position he wished to take, and was availing himself of all views expressed by men who were devoted to his candidacy. Only honest confusion and indecision explain the awkward events that subsequently occurred in Chicago.[14]

III

Sunday evening, 4 August, was surprisingly cool in Chicago. A breeze came off the lake to welcome the delegates as they arrived on the trains. They were immediately recognizable because of the red bandannas they wore. A few wore the new emblems correctly, indicating at least a nodding acquaintance with the West; most seemed to be wearing them incorrectly, usually hung straight down the back, because they had never seen a real cowboy. They knew only that they were expressing their loyalty to a man who was, in effect, an Eastern cowboy of the imagination far more than a genuine veteran of the North Dakota Badlands that he so often liked to evoke. Some bandannas were knotted around straw hats or sleeves; they added "dash and out-of-doorness, an air of brisk adventure" to the scene. As the delegates drifted about the lobby of the Congress Hotel or wandered about in the dusk, they looked "less like the characteristic fauna of political conventions than ordinary suburban Americans just in from the country club." Their faces appeared fresh, if not always youthful, and were noticeably "more alight with some sort of inner fire" than was normal for a political gathering. A "thrill of clean and unselfish enthusiasm" seemed to charge "the very air."[15]

At 8:55 the next morning, Roosevelt's train arrived ahead of schedule. The problem of how to handle the disputed Southern Negro delegates had caused him displeasure, and newspaper distortions of his views made things much worse, but nothing dampened his mood once the excitement of the occasion was upon him. A small crowd met the train and followed the progress of his car. At one stop a man thrust a red bandanna in his hand, and then scores of other supporters did the same. In his other hand he carried a steel-gray sombrero, which he waved repeatedly. Behind him were several carloads of women, waving their own red kerchiefs. He gave only a brief address before entering the hotel to make his final plans.

In front of the Art Museum, the next most famous delegate was surrounded by a brass band and about 500 admiring women. Jane Addams had agreed to second Roosevelt's nomination, and in the eyes of most of the newspaper reporters, she was the most influential of his supporters with the voting public as well as with socially conscious women. Carrying flags with "Votes for Women" prominently printed on them, the crowd impressed one observer as being made up of "young matrons who are 'up' on everything from Ibsen to the Montessori method, and embody the modern spirit in its quintessential and perhaps most terrifying form." Another person noted the presence of the old and the young as well, and remarked on the seeming classlessness of the group: "Women whose fortunes run into the millions walked with school ma'ams, who haven't a cent in the world except next month's salary, and society girls, social settlement workers, professional nurses, factory girls"; leaders of civic and labor groups "made up the rank and file." The Illinois Equal Suffrage Society had handed out yellow banners, but their supplies ran out at 300, and so the rest of the crowd waved handkerchiefs. Led by the band and a squad of police, they marched down Michigan Avenue to the convention hall.[16]

Inside, the crowd found a scene different from the one that had nominated William Howard Taft there in June. Some changes were physical, such as the huge sounding board that had been installed to project voices throughout the cavernous hall. Others were symbolic, such as the large, handsome moose head that adorned the guest box over the main entrance. On the stage, to the right of the platform, aging G.A.R. veterans with fifes and drums shared space with former Confederates. A band played patriotic tunes, including a version of the "Star-spangled Banner" that featured a revolver shot. Boy Scouts represented the young. Flags flew everywhere. "Purple row," where the wives of the Republican millionaires had sat in June, their faces frozen, now filled with women "in shirtwaists and young girls," who seemed "intent and earnest": "a different type, on the whole a type that was pleasanter to look at." At the front of the convention hall, almost immediately becoming a Progressive Protestant variety of the Virgin Mary, Jane Addams herself sat for an hour before the convention came to order. "Everyone who stopped spoke to her with an earnest face and manner. Miss Addams herself acted like a religious devotee," for most of the period sitting "with the same intent and almost reverent look on her face that the shouting Methodists and Presbyterians behind her wore."[17]

Throughout the morning observers noted a sense of puzzlement among the delegates: they "had not found themselves; the convention had not 'jelled'; it had not defined itself; it had not crystallized." Things were quieter and more orderly than in other conventions. "The hall was as still as a big church just before services begin." About half an hour before the convention opened, the Indiana delegation startled the other delegates by jumping on their chairs and bursting forth with "My Country, 'Tis of Thee." They sang "with tears in their eyes, and the look on their faces such as you might imagine on the faces of the men who backed Martin Luther after he had nailed his theses to the church door at Wittenburg, or the men who applauded Wendell Phillips when he indorsed Garrison's charge that the American Constitution was 'a league with death and a covenant with hell.' " Again and again observers connected the delegates with these two traditions of Protestant militancy and abolitionist idealism. "It slowly stole in on you that you were in a convention of fanatics, and that they knew it and were proud of it."[18]

In this supercharged atmosphere, former Senator Albert J. Beveridge of Indiana became the temporary chairman of the convention. He had been a warm and loyal supporter of progressive policies for many years, and had strongly favored the candidacy of Senator Robert La Follette during the year before Roosevelt decided to become a candidate. He had the reputation of being a man of enormous ego who had ambitions of his own to become President or at least vice-president. He was also a passionate nationalist, an imperialist with contempt for races not lucky enough to be Anglo-Saxon, and a staunch supporter of the rights of big business. He detested President Taft for his antitrust policies, his attempts to achieve Canadian reciprocity, and his clear opposition to Beveridge's own political progress. Despite all these reasons for supporting Roosevelt, however, he had held back during the days after the Republican convention. Only strong pressure from his closest friends—men like Perkins, Munsey, and John C. Shaffer—and his sense that if he had any political future at all it lay with Roosevelt and the Progressives, pushed him into joining up. Once he accepted a role, his oratorical skills and his national prominence made him an obvious choice for keynote speaker, since few established politicians had been willing to leave the parties that had provided them with careers. He finally agreed, and on Sunday night, had delivered the speech to a small group in his room. According to one observer, Roosevelt had been "visibly affected," while George Perkins had been so moved that he began to cry and had to leave the room to compose himself.[19]

Dressed "in a blue serge suit that fitted him neatly, a white vest that had two rows of buttons on it, a blue bow tie and a collar that came well up under his chin," Beveridge gave what many thought the speech of his career. It remains a curious document. In many ways it violates all kinds of presuppositions that students of later generations might have about the Progressive Party program. On the one hand, it furiously smites the bosses, the politicians, and the invisible government that cheats the masses out of direct control of their country. It advocates the initiative, referendum, and recall. It asks for an end to child labor, and requests that the chivalry of the state protect women workers. It demands minimum wages and state aid for the aged, maimed, and crippled. It insists on the conservation of natural resources and votes for women. It urges the Constitution to change and grow with the needs of the people. American institutions imply all these demands, Beveridge continued with imagery not lost on his audience, for "if they do not mean these things they are as a sounding brass and tinkling cymbal." His audience should "never doubt that we are indeed a nation whose God is the Lord." Critics "who now scoff soon will pray. Those who now doubt will believe."

On the other hand, Beveridge found the approval of God in some peculiar places. When he tickled a trust, it always seemed to respond "T.R., T.R." Small business was somehow dishonest in Beveridge's world, or at best in a primitive level of evolutionary development. He argued that "great business concerns grew because natural laws made them grow" and that antitrust legislation would thus be "at war with natural law." In a passage replete with the imagery of prisons, he said that antitrust efforts manacled big business while ignoring real corruption, and insisted that other nations wisely had no such laws. He seemed to feel that decent, law-abiding business leaders lived in constant fear of the penitentiary because they might allow natural law to put them in charge of a trust only to find Taft's attorney general imprisoning them. A nonpartisan tariff commission could easily handle what few wrongs a strong tariff might produce. This would protect the nation and its businesses and keep wicked politicians out of problems that they did not understand or wished to use for corrupt purposes. To anyone schooled in economics or remotely familiar with conventional European modes of class analysis, the result was sheerest hypocrisy: half the speech contradicted the other half, creating a muddle even worse than that encountered in the thought of Roosevelt himself. As William Allen White wrote to Claude Bowers, Beveridge's biographer, twenty years after the speech: "I am persuaded that he didn't just happen, that I dreamed him. Earth doesn't hold such contradictions, such spiritual grotesques, such elemental spirits all twisted and woven into one as appeared in Albert." Yet in its sincere attempt to unify business, labor, social reform, and the interests of the nation as a whole, the speech did sum up the aspirations of most delegates to the convention; it did not deceive the delegates so much as reflect their own genuine confusions.[20]

Certainly Beveridge put his heart and his future political career into the speech; by the end his beautiful high collar "lay like a limp rag around his neck." The convention that had so obviously lacked cohesion at first, had found its first leader. With a magnificent sense of his audience, Beveridge closed by intoning the words from the abolitionist hymn of Julia Ward Howe, the famous "Battle Hymn of the Republic":

> He hath sounded forth a trumpet that shall never call retreat,
> He is sifting out the hearts of men before his judgment seat.
> Oh, be swift our souls to answer Him! Be jubilant our feet!
> Our God is marching on!

The audience began to sing. Senator Dixon frantically signaled for the band to start playing the hymn to keep the enthusiasm going. They burst into "The Battle Cry of Freedom," continuing unaware that everyone was singing something else. It hardly mattered, since only the drum could be heard throbbing in the racket. Finally, the band gave its support to the symbolic union of abolitionism, Protestantism, nationalism, and Progressivism. As the *New York Times* noted, "nobody had to tell them the words":

> Mine eyes have seen the glory of the coming of the Lord;
> He is tramping out the vintage where the grapes of wrath are stored;
> He has loosed the fateful lightning of His terrible swift sword:
> His truth is marching on.
> I have seen Him in the watchfires of a hundred circling camps;
> They have builded Him an altar in the evening dews and damps;
> I can read his righteous sentence by the dim and flaring lamps,
> His day is marching on.
> I have read a fiery gospel writ in burnished rows of steel;
> "As ye deal with My condemners so with you My grace shall deal;
> "Let the Hero, born of woman, crush the serpent with his heel,
> Since God is marching on."

The *Times* inadvertently provided a useful vantage point from which to analyze the proceedings. Its reporters were clearly skeptical, possibly sent by the paper to report hostile material. Yet the articles that came back paid their respects to the crowd and its spirit—often appearing beneath headlines written by hostile staffers in New York who were safely isolated from the spirit of the convention. The lead story stressed the sincerity and strength of conviction of the audience, and found the scene "strange, moving, and compelling." Indeed, the writer continued, "it was not a convention at all. It was an assemblage of religious enthusiasts. It was such a convention as Peter the Hermit held. It was a Methodist camp meeting done over into political terms." Every face seemed to wear an expression of "fanatical and religious enthusiasm." Those who had listened to Beveridge obviously believed "that they were enlisted in a contest with the Powers of Darkness." The speech itself was "the best speech he ever delivered in his life. He knew the Lutheran and Garrisonian strain in his audience, and he skillfully bent his speech to that element." Not content with simply singing Howe's hymn, the delegates also went through such stalwarts as "Onward Christian Soldiers," and no one seems to have thought it strange that the leader of the New York delegation, as it marched off to war around the convention hall, was Oscar S. Straus, one of the most prominent Jewish political figures in the nation.[21]

IV

Tuesday's session began at 12:35 P.M. The crowd was excited with expectation of Roosevelt's imminent appearance. Strange mooing noises swept over it, "like a storm wind through the pines"; the call of the bull moose had become as common as red bandannas and religious hymns. After the customary prayer, Roosevelt "burst through the curtain of cops that veiled the entrance to the stage like a comic opera tenor coming to the rescue of the enslaved soprano." The band instantly began to play "America" and "with pistol shot accompaniment," the crowd came to its feet and went wild. Roosevelt himself was so excited that he had forgotten to eat lunch at all, but he had managed to change from his usual outfit of sack suit and cowboy hat into a frock coat and a silk hat—nothing was too fine to wear for his "Confession of Faith." Roosevelt was in his element, frantically waving to familiar faces as he spotted them in the galleries and punctuating his gestures with "such staccato snappings of the famous teeth that you could seem to hear each note." People came up to the platform eager for a handshake. Two Negroes were among them, and as if to atone for the shabby way the credentials committee had treated Southern Negroes, Roosevelt "grabbed these two colored brothers as if they both just came back from discovering a new Pole." Many women came forward, and in the process of hauling them up onto the stage Roosevelt "had to convert himself into a human winch." The coliseum became "one joyous field of noise incarnadined." One delegation suddenly stood up and began to sing the old revival hymn "Follow, follow, we will follow Jesus." They changed the words slightly:

> Follow, follow, we will follow Roosevelt,
> Anywhere, everywhere, we will follow him.

The delegates began marching around, and soon were back to singing "Onward Christian Soldiers." Even Jane Addams marched down the aisles with her delegation, although she seemed "a trifle dubious."[22]

The din lasted for 55 minutes. The enthusiasm was genuine, and the appearance of the candidate at all was a sharp break with precedent. Some of the activity was simply antic good spirits, like the huge golden teddy bear that the California delegation insisted on bringing up to the platform. Much of it was religious. The skeptical *New York Times* reported that Roosevelt seemed puzzled at all the revivalism and was himself not a crusader, a comment that seems doubtful in view of his persistent moral rhetoric, the labeling of his "Confession of Faith," and the long record of his own attitudes to political problems. Perhaps his nearsightedness or the lighting caused him to look puzzled, or perhaps the comment was simply a snide aside from a hostile source. Even the *Times* noted "the intense Christian feeling in that crowd all over the hall."

Certainly, Roosevelt managed to tie religion, morality, and political positions together effectively in his address, from its announced title to its rousing if repetitious conclusion. He continued his attack on the Taft bosses, finding that "there is no health in them, and they cannot be trusted." Identifying his own personal cause with general moral principles, he insisted that the people had to rule: they needed primaries, direct election of senators, the short ballot, and a corrupt practices act. The courts had damaged laws that Roosevelt advocated, and so their decisions must be subject to recall. Labor required protection, better pay, and health assistance. The Germans had shown the utility of pension and insurance plans, and Roosevelt wanted to follow their precedents. He wanted an income tax, women's suffrage, and efficient commissions to regulate the growth and activities of large business. He wanted to pass prosperity around and advocated a scientifically administered tariff that would protect domestic workers. He wanted a stronger currency, more powerful armed services, and a more sensible conservation policy.

At this point voices from the hall intervened. Roosevelt had already deviated in places from his printed text and would do so again, and some question remains as to how spontaneous some of these interruptions really were. At least one, by a Southern Negro, may well have been planted to deal with the problems of the Negro delegates and the place of the Negro in the party. Roosevelt took on the question without hesitation, repeated his general position, and stressed that he would like to be judged by his approval of the place Negroes had in the northern delegations. They had earned their places in the North, and in time Southern Negroes would do the same. Above all he stressed his good intentions: "I will do anything for the people except what my conscience tells me is wrong, and that I can do for no man and no set of men; I hold that a man cannot serve the people well unless he serves his conscience." The progressive cause "is based on the eternal principles of righteousness," and even though that cause might fail in the short run, "in the end the cause itself shall triumph." He then repeated what he had told his supporters in June: "We stand at Armageddon, and we battle for the Lord."

After the cheering died down, the credentials committee finally made its report, denying both Florida delegations their seats, accepting the Fridge delegation from Mississippi but disavowing specifically the racist wording of its call for delegates, and establishing an essentially states' rights procedure for future cases. The report was adopted quickly, and the session adjourned at 3:55 P.M.[23]

Roosevelt's speech dominated public attention, but the real drama was the backstage war about that covenant with the people, the Progressive Party platform. A subcommittee that included Dean William Draper Lewis of the University of Pennsylvania Law School and Gifford Pinchot had been at work for weeks on proposals. Pinchot, in particular, was working hard for a radical document. The full committee first formally convened at 9:00 P.M. Monday, 5 August. It had a number of vexatious problems to settle. A general advocated better veterans' pensions, and a feminist advocated women's suffrage. A Minnesota group attacked Canadian reciprocity and little-known issues relating to a national bureau of health, the civil service, liquor, the currency, divorce, and the income tax all had an airing. The most significant of the early proceedings was the plea, quickly ignored, that Jane Addams and Henry Moskowitz made in support of Negro rights. No one was going to be permitted to drag the Negro issue out if most progressives could help it. That meeting adjourned at 1:30 A.M. Tuesday, the first of many long, numbing open and closed sessions.

In addition to Chairman Lewis and Gifford Pinchot, committee members at Chicago included William Allen White, Senator Dixon, Chester H. Rowell, Dean George W. Kirchwey of Columbia University Law School, and Professor Charles E. Merriam of the University of Chicago. Many other progressive leaders like Professor Charles McCarthy, the general counselor to the group, were in and out of the room continuously. Roosevelt closely monitored the proceedings, approving each plank as it appeared so that nothing would come out of the party that would conflict with his basic position in the Confession of Faith. Rowell, Kirchwey, McCarthy, White, Lewis, and Pinchot had all prepared full drafts of proposed platforms in advance. The squabbling among them delayed the presentation of the platform to the full convention and also helped cause one of the few genuine embarrassments of the proceedings. The group worked through Tuesday and well into the night. Individuals made strong efforts to adopt a single tax plank, some kind of prohibition, and a declaration in favor of Philippine independence. All were defeated and remained unmentioned in the platform. Roosevelt intervened to tone down the planks on popular democracy measures, rejecting the recall of senators, the amendment of the Constitution by referendum, and the recall of judges.

The most divisive plank concerned the trusts. All Progressives agreed that an antitrust law alone was not an answer to the problem of big business. Most favored federal regulation by commission. But on the breaking up of large corporations merely because they were large, disagreement was sharp. Over a dozen planks were proposed, most of them advocating a stronger Sherman Act or avoiding any mention of it at all. The majority of the committee apparently was the more radical group, led by McCarthy and the Pinchots. They produced a document of two paragraphs: the first endorsed a regulatory commission to guard against irresponsible power; the second and more important read:

> We favor strengthening the Sherman law by prohibiting agreements to divide territory or limit output; refusing to sell to customers who buy from business rivals; to sell below costs in certain areas while maintaining higher prices in other places; using the power of transportation to aid or injure special business concerns; and other unfair trade practices.

McCarthy regarded this as his favorite paragraph, and it symbolized the western, more radical, progressive attitude toward the East and big business.

George Perkins would not stand for it. "Spick-and-span, oiled and curled like an Assyrian bull, and a young one, trim and virile," in White's memory, he brought Roosevelt to the group and insisted that unacceptable provisions were "all through it." He then left the room and Roosevelt informed the committee that Perkins had done more for the party than anyone except possibly Senator Dixon, that he paid the bills, and that Roosevelt wanted the platform written so as to obtain Perkins' approval. McCarthy replied that he was there to represent the people. He did not care what Perkins felt in the slightest and Pinchot agreed with him. Pinchot,

White, and Perkins began to talk in the next room and suddenly White scurried back in to say that Pinchot and Perkins were arguing; Perkins was threatening to leave the convention. Roosevelt scurried after his friend, and for forty-five minutes shuttled back and forth trying to hammer out a compromise. Perkins not only objected to the trust plank, he objected to many of the preliminary direct democracy, tariff, and currency planks. Finally, the meeting broke up and during the evening Beveridge, McCormick, and Dixon added their support for toning down the more radical provisions.

Roosevelt sided with his financial supporters even while assuring his western friends of his good intentions. He insisted that the platform did not need the amplification provided by the McCarthy plank and said that he had dealt with the issue clearly enough in the Confession of Faith. He finally produced a plank that satisfied Perkins, Beveridge, and Dixon, and managed to get the grudging approval of the committee for it. What no one apparently noticed at the time, however, was that the committee had also endorsed the brief second paragraph of the Pinchot proposal, and so had glued together two contradictory positions. By this time everyone was exhausted. William Allen White lay down and voted languidly by raising his pudgy hand whenever he could manage it from his prone position. Presumably, the others were less than alert. Possibly, they were unaware of what they had done. Possibly, they were quite aware and were attempting to mollify Roosevelt and his pet banker while remaining true to their consciences. Certainly, Roosevelt had supplied any number of precedents for the holding of vague and contradictory positions on public issues. Moral vigor rather than intellectual consistency dominated the convention, and such a compromise would have been quite in character.[24]

V

The most important of the nominating speeches was that of Jane Addams. A vastly different kind of progressive from Roosevelt, she nevertheless, strove to find common ground to advance her causes. Without mentioning Roosevelt by name at all at first, she seconded the nomination of the convention "stirred by the splendid platform adopted by it." Measures of industrial reform and social justice that charity organizations had long sought "have here been considered in a great national convention, and are at last thrust into the stern arena of political action." Even though, in the confusion of the deliberations, the convention as yet did not even have a platform, she spoke confidently about it. She especially approved the party's determination to protect children, care for the aged, relieve over-worked women, and safeguard male workers. She said that such measures could not help but appeal to women, as well as rouse the moral energy they possessed in ways that no other party ever had. She found the platform especially attractive because of the women's suffrage pledge that would presumably be in it. Democracy could never be truly realized while any competent group remained unrepresented—she did not mention Negroes but her language could easily include them as well. Only at the end did she leave the subject of the platform for the candidate, praising Roosevelt as one of the few men in public life "who has been responsive to the social appeal and who has caught the significance of the modern movement." Only, it seemed, "because the program will require a leader" of courage and democratic sympathies, did she "heartily second the nomination so eloquently made by the gentleman from New York."[25]

Addams had grave doubts about Roosevelt and always would have. She entered the political arena, not because of Roosevelt or any illusions of victory, but because of the "wonderful opportunity for education" that the convention and the subsequent campaign presented. Again and again this theme reappears in the writings of the settlement group: Roosevelt might well be an undependable opportunist, but he had joined their side of the battle. His charismatic personality meant publicity for their ideas and votes for social justice. Roosevelt might secretly regard the social workers as "moonbeamers" or wild-eyed radicals, but they judged the situation accurately enough in their own way, and they were using him fully as much as he was using them. Addams' speech was thus something of a tactful warning that Roosevelt would misread at his peril. The delegates hardly noticed. They greeted the speech with enthusiasm, which developed into a roar of approval when she picked up a large yellow banner with "Votes for Women" on it and proceeded to lead the Illinois delegation around the hall. Roosevelt might not have his heart in that issue but he was certainly more favorable to women's suffrage than Woodrow Wilson.[26]

After the nominating speeches, Dean William Draper Lewis finally made a belated presentation of the platform. Carrying a pile of papers in no obvious order, he and the rest of the committee were still groggy from their marathon session of the night before. The final platform became the subject of fierce debate almost immediately, and the controversy continues even today.

The Progressive Party Platform, its "Covenant with the People," was in many ways the central statement of political progressivism. Other parties had produced and would produce policy statements, but never had so much energy from so many different sources come together with so many innovative political suggestions in all of American history. Rhetorically, it combined the rhythms of American Protestantism, Jefferson and the democratic radicalism of the early Republic, and Lincoln and the early Republicans. In effect the Eastern Progressives had their Lincoln and their emphasis on national responsibility, and the Western Progressives had their Jefferson and their emphasis on measures of popular democracy. Both had a common heritage of evangelical Protestantism with its emphasis on sin and redemption, on the virtuous conscience versus the corrupt bosses. The platform would be only so much puffery, however, if it did not also contain hard proposals for solving the problems facing the nation.

For the West the platform demanded direct primaries for state and national officials including the president, direct election of senators, the short ballot, the initiative, referendum, and recall. It supported women's suffrage and the limitation of campaign contributions. It asked that if a law were held unconstitutional by a state court, that the people be allowed to overrule that decision by vote; and that if a state appellate court held a law unconstitutional because it violated the Constitution, then that decision could be reviewed and reversed by the Supreme Court. It approved of the pending income-tax amendment, and suggested a graduated inheritance tax.

For the East Roosevelt's insistence on federal authority and his scorn for states' rights were evident both in his request for a national-health service and in the tender tone taken toward business. He asked for strong national regulation of interstate corporations, with commissions to supervise business activity, tariff legislation, and related economic activities. But the platform went out of its way to speak of business having grown large out of efficiency and necessity, and of how businessmen ought to be able to have clear legal guidelines within which they could operate. Capital had to be secure and dividends safe; if they were, then investments would be steady, and everyone would prosper. All that was necessary was publicity, supervision, and regulation; promiscuous antitrust activities were out. The tariff needed study and a general lowering of rates; the Payne-Aldrich rates were too high, while the Democratic request for a tariff for revenue only would lead to industrial disaster.

The platform forged new ground. The policies that meant so much to Jane Addams and her social workers were, in fact, included. The party pledged to work for laws to prevent industrial accidents, occupational diseases, overwork, unemployment, child labor, night work for women, and abolition of convict contract labor. It wanted industrial safety and health standards, a decent minimum wage for women, an eight-hour day and six-day week, publicity about wages, hours, and labor conditions. It wanted full reports on industrial accidents, inspection of weights and measures, compensation for accidents and death, unemployment insurance, better industrial education, and labor unions. While none of these could be termed radical by anyone with his wits about him, these positions were all threats to established ways of thinking, running businesses, and administering the government. But they were also ways of improving the system, not rooting it out; they would work to the strengthening of capitalism rather than its overthrow. As the *San Francisco Examiner* noted, most of the ideas had been around for some time; a few years earlier, however, Roosevelt himself "would have denounced its each and every plank as incendiary and anarchistic." Progressivism, like Roosevelt, had come a long way.

The great controversy came when Lewis read the antitrust plank. As the committee had decided, he read both the Perkins plank tolerant of big business and the brief second paragraph of the McCarthy plank. Most of the people in the audience were not listening, but for George Perkins this plank was important: a look of blank amazement came over his face. He was under the impression that his pressure on Roosevelt had caused the removal of that plank. He hurried out of the hall in great distress to confer with Roosevelt, Beveridge, and Dixon. He persuaded them that the paragraph could not be included in the platform. They conferred with Lewis, who along with a number of others, was apparently in genuine confusion about what had been voted. Then O.K. Davis went to work to make sure that the newspapers did not carry the offending paragraph in their stories. He met some opposition from the Associated Press, but managed to convince the majority of newspapers to omit the offending paragraph. The result was a controversy in which even now some of the basic facts remain in dispute. Even the documentary evidence conflicts. The typescript record of the convention, presumably an official document, makes no mention of the paragraph or the confusion. Yet the most consulted collection of platform documents includes the paragraph without a word as to its authenticity. The editors are correct in doing so, but anyone studying the convention might find it helpful to have at least a footnote explaining that the paragraph was not the one presented to the American people at the time.[27]

VI

The convention was in its usual unanimous mood. It adopted the platform without dissent and without most of the delegates having heard a word of it. Roosevelt was formally nominated, and the convention turned to the vice-presidential nomination. Governor Hiram Johnson was clearly the strongest available candidate for the job. The California Progressives had been among the most enthusiastic Progressives from the beginning, and Johnson had supported Roosevelt longer than almost anyone else in public life. A man of intense moral feeling with a rigid personality, Johnson had entered public life reluctantly as the man most competent to end the dominance of California politics by the Southern Pacific Railroad. A lawyer with wide appeal in prohibitionist and agrarian areas of the state, he not only proved successful in his initial term as governor, he used the position to become a national influence for progressive ideas. Johnson consulted Lincoln Steffens and Robert La Follette, but he always felt closest to Roosevelt and was never shy about expressing his admiration for the former president. The relationship had begun during the 1910 campaign for governor. Theodore Roosevelt, Jr., was living in San Francisco at the time, and his father addressed a letter to him that was full of praise for Johnson. With Roosevelt's cooperation about one million copies of the letter were distributed in support of Johnson, thus helping his candidacy materially. Johnson expressed his gratitude, and in 1911 Roosevelt made an extended tour of California. He stayed with his son, exchanged visits with the Governor, and shared a podium with him several times. In October the California progressives concluded that another Taft candidacy would be a party disaster. Johnson wrote his "political father-confessor," urging him to run for the Presidency. Roosevelt had not made the decision to run, but heartfelt adoration and honest praise always had a rejuvenating effect on him, and he was unlikely to forget who his friends were.[28]

California progressives felt they had a man of national stature in Johnson. They were wrong, but the limits of his personality were not as clear in 1912 as they became during his long career in the Senate. In February Edwin Earl, owner of the *Los Angeles Tribune* and *Express,* began a strong campaign to win the vice-presidential nomination for Johnson on a Republican ticket headed by Theodore Roosevelt. At the time Roosevelt was at the start of his campaign, and the ticket made a great deal of sense. On 16 February, the front page of the *Express* had pictures of both men, a supporting editorial, and this caption from Kipling:

> For there is neither East or West,
> Border nor breed nor birth
> When two strong men stand face to face,
> Though they come from the ends of the earth.

Johnson was eager for the spot, and such a Republican ticket could have swept the country in the fall. But the Taft steamroller did its work, the

California delegation led the exodus out of the convention, and Californians were prominent in pushing the new party. Johnson had no illusions about the chance for a Progressive Party victory in November, but his previous relationship to Roosevelt made it all but impossible for him to refuse. Both men had convinced themselves that they had a moral duty to form the new party. Since Johnson was the most prominent office-holding politician in the party, he was stuck with the honor. The brief campaign to put a Southerner on the ticket, most obviously John M. Parker of Louisiana, collapsed from lack of stature in the candidate and lack of support in the South. The unity of East and West seemed more fruitful than that of North and South.

At the conclusion of the convention, the California delegation unfurled a huge 20-by-40-foot banner, with the inscription:

> Hands across the Continent
> Roosevelt and Johnson
> New York and California .
> For there is neither East nor West.

Hiram Johnson stepped out after Roosevelt to accept the vice-presidential nomination. The delegates then reverently sang the "Doxology," and at 7:24 P.M., the convention adjourned.[29]

The great revival was over; it would never come again. Even the young Republican Party that had nominated Abraham Lincoln had not approached this level of intensity. No subsequent convention would. The country was soon too pluralistic and the force of Protestant religion too weak ever for such a thing to happen again. When the Christian soldiers marched off to war, they marked the end of a great tradition in America as well as the coming of something new. As religion charged out, social justice tiptoed in.

Chapter 8

A Presbyterian Foreign Policy

Mored than any other aspect of progressivism, foreign policy remained a subject of contemporary relevance to succeeding generations. Scholars have tried to place blame and assign guilt. They have sought to determine whether the United States acted wisely when it entered the European war and which statesmen deserved praise or blame for American participation. During the succeeding three decades, revulsion at the results of intervention gave way to an almost desperate need to justify it because of presumed parallels to World War II. Many early students made their judgments on the basis of tendentious views of figures such as Walter Hines Page and Edward M. House, who reached print first. Others based their assumptions about Woodrow Wilson's abilities on erratic memoirs, or judged William Jennings Bryan's skills as if they were extensions of his earlier agricultural radicalism or his later religious fundamentalism. The variety of materials now available and the passage of time permit a new assessment in which Bryan recovers his proper role as a significant spokesman for progressive foreign policy views and House sinks back into the obscurity from which he so mysteriously emerged. Wilsonian policy as a whole becomes a genuine group effort, with three key men—Wilson, Bryan, and Robert Lansing—all in agreement about the portent of that policy regardless of later disagreements about specific applications of it. They formulated a Presbyterian foreign policy that asserted the virtue of American motives and the superiority of American ideals and insisted on the right to export those ideals to any unstable area in the world.

Contrary to many accounts, Secretary of State Bryan was the most important foreign policy figure in the administration throughout most of his tenure. Although he had never held an official job relevant to foreign policy, he had more experience in this area than other possible Democrats. Bryan received his place in the cabinet because of this experience as well as for his vital support of Wilson's candidacy and his predominant role in the Democratic Party machinery. Three times nominated for president and representative of Midwestern and Southern agrarian interests that Wilson had difficulty inspiring, Bryan was too significant a figure to ignore. Within the administration, he became the lobbyist for the most important measures of the New Freedom. Bryan was frequently away from Washington during his tenure, usually on trips that seemed important to the survival of the Democratic Party in office. In Washington, he worked hard and controlled the vast bulk of the correspondence of his office. His judgments were often sensible, reasonably consistent, and for a long period remarkably close to those of the president. Although an inefficient administrator unfamiliar with the niceties of international law or diplomatic protocol, Bryan was an abler man, all things considered, than his immediate predecessor and successor and hardly deserved the savage treatment that he received from Republicans, newspapermen, and many "liberal" historians.[1]

Bryan became a scapegoat for many reasons. Although he represented the broad middle range of public opinion away from the Atlantic Coast better than any other public figure, everything about him offended politicians suddenly out of power and the journalists scaring themselves for twenty years about the raving radical from the West. At the time, few were willing to acknowledge Bryan's role in rebuilding the Democratic Party or in developing a public audience for many of the reforms for which Woodrow Wilson and Franklin D. Roosevelt later took credit. Critics disliked Bryan's clothes, his obesity, his fundamentalism, his naiveté, his churchgoing, and his optimism. He refused to smoke, drink, dance, or tell off-color stories. He was relentlessly good-humored and seemed immune to the pressures of fashionable Washington society. Important members of the administration underestimated him, deceived him, and expressed contempt behind his back while professing to like him socially. Colonel House, who had been a friend of Bryan for almost twenty years, demonstrated this when he wrote Wilson in the opening days of the World War that if the president made a public statement, he should emphatically make it in his own name: "If the public either here or in Europe thought that Mr. Bryan instigated it, they would conclude it was done in an impracticable way and was doomed to failure from the start." He said that he hated "to harp upon Mr. Bryan," but that the president could hardly imagine how little respect the Allies had for him. He then added, with the air of unctuous superiority of which he was a master, that Wilson and he "understand better and know that the grossest sort of injustice is done him." When it came to gross injustice to Bryan, House was himself in the lead, closely followed by other highly visible men in the administration ranging from Ambassador Walter Hines Page to State Department Counsellor Robert Lansing. As a person, if not as a policy maker, Bryan hardly had the opportunity to do a good job.[2]

House managed to get his self-serving views into print first and did incalculable damage to Bryan's historical reputation. Other figures who observed Bryan day by day formed higher opinions. Secretary of the Interior Franklin K. Lane assured a friend early in 1915 that "Bryan is a very much larger man and more competent than the papers credit him with being," while Navy Secretary Josephus Daniels pointed out that the cabinet was usually unified in its policy decisions, and that Bryan's views on peace treaties and alcohol were acceptable not only to the president but also to most of the others in authority. Wilson himself changed his estimate of Bryan as the two men collaborated over time. He grew to have a genuine affection for a man who could cooperate with such cheer and humility. Some of this approval could perhaps be dismissed as signs of friendship or gentility, but the case of John Bassett Moore sheds light on the matter. The most distinguished scholar of international law in the United States, Moore was the apolitical Counsellor to the Department of State during the early part of Bryan's tenure. Reluctant to join the administration, he accepted his position expecting to agree with Wilson and disagree with Bryan on important issues. He found to his surprise that he agreed more and more with Bryan and less and less with Wilson. His opinion had so changed by the time he left office that he retained Bryan's friendship. In view of the different backgrounds and outlooks of the two men, this was a valuable outside comment on Bryan's performance in high office.[3]

Foreign diplomats provide an external view of Bryan and the other members of the administration. Sir Cecil Spring Rice, the highly irritable British Ambassador, proved to be as moralistic as Bryan; the men established ties in 1913 because of their mutual interest in the Bible. They maintained a personal warmth despite Bryan's unwillingness to adopt

MINISTERS OF REFORM

Spring Rice's militantly pro-British stance on the war and the ambassador's firm belief that Bryan was "incapable of forming a settled judgment on anything outside party politics." The ties between the men perhaps come through most clearly in the letter that Spring Rice wrote to Mrs. Bryan after her husband resigned: "I can't say how grateful I am to you both for your invariable kindness and good humour. When I think how aggravating I have often been, I am simply appalled at Mr. Bryan's imperturbable good humour." Spring Rice's long-time friend, Austro-Hungarian Ambassador Constantin Dumba, had a similar opinion. Although he judged Bryan to be "a very indifferent Secretary of State," he also found him "the most congenial personality among all the politicians whom I met in America." Bryan seemed "exceptionally open and sincere, kindly and dignified. . . . His clean-cut, expressive features, his tall and somewhat portly figure, and his beautiful baritone voice immediately made a favorable impression." Memoirs and public statements may frequently conceal the true sentiments of observers, but it is highly unlikely that German Ambassador Count Johann Heinrich von Bernstorff felt the need to be anything but honest when he informed Theobald von Bethmann Hollweg that Bryan's first appearance as Secretary of State "made a favorable impression on me and all colleagues with whom I spoke. He appears to want to be more approachable than his predecessor who, as is well known, had little interest in foreign countries." Where outsiders were critical, as when Bernstorff found Bryan "an honest visionary and fanatic," the context frequently made it clear that the writer had an even lower opinion of the president or other figures in the administration.[4]

The world may have assumed that Wilson dominated his own foreign policy and his provincial Secretary of State, but except in rare periods of crisis he did not do so until European military affairs demanded his attention late in 1914 and early in 1915. Instead, Wilson concentrated on New Freedom issues such as tariff, antitrust, and currency reform, and was increasingly distracted in his personal life by the illness and death of his wife during the first eight months of 1914. Even had Wilson paid closer attention to foreign policy, his views and attitudes were so similar to Bryan's that his attention hardly mattered. The two men were both Presbyterian elders who believed in the omnipotence of God and the ultimate triumph of righteousness; they shared a Scots-Irish heritage of industry, sobriety, and religious zeal inherited from fathers whom they idolized; and neither man knew anything significant about diplomacy when he took office. Whereas Bryan could be accommodating and flexible in new diplomatic situations, Wilson had what his chief biographer called "an astounding array of prejudices" and personal hatreds. If Bryan was all too often uncritical and trusting, Wilson went to the opposite extreme "to equate political opposition with personal antagonism and to doubt the integrity of any man who disagreed." As Senator John Sharp Williams of Mississippi wrote of the President: "He was the best judge of measures and the poorest of men I ever knew."[5]

Wilson was certainly less popular with foreign diplomats than Bryan. Spring Rice found it hard to meet with him at all, although he sensed that the President was an Anglophile. To the Englishman Wilson was a "hardened saint." Bernstorff found him reclusive and stressed Wilson's tendency to take all criticisms of his policies as personal slights. The President was also ignorant of Europe. "Nature had equipped him with brilliant gifts, but they were not the gifts called for by his position at that time, and were rendered even more ineffective by the President's incapacity for personal negotiation." Dumba repeated these criticisms, finding Wilson "stiff and repellent," "unskilled socially," and "lacking in common sense." "He was doctrinaire and, above all, a man of words" with "a considerable meed of conceit and presumption." Some of these criticisms may be discounted because of the war atmosphere and Dumba's own awkward role as America moved toward intervention, but he was not the only observer to find in Wilson "a far-reaching lack of decisiveness," or to have "the feeling that Wilson considered a well-turned speech or message to Congress as itself the solution of a problem."[6]

The best way to understand the evolution of American foreign policy while Bryan was Secretary is to visualize it as an effort by two groups for control of Wilson's allegiance, and thus, of his diplomacy. Working reasonably well with Bryan were Navy Secretary Josephus Daniels and Bryan's Solicitor in the State Department, Cone Johnson; they had widespread support in both houses of Congress. Often contemptuous of Bryan were House and his Texas clique, which included Secretary of Agriculture David F. Houston and Postmaster General Albert Sidney Burleson; Robert Lansing, who was the Acting Secretary of State during Bryan's last year whenever Bryan was out of town; and the traditionalists, usually from the East Coast, who remained influential in the State Department or in the media. The two groups agreed with surprising frequency until the issues of the first six months of the World War demonstrated the incompatibility of various strands of the progressive mentality.[7]

The most important figure to oppose Bryan's influence was Colonel Edward M. House. Because House never took an official position in the administration, and because he kept a voluminous diary, he has been a difficult person to assess. Too much of the material available for analysis came directly from him or from his close friend and editor, Charles Seymour of Yale University. Now that far more material has come to light, it is necessary to distinguish between two aspects of House: his actual accomplishments and his own sense of importance. House had been a prominent supporter of Texas Governor James Hogg, a mildly progressive Democrat in one of the key states for obtaining power within the party. Plagued by delicate health, House nevertheless had a rich fantasy life and yearned for an appropriate candidate to carry him to power. After several failures he found Woodrow Wilson. The two men became close, as much because of Wilson's need for flattery as for House's ambition. House proved valuable as a liaison between the Eastern and relatively conservative Wilson people and the Southern, Western, and more radical Bryanites. Once Wilson assumed office, House provided a much-needed European perspective to an inexperienced and provincial administration. He had lived for an extended period in England and had the ability to get along with Europeans; he also valued a professional foreign service. He recommended men of the quality of John Bassett Moore and William Phillips in the hope that they would keep Bryan from embarrassment. House also proved invaluable for years as Wilson's unofficial ambassador to Europe. The level of ambassadorial appointments proved low, and men like James Gerard in Berlin and Walter Hines Page in London turned out to have severe if different limitations. The British Ambassador in Washington, Cecil Spring Rice, was also difficult to work with for reasons of health and temperament. Only House provided Wilson with a trusted line of communication to the warring European powers.[8]

Less attractive features balanced House's virtues and services. Some, like his passion for codes in his messages, his fawning over Wilson and others, and his going about in what Ray Baker liked to call a "blaze of mystery" were harmless. Ellen Maury Slayden, wife of a Texas Congressman and a woman who had disliked the Austin clique for years, sneered at House as "the Veiled Prophet of Austin"; many writers have questioned the actual influence House may have had. The tendency to agree with the person to whom one is talking could have a pernicious effect in serious diplomacy. Foreign diplomats could seriously misperceive the position of the American government, all the more when they assumed that the voice of House was the voice of Wilson. House charmed many people, including Bernstorff, Dumba, and many of the British. Yet when Bernstorff read the *Intimate Papers* he realized that House had deceived him about American attitudes. Josephus Daniels was convinced that House had misinterpreted Wilson's position during the Mexican involvements and had misled Sir Edward Grey. House proved obtuse in certain later dealings with Grey involving key war issues, and Daniels referred bluntly in one place to

House's "ineptness." At least one French writer noted acidly that House was not so much a Talleyrand from Texas as one who would pass in Texas as a Talleyrand, and the distinction implies something important about the quality of American foreign policy and its formulators in this era. Even those like House who regarded themselves as worldly, sophisticated, and cosmopolitan could be as innocent as Bryan seemed to be when it came to important negotiations, assumptions, and biases. House and Bryan might seem to represent the extremes in foreign policy formation, but viewed from afar, both men had severe limitations, lived rich fantasy lives, and misunderstood the Europeans they patronized. Wilson himself was more intelligent and articulate, but in these matters, he was no wiser. The results of this smug ineptitude had incalculable effects on the war and the peace. What remained beyond question was that Wilson was strongly devoted to House until late in his Presidency, and that House did his best to influence Wilson to intervene in the war on the side of the British.[9]

Equally pro-British and usually more influential was Robert Lansing. While House traveled with romantic ineffectuality, Lansing stayed in Washington as Counsellor to the Secretary of State after the resignation of John Bassett Moore early in 1914, and then became Bryan's successor as Secretary of State. On the surface Lansing was the perfect choice for such offices. In an administration unblessed by much diplomatic and legal talent, he was the son-in-law of former Secretary of State John W. Foster and a man with what seemed to be an impressive legal background. He had been a conservative Gold Democrat, and his presence lent an air of balance to a department headed by Bryan. These impressions were deceptive. His appearance might be as neat and distinguished as any in Washington, but his mind tended to mire in legalistic detail, and he regularly lost sight of long-term goals in moments of anxiety. As Ernest May has said: "At no time during his temporary occupancy of the Secretaryship did Lansing betray a single gift for statesmanship." At the same time, he shared completely the religious biases already strong in American policy. As much a fundamentalist as Bryan, he was a Presbyterian elder who devoted most evenings to Biblical study. His legal and governmental background might make him look knowledgeable in the realities of foreign policy and the necessities of the balance of power, but his mind ran in the same narrow rut as those of Wilson and Bryan. Wilson had a certain measure of contempt for Lansing and regarded him mostly as law clerk and errand boy. Certainly Lansing never had the control over foreign policy that Bryan exercised during most of 1913–14. Nevertheless, Lansing agreed so closely with House on certain matters of policy that he could lend support on important issues, and as the man who shaped many official messages, he could sharpen or blur key points or leave out important issues entirely. His alliance with House on the Allied side ultimately contributed materially to drawing the President away from Bryan's neutrality.[10]

At the beginning and at the end of the administration, one issue provides a convenient focus within which to examine the ways in which progressivism produced foreign policy. The first test for the Presbyterians in Washington was in Mexico; preconceptions and patterns of behavior established in dealing with Mexico remained crucial to more significant issues of European wartime diplomacy. That European diplomacy, in turn, produced the debate over the Covenant of the League of Nations. That Covenant was the single most important product of progressivism in foreign policy. Its defeat in the Senate ended progressivism as a dominant climate of creativity.

II

In May 1910 President William Howard Taft had defended the business orientation of his foreign policy. While he was devoted to justice throughout the world, he thought the American government had the right to "include active intervention to secure for our merchandise and our capitalists opportunity for profitable investment which shall insure to the benefit of both countries concerned." Peaceful relations should go hand in hand with trade relations "and if the protection which the United States shall assure to her citizens in the assertion of just rights under investment made in foreign countries, shall promote the amount of such trade, it is a result to be commended." He thought it a shame that people should sneer at this as "dollar diplomacy," and thought critics ignored "entirely a most useful office to be performed by a government in its dealings with foreign governments." Pursued chiefly in Latin America, the dollar diplomacy of Secretary of State Philander C. Knox sought to ensure stable democratic regimes in the countries concerned. These facilitated the loaning of money by American banks and the production of raw materials valuable to American business. They also worked to the diplomatic advantage of American power and prestige, and helped keep out European powers. If political or financial instability threatened, then American troops assisted Latin countries by keeping order and the businessmen by maintaining the payment of interest and legitimate bills.[11]

Bryan and Wilson shared a great distaste for dollar diplomacy. Bryan had spent his career worrying about the power of money in American business; he thought capitalists exploited their workers and always seemed to enjoy the protection of the government and the courts. His campaigns for free coinage of silver and his desire to reform the national banking system gave him experience in this field and a monumental set of prejudices. Bryan perceived the Taft-Knox policies as being external versions of internal capitalist exploitation, and he reacted in the same way to both. He recorded his sympathy for the people of the exploited nations, and wanted to assure them that a concerned American government would not permit rapacious American businesses to exploit them. In his scenario American banks charged exorbitantly high interest rates for loans to weak foreign governments, receiving the excess money because of the risk involved. They then asked the American government to protect their investments and thus remove the risk that justified the interest. He saw American foreign investment as simply another version of the trust using the government to take money from the poor.[12]

The classic statement of the administration attitude to dollar diplomacy was President Wilson's Mobile speech on Latin American policy, delivered 27 October 1913. It provided a case study of the goals and limitations of Wilsonian policy. The future would be different, Wilson said. The countries of the Western Hemisphere would be drawn together in many ways, "chief of all, by the tie of a common understanding of each other." Economic self-interest was inadequate, but "sympathy and understanding" could help. "It is a spiritual union which we seek." Capitalists had no right to ask for concessions in Latin America, only the right to invest and to take legitimate risks and profits. Wilson sympathized with the exploited peoples, flattered their achievements, and reverted to Lincolnian language to speak of their "emancipation" from the policies of dollar diplomacy. Friends had to be equal, and Wilson wanted America to be friends with her sister republics to the south. He thought it "a very perilous thing to determine the foreign policy of a nation in the terms of material interest," and hoped that Latin America would have a place in the "development of constitutional liberty in the world. Human rights, national integrity, and opportunity as against material interests—that, ladies and gentlemen, is the issue which we now have to face." He promised that America would "never again seek one additional foot of territory by conquest." The true relationship of the United States with the rest of America was "the relationship of a family of mankind devoted to the development of true constitutional liberty. We know that that is the soil out of which the best enterprise springs." Always, spiritual values took precedence over material ones. American history demonstrated the spiritual value of the democratic process and the legitimacy of constitutions.

Wilson then invoked the most ancient American tradition. John Winthrop had settled Massachusetts Bay secure in the feeling that his community would be "a city on a hill" that the rest of the world would admire and emulate. For him America enacted the will of God in the forms of its government and the manner of its citizens' political and religious actions. America, Wilson said in closing, did not become a great nation "because it is rich. . . . America is a name which sounds in the ears of men everywhere as a synonym with individual opportunity because a synonym of individual liberty." He then made the statement that best marked the progressive nature of his foreign policy and its failure: "In emphasizing the points which must unite us in sympathy and in spiritual interest with the Latin American people we are only emphasizing the points of our own life, and we should prove ourselves untrue to our own traditions if we proved ourselves untrue friends to them." Morality and not expediency guided American policy, and in only a few generations, Wilson looked forward to being "upon those great heights where there shines unobstructed the light of the justice of God."[13]

The hallmarks of the Wilson-Bryan policy outlined at Mobile never changed. Both men were constantly aware of past diplomatic acts of arrogance and greed, and determined that, above all, they would not perpetuate dollar diplomacy. Firmly convinced that God guided their hearts and hands, they acted to purify the political and business practices of Latin America. Genuinely selfless in many ways, they elevated personal morals into government policy to benefit the poor of Latin America. It went without saying that good intentions produced good results. Countries got to heaven the same way people did, because their hearts were pure and their acts worthy.

Any number of criticisms can be made of this progressive outlook and its consequences. American corporations might well have been exploiting both their own government and the governments of Latin America, but the interests of the local people were best served by the creation of jobs and an environment in which economic security made genuine freedom financially possible. A word like "exploit" had several meanings, some of them beneficent. To exploit and develop the natural resources of a new, poor land might have salutary effects on the host. Capitalist "greed" might produce more stable governments and happier people. Wilsonian rhetoric, by contrast, seemed to assume that people preferred words to meals, and that the will of a gringo God should somehow be of concern to a people who had never known such a God, any comparable religion, or economy. Philander Knox may well have been an indolent lawyer whose real talents lay on the golf course, but he could never have been accused of hypocrisy. He spoke a language of economic self-interest that was cosmopolitan and imbued with the harsh honesty of the marketplace. Above all Taft and Knox did not pretend to direct the lives, work, worship, and votes of others. In the long run, everyone would do better working for money than for abstractions.

The real problem in Wilson's foreign policy was its provincial, culture-bound, intolerant, and arrogant ways, known to Christians since the Pharisees. Like so many other Presidents, Wilson assumed the superiority of Anglo-Saxon peoples and institutions. He used familial imagery in which no one could possibly doubt the identity of the father. He assumed, as American Protestants had assumed so often since 1630, that the political processes of voting for the minister or for the magistrate implied divine sanction. He and Bryan were textual literalists, and it did not matter to them whether the texts were the Bible, a law, a treaty, or a constitution. A good Presbyterian read his Bible and regulated his private life accordingly. He read his constitution and voted accordingly. He comprehended a law or treaty, and he obeyed. Petty violations of these codes were fraught with large implications, because every infraction implied disobedience to the divine plan. Actual results hardly mattered because such a God was interested only in intentions. Every pain, every disaster, was somehow a necessary obstacle on the path to the divine goal. If men did not understand they could rest assured that God knew, and men should attempt to execute his will. Neither Wilson nor Bryan could imagine a society that did without a constitution, honest voting, and liberal democrats in office. Such political processes were so much a part of their American environment that they were incapable of calling them into question.

Wilson and Bryan were also cultural absolutists, and while they might acknowledge the existence of decency among Roman Catholics, Jews, Muslims, or Buddhists, they always seemed to assume that people or nations would ultimately realize the degraded nature of their allegiance and convert to the American Way. They could not conceive of a people who did not share their moral principles. They could not conceive of a people who might prefer dictatorship, aristocracy, monarchy, theocracy, or communism to liberal democracy. They could not conceive of a rewarding national life that was not bound by covenants just like colonial Massachusetts Bay and modern American democracy. The language of democracy became reified for them; the process of voting had divine sanction. Those foreign statesmen who could speak this language received their support, even when the statesmen were themselves impotent, incompetent, or corrupt; even when they were far wiser in the ways of the world and cynically using these words and American presuppositions to obtain goals of which Wilson and Bryan did not approve.

MINISTERS OF REFORM

III

Mexican history in the early twentieth century is an oft-told tale. Long under the benevolent dictatorship of Porfirio Díaz, an élite group called the Científicos had run the country to suit themselves, cooperating closely with foreign investors. In the face of the resulting poverty, Francesco I. Madero had been agitating against the dictatorship since 1904. Gentle and mystical, Madero spoke a language that resembled the language of American liberalism. Despite periods of jail and foreign exile, he became president in 1911. Mexico, however, remained unstable. Groups led by Victoriano Huerta, Emiliano Zapata, and Francisco Villa threatened both Madero's life and Mexican democracy. In February 1913, things came to a head. Madero was murdered, and Huerta replaced him, to become a symbol of Mexican corruption to the Wilson administration.

Taft had followed a quiet, sensible course in Mexico. He remained neutral and refused to allow American troops to intervene. Unfortunately, his ambassador was the temperamental and bibulous party stalwart Henry Lane Wilson, a man openly devoted to any American business wanting to exploit Mexican resources. Ambassador Wilson had no sympathy for Madero, since the Mexican president showed little eagerness to do his bidding or to perform favors for American companies. Aware of the hostility, Madero asked for Wilson's recall; Wilson, in turn, bombarded Washington with indignant telegrams critical of the régime. High officials in Washington learned to ignore Wilson, but it remains something of a mystery why such a man remained in office under the new Democratic administration.

Diplomatic relations worsened on 9 April 1914. Several American soldiers and their commander were taken at gunpoint from a vessel flying the American flag by Mexican soldiers under orders to permit no one in the area without a proper pass. The military governor soon released the men and apologized for the mistake. No one at the time thought the incident worthy of much note. Then Rear Admiral Henry T. Mayo took strong affront and ordered Mexico to make a formal act of contrition to assuage the injured dignity of his country. He demanded an apology, punishment of the officer responsible, and a twenty-one-gun salute to the American flag, which his ship would then return. Two days later, a mail courier from an American vessel was detained and roughed up by a Mexican soldier who mistook him for a fugitive from Mexican justice. The matter was cleared up at the police station, and the captain of the American ship thought the matter too insignificant to note in his log. That same day Mexican censorship briefly delayed a dispatch from Bryan to Nelson O'Shaughnessy, chargé d'affaires in Mexico City. Under orders to have all messages decoded before delivering them, the censor was apparently unaware that diplomatic messages were exempt. When O'Shaughnessy checked into the matter he received his message; the delay was 55 minutes. Woodrow Wilson used these three trivial incidents as a pretext to occupy Veracruz. As a result about 200 Mexicans died, and 300 were wounded; 19 Americans died, 47 were wounded.[14]

Both Bryan and Wilson set the basic outlines of the administration's foreign policy toward Mexico. The President proved narrow-minded and stubborn, while Bryan seemed fearful and indecisive, but they acted as one throughout most of the crisis. Colonel House and Robert Lansing said little and played next to no role in the intervention. Instead, Wilson and Bryan concentrated on their larger Latin American policy of enforcing American ideas of morality, contract, democracy, legality, and constitutionalism on reluctant beneficiaries. In statement after statement, they declared that material self-interest and imperialism had no role in their thinking, that they were working solely on the best of moral principles, and that they acted from deep concern for the Mexican people. The implications were insulting: both Wilson and Bryan portrayed themselves as Sunday school teachers giving lessons to the immature.[15]

The most coherent statement of what they wanted was "Our Purposes in Mexico," circulated to many embassies and legations on 24 November 1913. Their sole aim was "to secure peace and order" by seeing that "the processes of self-government there are not interrupted or set aside." They termed Huerta's coup a "usurpation" that menaced "the peace and development of America as nothing else could." Huerta had made legitimate self-government impossible, ignored the law, endangered the lives and property of foreigners and citizens, invalidated concessions and contracts, and impaired "both the national credit and all the foundations of business, domestic or foreign." The American government wished "to isolate General Huerta entirely," to "cut him off from foreign sympathy and aid and from domestic credit," and "so to force him out." It would try to be patient, but could not rule out use of force.[16]

Neither Wilson nor Bryan ever separated standards of private morality from public policy. What led to their disastrous policy was their tendency to personalize matters rather than to examine policies for their own sake, or to settle for results that were tolerable. The result involved a fascinating cast of characters, all working at cross-purposes; two countries that came needlessly into conflict; one of many military interventions in Latin Amer-

ica that set patterns still all too visible; and some hard lessons for the
Presbyterian diplomats in Washington, lessons taught by events but never
learned.

IV

The President mistrusted Henry Lane Wilson and had little confidence
in Secretary Bryan. Woodrow Wilson also mistrusted the diplomatic corps
as a den of Republicans insufficiently moral and excessively devoted to
dollar diplomacy. But instead of finding good men and building up an
efficient service, Wilson insisted on appointing special envoys, often with
no official status, to report what they found. As with Colonel House, he
preferred loyal Democrats who spoke his truth. Bryan deferred, softening
official statements and urging delay where he could. When one of Wilson's
own experts, such as John Bassett Moore, rightly objected to the course of
events, he found his position uncomfortable.

The first of Wilson's amateur agents was William Bayard Hale, a social
gospel clergyman who had left the Episcopal Church for a career in journalism. After having campaigned for Bryan in 1900, Hale had worked for
publications ranging from the *New York Times* to *The World's Work.* His
writing attracted Walter Hines Page, then an editor, and Page commissioned a biography of Wilson that turned out to be flattering enough for
use as a campaign document in 1912. Hale then went on to edit the official
version of Wilson's speeches, *The New Freedom.* His enthusiasm and loyalty
were thus unquestioned, but except for occasional travel in Latin America,
he had no competence for dealing with the situation. He spoke no Spanish
and was committed to Anglo-Saxon racial superiority. In other words he
shared Wilson's general prejudices as well as his ignorance of Mexico.
Wilson asked Hale if he would travel about "ostensibly on your own
hook" to "find out just what is going on down there." During the summer
of 1913, Hale filed several reports, among them one that reported Madero
"unequal to the task to which he had been chosen." Despite good intentions, he had been essentially "a little man, of unimpressive presence and
manner, highly nervous, overwhelmed by his troubles, surrounded by
incompetents, trying to be severe but yielding to his merciful instincts just
when he should have been unyielding."

The revolution that broke out against Madero, however, found no sanction with Hale. He made a characteristically progressive class analysis of

the enemies of democracy. He termed one leader "the Pierpont Morgan of Mexico," and characterized the revolutionaries as "a conspiracy of army officers, financed by a few Spanish reactionaries, in conjunction with Cientifico exiles in Paris and Madrid." He repeatedly found Huerta to be "an habitual drunkard," a remark that could only be termed loaded in view of the moral preoccupations of the secretaries of State and the Navy. Hale found that Henry Lane Wilson had displayed undisguised contempt toward Madero; "the Ambassador became more and more outspoken in the dislike of the President," expressing his hostility "to those who, even socially, consorted with him or his family" and making "predictions of his early fall." Hale's report finished H. L. Wilson's career. Specifying several more of the ambassador's prejudicial activities, Hale remarked that Wilson "had thus reached the point where he admonished the legal government as if it were a revolt, and treated the mutineers as if they were the Government *de jure* and *de facto.*" A number of plans for conducting the revolution apparently were made inside the American embassy. Hale also noted that on at least one occasion, Wilson had not even conveyed to Huerta a message from the State Department about possible mistreatment of Madero.

As he devastated Ambassador Wilson, Hale also laid a sturdy foundation for the moral foreign policy of Wilson and Bryan. He assured his superiors that Huerta had no democratic or constitutional right to power: "The betrayal of the President by his generals was mercenary treachery and was not in the slightest degree a response to the sentiments of a nation, or even of the city." Furthermore, America was itself hopelessly implicated: "Without the countenance of the American Ambassador given to Huerta's proposal to betray the President, the revolt would have failed." The Mexicans thus had legitimate cause to believe "that the Ambassador acted on instructions from Washington and look upon his retention under the new American President as a mark of approval and blame the United States Government for the chaos into which the country has fallen."[17] Thus, the scenario as seen from Washington: a weak, ineffectual but legitimate leader had been overthrown by a drunkard with the help of an American representative closely allied to the very business forces who most despised Woodrow Wilson and most vigorously supported dollar diplomacy. America was implicated and had to act; the enemies abroad in effect were the same as the enemies at home.

Wilson and Bryan recalled Henry Lane Wilson, ostensibly for consultation but really to remove him from the scene. They acknowledged their duty to offer their services as mediators in the Mexican revolution and decided once again to circumvent regular diplomatic channels by appointing another special envoy. With their genius for the inappropriate, they fastened on former Governor John Lind of Minnesota, appointed him "advisor to the American Embassy in the City of Mexico," and asserted their right to intervene "not only because of our genuine desire to play the part of a friend, but also because we are expected by the powers of the world to act as Mexico's nearest friend." Convinced of their purity, Wilson and Bryan bent over backward to dissociate themselves from intervention-ists of the dollar diplomacy type. They intended "not only to pay the most scrupulous regard to the sovereignty and independence of Mexico," but also to prove that "we act in the interest of Mexico alone," with no concern for those with a merely pecuniary interest in political stability. "We are seeking to counsel Mexico for her own good," for the American govern-ment "would deem itself discredited if it had any selfish or ulterior pur-pose" in the matter. The document offered no evidence whatever that the world expected America to intervene or that Mexico welcomed such atten-tions. Not only was its tone condescending, informing an errant dependent of its "own good," but it also ignored the Shakespearean ironies that a man of Wilson's education should have recognized. Countries, like women, should not protest too much, or wise observers might well doubt the virtue of both.[18] American policy had four goals:

(a) An immediate cessation of fighting throughout Mexico, a definite armistice solemnly entered into and scrupulously observed;
(b) Security given for an early and free election in which all will agree to take part;
(c) The consent of General Huerta to bind himself not to be a candidate for election as President of the Republic at this election; and
(d) The agreement of all parties to abide by the results of the election and coo-perate in the most loyal way in organizing and supporting the new administration.

Wilson and Bryan also asserted, somewhat ominously, that they "can conceive of no reasons sufficient to justify those who are now attempting to shape the policy or exercise the authority of Mexico in declining the offices of friendship thus offered." In effect, the argument of the United States was that, because of its long-standing divine mission to liberate the world for democracy, it had the right and duty to intervene whenever and wherever it felt the moral urge. Mexico's right to elect its own government and run its own affairs did not mean the right to live in ways that the United States did not approve or to refuse correction from outside when it erred. Presbyterian political science lacked both a sense of humor and a sense of irony. Despite explicit warnings to the contrary, Wilson never realized that the vast majority of Mexicans on all sides preferred to settle their own affairs. Bryan, to his credit, feared "that intervention would unite the country against us and that we would have a long and difficult work upon our hands," but Wilson paid little attention and the Secretary of State reluctantly subsumed his own misgivings into the policies of the President.[19]

John Lind's qualifications for delicate revolutionary diplomacy were even slighter than Hale's. Swedish-born, a populist-progressive who had been a friend and supporter of Bryan since they served in Congress together in the early 1890s, he was as surprised as anyone when Wilson named him to the post. He had no knowledge of Spanish, diplomacy, or Mexican history; he was simply honest, unassuming, available, and loyal. In Washington he made an unforgettable impression on Ellen Slayden: "He has not traveled; he has stiff, awkward manners, and his moral standards are fixed by the Swedish Evangelical Church in Minnesota." She thought him "a perfectly square man in a perfectly round hole." When he and his wife came to dinner, "his face did not light up, and he couldn't smile because when talking he pursed his lower lip as if trying to hold something in his mouth. His solemnity would be incredible if his wife were not a present proof that someone could be solemner. She has two fervent enthusiasms happily combined in prohibition and Josephus Daniels."[20]

Lind, candid and blunt, might have performed certain duties admirably. But, like Hale, he represented the administration so closely that when he and Bryan corresponded, as they frequently did, each reinforced the other's prejudices and fears. Lind loathed Huerta, and he undercut the authority of chargé O'Shaughnessy when he decided that the Roman Catholic O'Shaughnessys were devoted to Huerta and hostile to Carranza and the Constitutionalists for religious reasons. He wrote his wife that O'Shaughnessy was a fool, impossible to work with. Lind tended to view European powers like Britain with populist hostility. He was convinced that the new British minister, Sir Lionel Carden, was dominated by greed for Mexican oil and related raw materials. Indeed, Lind had a large number of prejudices that indicated his many limitations. He disliked most of the resident Americans almost as much as he disliked Huerta and O'Shaughnessy, found the influence of Henry Lane Wilson "baneful and malevolent," and indicated that he was motivated by a casual and unthinking anti-Semitism with phrases like "every Jew banker" in his letters to Bryan.[21]

Lind was aware of the gulf between himself and the people he was observing. He informed Bryan that "we cannot expect to make them conform in any very great degree to our standards in the matter of government. They are incapable of understanding our viewpoints." They lacked a Teutonic sense of cooperation in both business and government, and were incapable of understanding patriotic self-sacrifice. "They recognize only two forces in politics, in religion, and, I might say, in business, namely power and 'pull.' Saint Peter holds the keys, the Padre has some 'pull' with him; appease the Padre, make him well disposed and the chances for enjoying the good will of Saint Peter are fair." Lind habitually put people into racial categories, comparing in one instance Mexicans to "the Irish in our cities," with undisguised revulsion. He thus displayed with clarity the reigning Protestant provincialism. But where Wilson and Bryan assumed that Mexicans could develop toward the higher status of American democrats, Lind washed his hands of them as biologically inferior. Yet his conclusion had a certain political validity: "Judged by our standards the elections here are a farce, nothing but the homage of a people to the forms of democracy."[22]

Two weeks later Lind wrote that politically, "the Mexicans have no standards. They seem more like children than men." Their only motives seemed to be "appetite" and "vanity," and all their talk about pride was "rot." "Their pride rarely compels action and is usually content with bluster and an exhibition on paper or in fiery speeches." Unlike Americans, they did not seek office to "make a record" or "to realize political or social ideals." He was sure that "Indian blood" in the country promised more for its future than "mongrel progeny of the early moorish spaniards." As a whole he found the people "wholly unfitted to cope with the complex situation that now confronts them." He recommended against intervention "unless forced upon us." President Wilson told Bryan that it was "a splendid letter and most instructive from every point of view."[23] The best part of Mexican public opinion and leadership had no way to get their views to the White House, and so remained incomprehensible and unrepresented to the leaders of American foreign policy.

Lind developed opinions that both reflected standard progressive thinking and reinforced it. He became convinced that Great Britain was supporting Huerta for its own economic advantage; he came to loathe Huerta as a hypocritical murderer whose assurances could never be taken seriously. He watched one of Huerta's farcical elections and wrote Bryan that "to hope to establish peace and to work the regeneration of a nation by men so utterly devoid of truth, honor, and decency as Huerta and his whole entourage is absurd." Having personalized all that he disliked about Mexico in the figure of Huerta, he also somehow managed to find virtues in those most likely to depose Huerta. He decided that Pancho Villa was intrepid and resourceful, though cruel and avaricious, perhaps the best that such a low country could produce. Venustiano Carranza he found honest if rigid; America should reach a definite understanding with him. Lind at first did not want to intervene directly and preferred to have America support those, like Villa and Carranza, who seemed most likely to dispose of Huerta and install the system most compatible with American standards. By the spring of 1914, however, Lind in exasperation was considering armed intervention to dispose of the problems. A few days before the occupation of Veracruz he left for Washington and was close at hand as the key decisions about intervention were being made. Insomuch as there was disagreement in Washington, it apparently was about which men to back: Lind preferred Carranza while Bryan preferred Villa. When Huerta finally fled and Carranza took over, Lind's advice may well have been decisive in causing Wilson, over many objections, to recognize Carranza's régime.[24]

V

One of the most disheartening things about the Mexican intervention was that it need never have happened. A sense of inevitability seems to loom over many tragic events, the thoughts and deeds of mere men but details to the impersonal course of history. The issues that the Wilson administration first faced were not of that sort. The Taft administration had made every effort to avoid Latin American interventions, and it is hard to imagine the genial former President personalizing his foreign policy to the point of invasion. Wilson's peculiar cast of mind, reinforced by Bryan's similar impulses, proved too inflexible to handle new and strange events in a sensible manner. The tragedy seems even more unnecessary when one examines the role of Counsellor John Bassett Moore.

Moore was the only man making foreign policy during 1913 who was able to make rational decisions based on law, precedent, and some sense of the nations with which he was dealing. Any President with a modicum of humility, especially one so untutored in foreign affairs and culture as Woodrow Wilson, should have valued Moore beyond price. His advice came through vigorously, and for a while, it helped restrain the belligerent impulses in the White House. Two of his memos deserve special mention. Most important was that of 14 May 1913, in which he argued that European nations were correct in recognizing *de facto* régimes such as Huerta's, and that such recognition was traditional American policy. He declared that the Mexican government was organized in accordance with the Mexican constitution. Even if it were not, it existed, exercised authority, and was able to discharge national obligations. For a revolutionary nation such as America, that should suffice. Americans might deplore the killing of Madero, but doing so "can not relieve us of the necessity of dealing with the governments" that might be formed in such violent fashion. "We cannot become the censors of the morals or conduct of other nations and make our approval or disapproval of their methods the test of our recognition of their governments without intervening in their affairs," he noted firmly, a year before Wilson's government did precisely that.

The Americans seemed innocent of their own history. They scarcely remembered that they had themselves once recognized five governments in Mexico within a few months. Recognition, Moore insisted, "is purely and simply the avowal of an apparent state of fact, and the advantage gained by it is that the country is held responsible for the acts of the authority so acknowledged." He pointed out that Ambassador Henry Lane Wilson, still at his post, favored recognition, seemingly spoke for Washington, and that his position was well-known to the Mexican government. By retaining him in this awkward position, America apparently kept him at his post "merely for the purpose of receiving and transmitting complaints against the attitude of his own government, with which attitude he does not himself agree." Moore felt the aimlessness that Wilsonian moral diplomacy had created, posturing from principle yet unable to act in a recalcitrant situation. He warned again: "It is also possible as a result of inaction to drift into intervention." He could not know at the time not only that his sense of impending disaster held true for Mexico, but that he had succinctly characterized the way Wilson, unhindered either by Moore's realism or Bryan's idealism, later drifted into World War I.[25]

In October, when the personalized, anti-Huerta bias of Wilson–Bryan diplomacy was evident to all, Moore also warned his superiors about their characterizations of European powers. He was opposed to the casual use of moral frameworks in diplomacy. The habit of seeing England, for example, as solely interested in its own economic advantage in Mexico and as conducting an immoral policy seemed fraught with danger to Moore. European countries and business people should not be viewed as American trust barons, exploiting an impoverished people. Moore wrote: "There is nothing on the record to show that the governments that recognized the Administration at the City of Mexico in May, June, and July last felt that they were actuated by any other design than that of recognizing, in conformity with practice, what appeared to them to be the only governmental authority holding out the prospect of being able to re-establish order in the country. Nor had the United States said anything to indicate to them

that it entertained a different view of their conduct." Britain, in particular, had a number of grievances against the United States on other issues and deserved careful treatment. Nothing "short of the clearest proof would at this juncture justify us in attributing to other governments, by means of a direct diplomatic communication, motives the imputation of which they would necessarily repel and resent." Wilson respected Moore, paid attention to him, but proved constitutionally incapable of responding to arguments not couched in moral terms. Moore briefly slowed the administration's course, and on this issue, the British placated Wilson and Bryan, but Moore received the reward of an honest advisor in succeeding months when he found himself outside Wilson's inner circle. He resigned on 2 February 1914, unable to tolerate administration Mexican policy any longer.[26]

All too many unfortunate tendencies became evident in the Mexican involvement. Wilson's mistrust of the professional and often Republican diplomatic service; the incompetence of his own appointees; the assumption that to recognize a government was to approve it; the projection of domestic categories onto foreign problems; the congenital use of inexperienced but trusted agents—these habits soon distorted the handling of far more significant issues in Europe. The inability of anyone in Washington to understand strange cultures or to sympathize with their aspirations did not bode well either for Wilson's handling of Germany during the war or of his allies at Versailles. Above all, in view of the events of the next few years, the Mexican intervention showed that the administration suffered from a kind of righteous paralysis when events and people did not act as it wished them to. If Huerta or Carranza or Villa did not take the appropriate sides and act in appropriate ways, symbolizing good, virtuous democracy and wicked, immoral autocracy, then Wilson did not know what to do. He drifted, uttering platitudes, and let events get out of his control. In Mexico an Admiral Mayo could set policy, make demands, and virtually start a war without prior approval in Washington. Neither Wilson nor Bryan dared to contradict him. In World War I, a similar sense of drift led to American intervention based publicly on moral issues that could not be sustained in reality.

This harsh assessment was not impossible to make at the time. Henry Cabot Lodge had known Wilson casually for decades. He had accepted a Wilson article for the *International Review* in 1879, and the two men met later at a Harvard commencement. They had had dealings over the Panama Canal tolls issue before Mexico became an issue, and Lodge afterward claimed that these early contacts were friendly and in no way anticipated the enmity of 1919. When Wilson prepared his message on Veracruz, he consulted the Senate members most involved after he had printed the speech but before he had released it. Lodge reported some years later that he found it "almost incredible" that an American President seriously proposed to declare war "against an individual, a single man whom he wished to name in the resolution." Lodge said that, at the time, he did not grasp the significance of the issue or the light it cast on Wilson's character. But he did see "even then what I afterwards came clearly to perceive, that the reason for the extraordinary proposition to make General Huerta the subject of the resolution authorizing the President to seize a Mexican city was that the salute to the flag, which was never given, was a mere excuse." Huerta had refused to obey Wilson, and the purpose of the occupation "was to punish the recalcitrant Mexican." Interpreting Wilson's personalized world of moral categories as an enormous egotism, Lodge insisted in 1914 that a matter so important as occupation and war could not rest on personal vendetta. America could only intervene on a genuine issue.

Lodge later noted with some contempt Wilson's great shock at the casualties at Veracruz. Because he hated war and could not think in military terms, Wilson was helpless in preparing himself psychologically for the use of force and its inevitable consequences. Lodge was right to note that Wilson had not determined any clear policy. Being far more bloody minded, Lodge was better able to think about death. He never dreamed that he could send in troops without the strong possibility of casualties. But "it was only too obvious that the President had made no preparation in his own mind for this most probable event. All he seemed desirous of doing, the fighting having occurred, was to get out of the trouble in any way possible without continuing the war which he had himself begun."[27] Lodge had any number of motives, during the 1920s, to think ill of Wilson, the greatest enemy of his life. But in essence, Lodge was correct, and in view of Wilson's later behavior, ominously so. Wilson did want to make war on one man, because he symbolized evil. Compromise was impossible.

VI

As in Mexico, so in Europe. The Kaiser replaced Huerta, Germany replaced Mexico. The stage was farther away, the audience larger, and the issues more complicated, but in essence little changed. Immoral autocracies had attacked moral democracies, and America had to take sides and assist the Allies in preserving Christian values. Among other things American

entry into the war was an attempt to impose Wilson's vision on a world
that rarely accepted his goals and never understood his principles. Political
leaders like Wilson could be successful at home because they spoke the
language of their constituencies. Voters in most districts still shared evangelical values and endorsed their expression. The war cast a cruel, new light
on progressivism. As Christians in Europe slaughtered one another, European statesmen acted amorally. Long accustomed to thinking in terms of
the balance of power, they wished to reestablish a system of alliances that
would maintain the superiority of their nations through military partnerships. Christian morals, while given lip service in public, were essentially
private in nature and not applicable to nations or armies—certainly not to
victorious ones.

The great virtues of progressivism were domestic. Kind, liberal, and
pragmatic, progressives wanted to improve their environments and love
their neighbors. Progressives regarded foreigners as essentially immigrants
to assimilate; the superiority of American democracy, landscape, and moral
values was beyond question. America for three centuries had been a place
to which Europeans fled to lose their outmoded habits, choose new characters, and lead free lives in the middle class. By definition all problems in
America were benign and soluble, only temporary setbacks on the long
march to a perfect society on earth. Such a background produced a reflex-
ive Manicheanism: he who was not with America was against her; the
forces of light were here, so the forces of darkness must be out there;
democracy was American, autocracy foreign. Just as America had been as
a city on a hill in 1630, so in 1917 she was a nation on display: when
Europeans stopped their carnage, they could find an exemplary society
across the ocean. America, Wilson informed Europe, had no stake in the
war or the peace. She had rights to defend, to be sure, but more important
she had values to convey. European statesmen, mired in the amorality of
their post-Metternich maneuverings, no longer represented their own people; Wilson would.

Unfortunately, nothing in Wilson's background had prepared him for
the role of diplomat. An academic who was widely read in British, German,
and American social science, with a brilliant political career in Trenton and
in Washington behind him, Wilson nevertheless was a hopeless provincial
when it came to understanding other cultures. He had traveled occasionally in Great Britain, largely confirming his prejudices rather than enlarging
his sympathies, but his world was never cosmopolitan. He used to talk to
intimates about his "one track mind," and historians have often commented generously on this insight. But Wilson was genuinely proud of his
narrow-mindedness, and if he doubted for one minute the truths of his

early eighteenth-century religion or the superiority of its realization on American soil, it has escaped the editors of one of the most massive documentary records in history. Not all progressives shared his narrow-mindedness or his intolerance of those who disagreed with him, but they all tended to use Christian ethics as the basis of their perceptions of the world and to deal with nations as if they were individuals subject to sin, forgiveness, and redemption. If only nations would learn to turn the other cheek when confronted, to love one another, and to vote their consciences, then all would be well in the world, and the spirit of Christ would reign.

The bare facts of Wilson's venture into Europe, his failure to produce a treaty consonant with his own principles, his unwillingness to admit his failure, and his inability to win Senate ratification for his handiwork have been told many times and need no elaborate repetition. In its broad outlines, the story becomes the last great chapter in the history of progressivism as a creative environment. The League of Nations Covenant was the last significant achievement of progressivism; its failure killed Wilson and progressivism. As it did so, it offered observers a rare opportunity to see progressivism from the outside. The view is of more than historical interest, for Wilson and his opponents, both at home and abroad, set the terms of American foreign policy for the rest of the twentieth century.

VII

On the trip to Europe, Wilson made the first of many mistakes in dealing with the French. Speaking of them and the other allies, he said that people like Clemenceau and Lloyd George "are evidently planning to take what they can get frankly as a matter of spoils, regardless of either the ethics or the practical aspects of the proceedings." Several among those who heard him repeated the gist of his remarks when they got to Paris, and both delegates and newspapers soon spread versions of the story. House attempted to reassure Clemenceau, but the damage was done. French hostility to Wilson and his positions continued throughout the conference. Indeed, Clemenceau's witticisms, apocryphal or not, were soon everywhere. He took to referring to the President as "Jupiter," and developed the habit of touching his forehead in disbelief whenever Wilson began to talk about idealism, as if to say, "A good man, but not quite all there." Clemenceau even once confided to Colonel House: "You and I could settle anything; but I don't understand President Wilson very well; I could talk just as easily to Jesus Christ."[28]

This caustic tone was not at all unusual for French public opinion. Clemenceau enjoyed wide support and controlled several influential Paris newspapers. Thus, *L'Homme Enchaîné* could remark after Wilson's speech of 22 January 1917: "Never before has any political assembly heard so fine a sermon on what human beings might be capable of accomplishing if only they weren't human." The French wrote off Wilson's moral imperialism as simply another expression of American provincialism. "No matter what he does, in spite of himself, he is primarily a citizen of the United States before being a citizen of the world, since he wants the world to resemble the United States," wrote *Figaro* editor Alfred Capus. Diplomats were hardly more friendly. General Bliss reported to his wife that Ambassador Jusserand described Wilson to him as a man who, "had he lived a couple of centuries ago, would have been the *greatest tyrant in the world, because he does not seem to have the slightest conception that he can ever be wrong.*" Capable of giving Wilson a tumultuous welcome when he first arrived, the French were just as capable of ridiculing him out of town upon closer inspection. The chief French scholar of American diplomacy in modern times has repeatedly stressed the offending themes. One can only label as "extreme nationalism" Wilson's "obsession for affirming that one's own country is more moral than any other, and thus endowed with a 'mission,'" wrote Jean-Baptise Duroselle. "It was indeed ultranationalistic to believe, as did Wilson, that the interests of the whole world coincided with those of the United States." Both he and America were presumptuous "to think that the United States alone possessed the formula for a just and durable peace." In the final analysis, Wilson "lacked consideration for others."[29]

The British representatives at Versailles shared the French view. On the surface this critical stance seems odd: Wilson had revered British statesmen like Prime Minister Gladstone and political scientists like Walter Bagehot and Lord Bryce throughout his professional career. He thought the British parliamentary system superior to the American presidential one. Indeed, a number of his problems as President arose directly from trying to act like a prime minister with a parliamentary majority, instead of like a president with a fractious congress, a judiciary that might declare reform laws unconstitutional, and after early 1919, an opposition party not only in power but bitterly critical of his leadership. To the British, Wilson should have appeared as a John Morley figure, a liberal intellectual striving for humane values.

Certain British statesmen like Lord Robert Cecil apparently came to regard Wilson in this way, but they remained a minority. Splenetic comments about Wilson reappear in many British sources, but three serious critiques stand out for their length, their agreement, and the fame of their authors. The first appeared in 1920, written by an obscure economic expert of the British delegation, John Maynard Keynes. In a brilliant polemic, Keynes pinned Wilson to the wall like some mutant butterfly, fascinating yet repulsive. To Keynes, Wilson "had not much even of that culture of the world which marks M. Clemenceau"; Wilson "was not only insensitive to his surroundings in the external sense, he was not sensitive to his environment at all." He seemed instead a "blind and deaf Don Quixote" who was "entering a cavern where the swift and glittering blade was in the hands of the adversary." Wilson "was like a Nonconformist minister, perhaps a Presbyterian. His thought and his temperament were essentially theological not intellectual."

Any number of commentators found the American delegation unprepared for Versailles, and stated flatly that Wilson arrived with no clear plan of what he wanted or how to get it. Instead, he preferred to repeat slogans about the self-determination of nations or the freedom of the seas, which rang with altruism but were wholly impossible to accomplish. He knew nothing about Europe and was severely hampered by a mind that worked more slowly than those of the other major leaders. He ignored his advisors. Keynes spoke for many, including many Americans sitting around writing in their diaries or sending letters home, when he said that "the President had thought out nothing; when it came to practice his ideas were nebulous and incomplete. He had no plan, no scheme, no constructive ideas whatever for clothing with the flesh of life the commandments which he had thundered from the White House." As a good preacher, he could have given "a sermon on any of them or have addressed a stately prayer to the Almighty for their fulfilment," but he never could manage to "frame their concrete application to the actual state of Europe." Thus Clemenceau and Lloyd George could force him to compromise seriously his most cherished principles. He could not admit to himself that he was compromising, and became stubborn, irritable, and belligerent. "Then began the weaving of that web of sophistry and Jesuitical exegesis that was finally to clothe with insincerity the language and substance of the whole Treaty." The French won their revenge against Germany, and all Wilson had to show for his principles were the tattered vestiges of his pet rhetorical phrases. Wilson was "a really sincere man," and no one should attribute malice or intentional deceit to him; "to this day he is genuinely convinced that the Treaty contains practically nothing inconsistent with his former professions." General Jan Smuts later told him that his "portrait of Wilson was absolutely truthful."[30]

Almost twenty years passed before the publication of another of the most famous analyses of Versailles. By then Harold Nicholson was almost as well known as Keynes became and his book remains a standard primary as well as secondary source. Nicholson too fell instantly into religious imagery as soon as he began to discuss Wilson. The need at Versailles was to find "a middle way between the theology of President Wilson and the practical needs of a distracted Europe"; the essence of the theology was "the fourteen commandments." Nicholson was kind to House and complimentary to the American experts in The Inquiry; he was even kind to "Wilsonism," but insisted that Wilson was not "a philosopher. He was only a prophet," and that his presence at Versailles had been a great misfortune. Wilson's compromises, especially on Shantung and Italy, disillusioned those who, like Nicholson, had held great hopes. Wilson had been tested and had failed. "We were shocked by this failure. We ceased, from that moment, to believe that President Wilson was the Prophet whom we had followed. From that moment we saw in him no more than a presbyterian dominie."

Nicholson brooded over Wilson at length. He repeated the usual remarks about Wilson's "one track mind," but went on to insist that Wilson quickly became "profoundly bored" even by his own fourteen points and their various corollaries once he was on the scene. Nicholson felt that Wilson had somehow become derailed by the dumb, adoring crowds that greeted his arrival, and that given his mystical faith in the masses, he had come to a sincere belief that he was the representative of those people and that their voice was the voice of God. "It is not a sufficient explanation to contend that President Wilson was conceited, obstinate, nonconformist and reserved. He was also a man obsessed: possessed." To him, "the League Covenant was his own Revelation and the solution of all human difficulties." Yet those defects were certainly there, and "his rigidity and spiritual arrogance became pathologically enhanced after his arrival in Europe. They loomed as almost physical phenomena above the Conference of Paris." Spiritually arrogant, Wilson began to concentrate almost entirely on the League of Nations Covenant, "and before the Ark of the Covenant he sacrificed his Fourteen Points one by one." The covenant "became for him the boxroom in which he stored all inconvenient articles of furniture." He placed there every issue that he could not solve in hopes that somehow it would go away.[31]

At roughly the same time, the most famous critic of Wilson published his own version of the conference. David Lloyd George had been Prime Minister of England and the foreigner of importance who had cooperated most closely with Wilson. A generation later, he was still stressing the vast

common ground that should have united the British and American governments; his considered judgment was that only the issue of the freedom of
the seas was nonnegotiable to the British. But Wilson gave no indication
either to Lloyd George or to the French that he was aware of the great
suffering that the war had caused. Wilson "was indeed an incomprehensible character." He also seemed badly prepared and uncertain of his goals,
except for a League of Nations Covenant, which would somehow solve
everything later. The word "vague" keeps popping up in the text, as if even
after twenty years of ruminating, Lloyd George still did not know what
Wilson really wanted. He was not alone. It was "intolerable," he quoted
Australia's outspoken Prime Minister William Morris Hughes as having
spluttered, "for President Wilson to dictate to us how the world was to be
governed." If civilization had depended on the United States, "it would
have been in tears and chains today." As for the League of Nations, Wilson
"had no practical scheme at all, and no proposals that would bear the test
of experience. The League of Nations was to him what a toy was to a child
—he would not be happy till he got it." Lloyd George clearly agreed.

Early on, Wilson had "regarded himself as a missionary whose function
it was to rescue the poor European heathen from their age-long worship
of false and fiery gods." The Allies soon became "impatient at having little
sermonettes delivered to them, full of rudimentary sentences about things
which they had fought for years to vindicate when the President was
proclaiming that he was too proud to fight for them." The memory may
be apocryphal, but Lloyd George even remembered an occasion when
Wilson went so far as to ask why Christ had failed to convert the world.
It was "because He taught the ideal without devising any practical means
of attaining it," and that was why Wilson was pushing so hard for a
League. "Clemenceau slowly opened his dark eyes to their widest dimensions and swept them round the Assembly to see how the Christians
gathered around the table enjoyed this exposure of the futility of their
Master." Yet despite all this, Lloyd George also included several pages
about his growing areas of agreement with Wilson and his increasing
respect and affection for the man. "When I criticise Wilson it will be with
genuine personal regret."

Like so many progressives, "Wilson copied Lincoln," but he lacked
Lincoln's wit, balance, and his common touch; Wilson was "highly cultivated and polished; Lincoln was a man of genius." For Lloyd George the
explanation lay in some sense of Wilson's dual nature. "He was the most
extraordinary compound I have ever encountered of the noble visionary,
the implacable and unscrupulous partisan, the exalted idealist and the man
of rather petty personal rancours." A man with "rather an ecclesiastical

than a political type of mind," Wilson "believed all he preached about human brotherhood and charity towards all men." But where Lincoln had felt brotherhood and charity in his bones, for Wilson it was all rhetoric. He "was a bigoted sectarian who placed in the category of the damned all those who belonged to a different political creed and excluded them for ever from charitable thought or destiny."[32]

Delegates to the conference and historians ever since have been debating the truth of this composite picture of Wilson as rigid, theological, unprepared, vague, and malicious. An equal number of quotes could be chosen, chiefly from American books but from a sprinkling of foreign ones as well, outlining Wilson's charm, the inspiring nature of his words and acts, his vision, and decency. Indeed, depending on the time and place, both the critics and the supporters were correct; surely, if any man were a different person in private from what he was in public, Wilson was that man. But this misses the point. As a delegate, Wilson was a public man in a public arena deciding the fate of nations. If he appeared to be all these unpleasant things to his fellow delegates, then pragmatically he *was.* His intentions or nobility of character were no longer relevant. Indeed, he failed in large part because of his virtues; they were the virtues of private life. In public life it is often impossible to tell the difference between a pharisee and Christ. It is quite impossible to treat nations as if they could sin and repent; neither theology nor politics can cope with the suggestion. Wilson's failure to understand this was in many ways a failure of progressivism and of the American people as a whole. They knew not what they did.

VIII

Indeed, historians have slowly come to the realization that Wilson quite literally knew not what he was doing. The all-but-missing element both in foreign assessments of Wilson and in the judgments historians have made of his work is illness. Omitted in many accounts, noted in passing if noted at all, illness, in fact, pervaded Versailles. Wilson was only its most illustrious victim.

No American at Versailles felt well for long. In mid-November, Joseph C. Grew, Walter Lippmann, and Willard Straight all contracted the influenza then infecting the world; within two weeks, Straight was dead. Almost simultaneously, Colonel House fell ill for so long that he lost track of events. His influenza was complicated by a kidney stone, and at the end of January 1919, he was recovering only slowly and was still having trouble during the battle over ratification months later. Charles Seymour came down with what he called a cold, but he compared its impact on him to that of typhoid. James T. Shotwell caught the flu; Norman Davis got pneumonia; and Mrs. David Hunter Miller somehow managed to fall victim to both. On the worst day, doctors made 125 sick calls on various members of the American delegation. Misfortune proved to be nonpartisan and international. On 19 February 1919, an assassin shot and almost killed Clemenceau; for the rest of the conference, he seemed diminished in intellectual power, physically less resilient and subject to racking fits of coughing that scared those around him. On one occasion House suggested, presumably in jest, that some of Clemenceau's germs ought by rights to take refuge with Lloyd George; soon the prime minister was in bed with a sore throat. John Maynard Keynes and other members of the British delegation had flu, several so severely that they left Paris and never returned. Jan C. Smuts, the South African leader who was part of the British delegation, spent most of February and March unable to function. One could go on indefinitely.[33]

The greatest casualty was Woodrow Wilson. The history of Wilson's health, both physical and psychological, has generated much controversy, although most of the literature concentrates on his early life and on the period in late 1919 when he lay near death. Much of this material remains hypothetical, since adequate records do not survive, and physicians then and later have reached no definitive diagnosis. But what no one has pointed out is the way in which Wilson's disease(s) at Versailles weakened his ability to negotiate. This weakening went far beyond simply "not feeling well" or being "sick" for a certain period of time, which many have noticed. Anyone comparing the accounts by those in constant attendance —Mrs. Wilson, Dr. Cary Grayson, Ike Hoover, Edmund W. Starling—with accounts by those who, like Ray Baker, were on the scene but not regularly at Wilson's bedside, cannot help but see that illness tended to freeze Wilson in certain patterns of behavior and thought. Suspicion became paranoia; lack of patience became irascibility; custom became rigid habit; and set ideas, such as that of the League of Nations, became obsessions that could not be questioned or negotiated. Because of illness, Wilson slowly froze into positions that represented caricatures of progressivism. The resilience that he had displayed in Trenton and in Washington left him, and in a basic sense, he was no longer "himself."

Wilson had had a long history of cerebral vascular disease that went back to his Princeton days. While young he was also prone to psychosomatic illnesses, like nervous stomach problems and debilitating headaches that seemed to set in during periods of stress and then vanish when circumstances changed. A workaholic, he tended to compensate for illness and depression with still more work. He suffered extraordinary problems in 1896 and 1906, and the second attack left him blind in his left eye. The assumption of modern medical experts is that he had high blood pressure, and that a blood vessel had burst in his eye. The most common behavioral changes that occur after such attacks include irritability, impulsiveness, aggressiveness, inconsistency, and intolerance of criticism; even close friends noticed these traits in Wilson after 1906. Yet to many observers, his tenure as Governor of New Jersey and his postinaugural months as President were periods of relatively good health, and a word like "robust" even occurs on occasion. Wilson clearly felt well when things were going well, when he rested adequately, exercised, and motored frequently. Yet even at the outset of Wilson's administration, Colonel House noted in his diary for 25 March 1914 that Dr. Grayson had told him disturbing news, via Dr. George E. de Schweinitz, that "there was some indication of the hardening of the arteries." Dr. de Schweinitz had treated Wilson for ten years and seemed qualified to make the diagnosis. In the spring of 1915, the blinding headaches returned, recurring intermittently until his final massive stroke. By the end of 1917, he was difficult to manage and capable of expressing irritation at the slightest setback. His physician, Dr. Cary Grayson, was so much a personal friend and "family member" that he was not as professional in his diagnoses and treatments as he should have been. Indeed, in view of what happened to Wilson in the fall of 1919, Grayson does not seem to have been overburdened either with medical judgment or simple intelligence.

Wilson went to Europe tired but well. He soon sustained a bad burn on his hand, caught a severe cold, and may well have suffered bladder and prostate problems. The flu was everywhere. Wilson wore his body out and refused to stop. He scarcely relaxed at all and took almost no exercise. The crisis came on 3 April 1919, when he suffered a sudden attack that included "violent paroxysms of coughing which were so severe and frequent that it interfered with his breathing." He suffered from a high fever, vomiting, and diarrhea; Grayson suspected poisoning. Wilson's face had twitched off and on in moments of stress for years, but the symptom became far more severe; he had bloody urine. Grayson and the chief historian of the flu epidemic believe that the flu was chief among Wilson's problems. The most informed scholar of Wilson's medical history has suggested the possibility of a cerebral vascular occlusion, possibly complicated by a viral inflammation of the brain. It hardly matters which diagnosis one accepts. Both diseases are capable of changing the mental balance of the patient, and Wilson was a changed man after early April.

He became forgetful, in ways no one had noticed before. He had a tendency to stumble. He became obsessed with the notion that all the French servants were really spies fluent in English; no one else thought they could understand English at all. He became convinced that thieves were stealing his furniture; a few pieces, in fact, had been moved around by order of the custodian of the place. Before his illness he had made free offers of automobiles to divert overworked delegates; after his illness he became rigid and strict with anyone who tried to act on his own earlier suggestion. When he went for rides himself, he demanded the arrest and trial of anyone who dared to pass the Presidential vehicle. He began to grope for ideas and worry needlessly about trivialities. Previously able to sleep soundly, he became an insomniac. Worst of all, he began to identify himself in some strange way with his own words. He got to the point where he could not tolerate the slightest change in documents that were important to him. Wilson had always been a stickler for style and usage; he became unbalanced. He could not believe that he had compromised on key issues, and would not agree on verbal compromises in anything so vital as the League Covenant. The Senate would *have* to adopt it; it would have to "take its medicine." The last creative achievement of progressivism thus became calcified in a form unacceptable to delegates and senators. Neither Wilson nor his handiwork would bend in a storm, so they broke. Progressivism collapsed in the heap with them.[34]

IX

Henry Cabot Lodge and other influential senators and Republicans had been warning against a league entangled with a treaty for months. In constant contact with Henry White, the one nominal Republican on the peace delegation, they had sent up one warning signal after another, always with the same message: settle the war quickly, get Europe functioning normally again, and then we can debate a league. The traditional American policies enshrined in President Washington's Farewell Address and the Monroe Doctrine were too important to change in a moment. The Senate would want long, hard consultation before agreeing to such a thing. Wilson ignored these warnings, even when repeated to him by foreigners whom he respected. Convinced that the course of history and the will of all peoples supported his cause, he had contempt for the Senate, and he let it show.

The idea of a league of nations dated back to classical Greece and medieval Europe. Americans since William Penn had toyed with the idea; Benjamin Franklin and Tom Paine were only two among many eighteenth-century thinkers, chiefly European, who made suggestions for one. By the second half of the nineteenth century, some European powers were developing machinery that could supply vital components for such an organization. By the end of the century, over 350 international conferences had met to discuss related issues; the Pan-American Conference of 1889 perhaps came the closest to touching American interests. Some people advocated a world-wide league of an indefinite sort, while others preferred more restricted groups unified by language, a common colonial past, or geographical proximity. At the same time, peace societies gained in influence, and many statesmen became convinced that human nature and social progress were evolving to the point where cooperation had to replace conflict as the primary means of settling disputes.

Politicians in both major American parties participated in these movements. William Jennings Bryan and Representative John Sharp Williams had attended the Interparliamentary Union sessions in London in 1906, the ideas developing in less than a decade into Bryan's famous "cooling off" treaties. John Bassett Moore, the leading Democratic expert on international law, wrote articles and attended conferences on the subject throughout the period. For the Republicans, Elihu Root was the leading legal mind, widely respected abroad and the recipient of a Nobel Prize. Others of comparable stature were Nicholas Murray Butler, who spent most of his mature life as president of Columbia University; Theodore Roosevelt, who toyed publicly with league-type ideas before having severe second thoughts; and William Howard Taft. Under such prestigious sponsorship, some sort of league was clearly going to have its opportunity. By the time of the Hague Conference of 1907, significant numbers of Americans were involved. They found and continued to find that Europeans were eager for American participation. The subject was on the minds of men in both parties, but especially the Republican, throughout the war.[35]

Woodrow Wilson had come late to most progressive reform ideas, and the idea of a League of Nations was no exception. Wilson had given little thought to the subject before the war, and with only occasional, vague exceptions, he ignored the subject for two additional years. But Wilson was president, and once an American president adopted an idea, it quickly became associated with his name, even if he devoted relatively little time to it. Colonel House was shocked, late in 1917, to discover how little Wilson knew about current internationalist ideas. Taft was appalled, shortly thereafter, at the haughty tone Wilson adopted when dealing with anyone who might seem to be his equal on this or other subjects. Wilson's congenital inability to develop applications of his general principles was especially evident in all matters relating to the proposed league. Not only was he ill-informed and essentially without constructive suggestions himself, but he discouraged others from studying it.[36]

Wilson's unwillingness to seek counsel seemed especially odd since he produced not a single idea that led to the league. No American did. The true birthplace of the league was England. Lord Robert Cecil, Lord James Bryce, and the group that produced the paper known as Lord Phillimore's Report deserved chief credit. They had assistance from the South African General Jan C. Smuts when that Boer officer came to London. The chief non-British influence came from Léon Bourgeois and his circle in France. Like many American senators, these figures preferred to work out some practical form of league that might actually function effectively in Europe, rather than ram through the league in the same document that ended the war. Although not all these Europeans agreed among themselves about specific clauses, they believed that America should play a larger role in European and world affairs. The French were especially eager to join in mutual defense treaties that would keep Germany under control.[37]

The British government was, in all likelihood, committed to some form of league even before Wilson became closely involved, as Lloyd George later claimed. He delegated responsibility to Cecil and Smuts, and both soon were working with Wilson. On 16 December 1918, Smuts submitted a major state paper on the subject to the cabinet, with the expressed intention of producing a document that would draw America into the search for some viable balance of power. Wilson absorbed Smuts' work as he absorbed so many progressive ideas, and soon behaved as though the ideas had always been his. Smuts noted sardonically in his private correspondence that Wilson's own draft "is practically my twenty-one paragraphs with some alterations, which are most unfortunately not improvements but much the reverse." Wilson even appropriated "my mistakes," something that "seriously alarms me, as the paper was very hurriedly written" and contained "many things I would now rather put differently. Not so Wilson; for him the first fine rapture is enough." Wilson even refused to accept Smuts' second thoughts, so much did he like the first version. "The idea is to work out the Convention or 'Covenant,' as he calls it in remembrance of his Covenanter descent, in full and then get the Conference to pass it formally."[38]

Wilson's "first fine rapture" not only made him adopt the work of Smuts, Cecil, Phillimore, and Bourgeois uncritically, it made him impervious to criticisms and warnings about the import of what he was doing. Many European governments were reluctant to accept a league with powers of enforcement, although they were not opposed to a tentative first step. But Wilson only seemed stimulated by continental apathy, even as accidental circumstances made for a peculiar diplomatic situation. The original planners of the Versailles conference had assumed that two conferences would be needed to frame the peace. A preliminary conference would end the war and start the various economies on the path to recovery. A more formal conference would settle the intricate problems of reparations, national borders, colonies, and so on. Wilson internalized this assumption in his conference strategy. Knowing that the Senate would probably not be happy with so abrupt a departure from American traditions, he apparently hoped to embed the league in the documents of the first conference. Such documents would be sanctioned by executive order, thereby evading the incertitudes of ratification. Wilson thus openly ignored the Senate and made no attempt to consult or conciliate it. Only when his own legal advisors told him that Senate confirmation would indeed be necessary, and when the first treaty sessions moved to final settlement, did Wilson abandon his scheme and make grudging, tentative gestures toward the Senate.[39]

Wilson's own Secretary of State, not to mention senators from both parties and Republican theorists outside the Senate, never made that mistake. Robert Lansing certainly had his limitations, but on these issues, he proved far more sensible than his leader. As early as December 1918, Lansing was warning Wilson that both foreign governments and American congressmen would object to military and economic sanctions; that the Monroe Doctrine was a document of great symbolic value to Americans and flouting it would cause trouble; and that the Senate, jealous of its own prerogatives, might defeat any treaty. Subsequent letters predicted that including the League of Nations in a treaty would delay its ratification and increase public criticisms of it, while pointing out that bolshevism and starvation were posing serious threats to the future of all Europe. Wilson was playing a dangerous game, and his cautious, legalistic secretary did not endear himself by pointing it out.[40]

Objections from the Senate reached Wilson directly and indirectly. Henry Cabot Lodge was in regular contact with Henry White, and that elder statesman did everything in his power to press Wilson to consider the objections. Lodge was especially upset at Wilson's refusal even to notice the Senate in his negotiations, and he repeated to all who would listen that that body had constitutional prerogatives at stake, and that, regardless of specific clauses, it intended to play a role in making any treaty. Acting Democratic Minority Leader Gilbert Hitchcock discussed Lodge's activities with Wilson directly and wrote that certain amendments were essential, among them the need for the contracting parties to have "exclusive control over domestic subjects," "a reservation on the Monroe Doctrine," and some means by which "a member state can withdraw from the union." He thought it "important that definite assurance should be given that it is optional for any nation to accept or reject the burden of a mandate." Wilson's partisanship was never so obvious as in his reactions to these objections. He compromised reluctantly on the basis of Hitchcock's recommendations, but the mention of Lodge's name or the Senate's prerogatives closed his mind, and he insisted to incredulous Europeans that he would only make changes if those changes were in the best interests of the treaty.[41]

The chief focus of criticism proved to be Article X. In the original draft of 10 January 1919, Article X read:

> If hostilities should be begun or any hostile action taken against the Contracting power by the Power not a party to this Covenant before a decision of the dispute by arbitrators or before investigation, report and recommendation by the Executive Council in regard to the dispute, or contrary to such recommendation, the Contracting Powers shall thereupon cease all commerce and communication with that Power and shall also unite in blockading and closing the frontiers of that Power to all commerce or intercourse with any part of the world, employing jointly any force that may be necessary to accomplish that object. The Contracting Powers shall also unite in coming to the assistance of the Contracting Power against which hostile action has been taken, combining their armed forces in its behalf.

As the Article appeared in the final version of the Treaty, it read:

> The Members of the League undertake to respect and preserve as against external aggression the territorial integrity and existing political independence of all Members of the League. In case of any such aggression or in case of any threat of danger of such aggression the Council shall advise upon the means by which this obligation shall be fulfilled.[42]

For Republicans of all persuasions, Article X was a can of worms. To a progressive like Herbert Hoover, supportive as he was of Wilson and his larger policies, "the Covenant would be more effective without Article X." Charles Evans Hughes and William Howard Taft, two of the most significant Republican party leaders, agreed; neither was a senator nor especially friendly to Henry Cabot Lodge. But the most important voice among the Republicans was the very man whom Wilson had scorned as too hopelessly conservative to be a member of the peace conference delegation. Elihu Root was in many ways conservative, but in his case, the charge was the usual Wilson canard that came up whenever Wilson felt threatened by a strong personality who did not support his every move. Root was the ablest theorist of international law in the Republican Party, and his conservatism was scarcely of much import on those issues relevant to the treaty. Root also supplied important Republican senators with policy positions. Scorned when he could have been useful, he opted for opposition when Wilson refused to compromise. Having opposed the 1911 arbitration treaties of his own Republican President, Root believed that many of these international agreements were visionary promises that democracies could not keep. Statesmen tried to freeze the status quo and ignored the fickle nature of public opinion. Root opposed Article X because he insisted that only Congress could sanction a declaration of war. Either Article X meant nothing and should go, or it meant that Americans could be involved in war without specific congressional sanction. Once Root took his stand, most Republicans lined up behind him. His prestige was great; he had no self-interest to advance; he was lucid and sensible; and the Republican Party could gain partisan advantage by agreeing with him. Besides, he was right.[43]

Wilson was always able to dismiss Republican criticism of his work as partisan and inspired by personal animosity against himself. The charge had enough truth to appear plausible to the ill-informed. He could likewise ignore later criticisms from other countries like France on the grounds that the French desire for revenge bred greed and hatred. Clemenceau's final comment was that "there are probably few examples of such a misreading and disregarding of political experience in the maelstrom of abstract thought." He thought Wilson's unwillingness to compromise with the Senate catastrophic and Wilson's insistance on combining the treaty and the covenant a mistake. In this case, as in so many others, the views of sympathetic foreigners who worked with Wilson should have weighed heavily. Smuts was the most telling case. Almost as ill as Wilson for a month of the conference, he never lost his sense of what was possible and desirable. Twice in May Smuts pleaded with Wilson in writing to see the damage that the Allied negotiators had done and to rectify as much as possible before it was too late.

Smuts reminded Wilson of the Allied obligation to make a peace based on the Fourteen Points. He pointed out how the Germans had begun the war by casually dismissing the "scrap of paper" that guaranteed Belgian neutrality. If the Allies ignored the Fourteen Points with similar unconcern, then "the discredit on us will be so great that I shudder to think of its ultimate effect on public opinion. We would indeed have done a worse wrong than Germany because of all that has happened since August 1914." Wilson's name was on the treaty, and people tended to assume that the treaty was just on that basis alone. Both men knew how hard Wilson had fought some of the worst clauses and decisions, but that mattered little in the long run. Smuts argued that the Germans had a good case. He could find nothing in the Wilsonian platform that would cover "the one-sided internationalization of German rivers." He insisted that "reparation by way of coal" should not "cover the arrangements made in respect of the Saar Basin and its people." He even doubted "whether the occupation of the Rhine for fifteen years could be squared either with the letter or the spirit of your Points and Principles." Smuts left unspecified the "many other points" about which he also had doubts. "There will be a terrible disillusion if the peoples come to think that we are not concluding a Wilson Peace," he concluded, and "we shall be overwhelmed with the gravest discredit, and this Peace may well become an even greater disaster to the world than the war was."

The exhausted Wilson was no longer able to compromise, even with friends and colleagues. He brushed off the criticisms with a vague promise to "restudy" the issues, but never really took Smuts' criticisms seriously. Smuts was so upset that he came near to not signing the treaty at all, and probably did so only for personal and domestic political reasons. His views, by and large, found their way into his friend Keynes' devastating polemic. He himself concluded that Wilson "has failed Democracy—the man who was to make the world safe for Democracy."[44]

X

When Wilson returned to the White House from Paris, Alice Roosevelt Longworth was there. "I got out of my motor and stood on the curbstone to see the Presidential party pass, fingers crossed, making the sign of the evil eye, and saying 'A murrain on him, a murrain on him, a murrain on him!' "[45]

Not everyone was as outspoken as Theodore Roosevelt's irrepressible daughter, but her response was common among politicians welcome in their circle. The most significant was Chairman of the Foreign Relations Committee of the Senate Henry Cabot Lodge. A frequent visitor to the Roosevelt homes and a man in constant contact with Elihu Root, Lodge proved to be the key man, at least in Wilson's eyes, who stood between himself and his League of Nations. As such Lodge for many years received much criticism. Because he and Wilson loathed each other and because each was an uncompromising partisan, this great departure in American foreign policy never had a chance. Or so two generations of students were led to believe.

Lodge has been the subject of considerable rehabilitation in recent years. His ideas contributed to the framing of the United Nations Charter after World War II, and the failure even of that organization to do much to stop war or the arms race has apparently led a number of scholars to resuscitate Lodge. Once his enmity is filtered from the ideas and documents, Lodge turns out to have been intelligent and consistent in his thinking and to have consulted with the best minds of his party in addition to following many of their suggestions. In fact Lodge held views that, had Wilson not forbidden his followers to accept them, could easily have been the basis for American participation in the league. Lodge's tory outlook on foreign affairs may not have appealed to the progressive imagination, but Lodge turned out to have been far more prescient than Wilson in his estimation of human nature and the place of America in the world community.[46]

Lodge was a Boston patrician of a peculiarly irritating sort. Highly educated, well-read, a fine companion in private life, he appeared snobbish and supercilious in public and often offended people without realizing it. Never progressive or evangelical, he entered politics seeking honesty and efficiency in government and a large and respected place for America in the family of nations. His family had long had trade connections around the world, and he combined a fierce patriotism with a willingness to use force in world affairs. Like his friend Theodore Roosevelt and his favorite naval theorist Alfred Thayer Mahan, Lodge focused these interests on the need for an American navy that could enforce the rights of American business and exact authority for the flag. Like many Americans of his background, he admired England even though he suspected her economic and political motives.

Lodge had a deep suspicion of Democrats and academics. A professional Republican and proud of it, he had also been an instructor at Harvard, earned his Ph.D. under Henry Adams, and come to dislike universities. Wilson thus represented many things that Lodge disliked long before the League of Nations. Lodge also had been a personal friend of Andrew F. West, Wilson's most visible enemy at Princeton, and so probably had little use for Wilson for personal reasons even before Wilson entered New Jersey politics. Still, Lodge displayed no obvious animosity when Wilson took over in 1913, and even made it clear that he thought that any President ought to have the full support of legislators when he conducted foreign policy. Partisanship should not play a role, and only matters of principle should provoke attack. Privately, however, Lodge was contemptuous of Wilson's opportunism in shifting from conservative to progressive positions in his search for public office; he moved to public hostility over Wilson's inept handling of the Mexican crisis.[47]

Like many expansionists Lodge nevertheless remained skeptical about entangling alliances either in Europe or Asia. The Monroe Doctrine had established the principle of America avoiding European quarrels and had prevented Europe from meddling in Western hemispheric affairs. Lodge thought the lessons of that doctrine had been proved by a century of history. America should be a power that Europe respected, not a power that Europe used in its interminable feuds. Lodge's policy in this area was so long held and so deeply ingrained that it transcended his personal friendship with Roosevelt and his attachments to the Republican Party. Like Elihu Root he was willing to amend or destroy an arbitration treaty negotiated by Roosevelt or Taft if he thought it involved America too closely in affairs that did not properly concern it. In honing his arguments, Lodge also established a tactical precedent: unlike Wilson, who always preferred ideals to application, Lodge insisted on clear, limited, specific language that he and his country would then be willing to support with force. He preferred no treaty at all to a vague one.

All these emotions and policies came together when Wilson began discussing his league in public and then brought in his treaty. Lodge proclaimed sincerely that he was not opposed to any league; he preferred a league that included chiefly those powers whom America had joined in fighting the war, a league that could secure the peace of the world by force of arms without involving America in quarrels that did not directly concern her. He expressed contempt for any league that was merely an "assemblage of words, an exposition of vague ideals and encouraging hopes." "If such a league is to be practical and effective, it cannot possibly be either unless it has authority to issue decrees and force to sustain them. It is at this point that the questions of great moment arise." Did Article X require the U.S. to enforce the decrees of the league or did it not? If it did, then it infringed the rights of Congress and the hundred-year heritage of the Monroe Doctrine, and it should be opposed. If it did not, then America should insist in clear language that it did not; if such language were included, then Lodge could support it even if he did not have high hopes for its impact. Wilson refused to answer the question, preferring instead to have Article X mean all things to all men.[48]

XI

Few at the time failed to notice that Wilson and Lodge hated each other. Each tended to personify for the other all that was wrong with the treaty.

Because Wilson was a progressive and Lodge a conservative, the implicit assumption was that the fight over the treaty was at heart a battle over a progressive issue; that somehow the wily, partisan Lodge set back history by a generation, took advantage of a sick leader, and paved the way for Hitler because of personal malice and partisan ambitions. Such a version of the tale makes for exciting reading and poor history. Progressives did not agree on the treaty, the league, or on any other issue. Quite apart from the worth of Lodge's arguments, he was not Wilson's chief opponent in the battle. He was simply the Republican Chairman of the Senate Foreign Relations Committee and as such held the spotlight. Wilson's true enemies were the group of senators called the Irreconcilables, or the Battalion of Death. Fourteen Republicans and two Democrats, they fought Wilson and his schemes to the end and successfully prevented Lodge from making compromises to ratify any treaty. The most visible leaders of the group were two of the most progressive men in Washington. In fact, the fight over the Treaty of Versailles was also the last great battle within progressivism. Each side continued to have a moral vision of America's place in the world that depended on past attitudes, progressive attitudes. Lodge was only the man in middle; if even these two men left him—and both were quite capable of it—he lost his position in the Senate and Wilson would presumably triumph.[49]

William E. Borah led the fight inside the Senate, while Hiram Johnson followed Wilson around the country refuting his arguments. Both were western progressives in the idealistic tradition of William Jennings Bryan; Borah was one of the few men in the Senate whom Wilson genuinely respected. Each man had long concerned himself with domestic issues, and until the war forced a change in perspective, they ignored foreign problems unless they impinged on local issues. Intensely nationalistic, proudly patriotic, they were suspicious of Great Britain in particular. The West in which they had grown up did not concern itself with Europe; it concerned itself with "interests," with the role of giant mining companies and railway monopolies and the way such capitalistic giants could take over control of state legislatures. England to them was a land of wily bankers, anxiously searching the globe for investment opportunities. American democracy certainly had some kinship with British democracy, but it was far more a product of the West itself. The West wanted total democracy—the initiative, the referendum, the recall of politicians and even of judges, strict regulation of corporations, and heavy taxes on their excess profits. The West was egalitarian and independent; it did not like striped trousers, British accents, or imperialist adventures. Not knowing England, it could indulge in fantasies, much the same way Europeans could fantasize about the West. Above all, the progressive West hated war. It might, under extraordinary circumstances, work itself into a righteous frenzy for a good cause, but it much preferred individual, self-reliant violence to organized, national violence. Wars enriched bankers and hindered domestic reform.

Hiram Johnson had never been a constructive thinker. All his life he had pictured himself as an independent battler against giant forces out to enslave independent Americans. Sinister bankers and greedy railroads were an integral part of his imagination, and the history of California gave ample evidence that he had grounds for his obsessions. Fully as much as Woodrow Wilson, he developed a mystical faith in "the people," and he always favored legislation that would force diplomats to work out in the open, where the people could see them. That way, the enemies of the people could not function as well. Johnson had supported entry into the war with great reluctance. A new man in Washington and one with acute presidential ambitions, he supported Wilson out of a sense of national honor. He never shared Wilson's messianic ideas about reforming world government. Sure that the French and the British were just like distant railway and banking interests, he mistrusted them and any agreements made with them. He watched with horror the debates over the espionage and sedition laws, sure in his bones that war abroad led to reaction at home. As Theodore Roosevelt's vice-presidential running mate on the Progressive Party ticket in 1912, he had developed a strong dislike for Wilson, their common enemy. In August 1919, as Wilson expended his last energy on persuading the Foreign Relations Committee to go along with him, Johnson focused all his misgivings and frustrations on Wilson's face. "He is an uncanny thing to look at," he wrote his sons. When Johnson questioned him, "he was quite tense, and his whole expression, although not so intended, was quite wicked. His face in repose is hard, and cold, and cruel. When he smiles, he smiles like certain animals, curling his upper lip and wrinkling his nose." His laugh was not that of a "red-blooded individual. His ponderous lower jaw gives a very strange appearance to his ordinary talking, and his brow, which is like the receding brow of a vicious horse, has in connection with the lower part of his face a singular sort of fascination." As you watch him, "it is not of the intellectual man you think, but of some mysterious ill-defined monster. And yet he was very courteous and very pleasant, and I think extremely forebearing during the day."[50]

Many factors entered into Johnson's repulsive picture of Wilson. He felt lonely and insignificant in Washington after his years as the center of attention in Sacramento. Family tragedies, including a son severely gassed in France, weighed on his mind. Suspicious by nature, something of a paranoid facing real enemies, he watched Wilson do one unforgivable thing after another. As progressivism died at home, it became hypocritical abroad. Wilson had promised open covenants openly arrived at, and promptly went into secret session. He refused to consult the Senate or respect its advice. He spoke of the self-determination of nations, yet left colonies in the hands of Great Britain especially, rewarding imperialism further by giving it six votes in the proposed league to America's one. He gave German-speaking Tyrol to Italy. As racist as anyone in California, Johnson despised the Japanese, and he could never forgive a man who would hand over Shantung to them. In close contact with former social worker Raymond Robins, Johnson was privy to Robins' first-hand reports about American troops on Soviet soil—"a shame and disgrace" that was the handwriting on the wall for American interventions abroad. Many progressives were neutral or supportive of the Russian Revolution; few thought that America should actually send troops in support of the counterrevolutionary White forces. America had gone to war to save the world for democracy, and within two years, was trying to save Russia for the heirs of the czar. His worst fears confirmed, Johnson concluded that the League of Nations would be "a war against revolution in all countries, whether enemy or ally." America would become a guarantor of the British Empire, and that would be the "end of American idealism."[51]

If Johnson carried the battle to the country, Borah led the Irreconcilables in the Senate with such skill and fervor that on occasion he could bring patriotic tears to the eyes of Henry Cabot Lodge. Far from hating Wilson, Borah admired him as a giant among the pygmies in Washington, but his was the admiration of a warrior happy at last to face a worthy enemy. Fully as much as Johnson, Borah insisted on judging Wilson and his work by the yardstick of Wilson's own progressive principles. By such a standard, as expressed especially in the Fourteen Points, the treaty failed. Wilson had not brought home a document worthy of his own ideals, and the duty of a progressive in the Senate was to tell him so and refuse the result. Borah had despised the league from the day he first heard of it. It was an "evil thing with a holy name" that would end the sacred tradition of nonentanglement with Europe. He, too, was suspicious of bankers and Britishers; indeed, he was almost as suspicious of Henry Cabot Lodge, who seemed all too willing to compromise with such people. Borah had watched with grim disapproval the way Wilson had intervened in Mexico and elsewhere in Latin America, and had developed a rudimentary but useful theory about "the logic of events." If once America found what it believed to be a righteous cause and began even a tentative intervention, this logic would suck the country in more and more deeply, until a needless war resulted. Americans would find their soldiers fighting under foreign generals, on foreign soil, for imperialistic goals, and suffering needlessly high taxes to pay for it. Congress would lose its power to declare war, and American democracy would suffer. Wilson believed that moral forces would control selfishness and that nations could deal with each other as personal friends did. Borah snorted at such misplaced idealism. The logic of events was there for all to see.[52]

Other progressives who opposed Wilson tended to repeat these same basic themes but were less visible to the public. Senator La Follette had long feared that the President would sell out his own principles at Versailles; he expressly feared that Wilson would lapse into secret diplomacy and compromise on issues. Also in contact with Raymond Robins, La Follette deplored the presence of troops in Russia. Like many progressives who were as moralistic as Wilson, he deplored the President's inconsistencies, compromises, and vaguenesses. When he got a copy of the treaty, he told his family that "it is enough to chill the heart of the world." He was appalled at the treatment of the German people and thought it would "bind us to fight in ev[e]ry future world-war." Wilson had acted unconstitutionally in not consulting with the Senate, and the result was American endorsement of the terms of the secret treaties, especially in the case of the shabby Shantung settlement, as well as American complicity in European imperialism. La Follette mentioned China, India, Ireland, and Egypt by name. In all probability he was not more prominent as an Irreconcilable because for much of the period he suffered from influenza, bad teeth, and gallstones.[53]

The progressives who opposed Wilson, it seems clear, shared a broad area of agreement. Wilson had betrayed his own principles and brought back a treaty unworthy of support. Popular democratic control of national policy was at risk. Illiberal forces in Britain and France had outmaneuvered Wilson and were attempting to co-opt America into support of colonialism. The reparations clauses were unfair to Germany; national boundaries were badly drawn; a few areas such as Shantung were given to the wrong powers. The sending of American troops into post-war Russia was a prelude to countless seedy cases of meddling with other peoples, which would mean the end of progressive reform at home. The doctrine of nonintervention, dating back to Washington's Farewell Address and the Monroe Doctrine should still be the rule. Wilson wished to export progressivism to the world under the misguided assumption that everyone wished to become Americans. His opponents said that if he did so, he would undermine progressivism at home and be ineffective anywhere else. The quarrel, in short, was between members of the same family who disagreed about how to handle threats from next door.

XII

Wilson did not agree. Convinced of his own rectitude, convinced too that the treaty was the best one negotiable under difficult circumstances, he seemed to calcify. Willing before his April illness to compromise on issues when he had to, after it the only compromises he seemed willing to accept were those he had written. He was sure of public opinion. If necessary he could appeal to the country—as if he were a British prime minister seeking a vote of confidence. But the American system did not work that way. The Senate had just had one-third of its members elected, and another third would face the people in 1920. Nothing in the recent vote had much to comfort Wilson, but he ignored his party's loss of the Senate. He ignored the disastrous results of his attempts to appeal to the people of Europe. Instead, he began to see Henry Cabot Lodge as the fountainhead of all opposition. He began to snarl about the "Lodge reservations" and how he would never accept them, conveniently ignoring European willingness to accept those reservations and Lodge's own position as the head of a coalition in which he was far from being the most rigid. Wilson dismissed his critics as blind and provincial, motivated by personal spite against himself. He summed up his attitude when his daughter Jessie produced a grandson. The proud grandfather visited the new arrival, only to find a baby that kept its eyes closed and its mouth wide open. The president said: "I think from appearances that he will make a United States senator."[54]

The contempt that Wilson and Lodge felt toward each other became a subject of great concern to people around the country. At Harvard, it became fashionable to repeat the remark of former President Charles Eliot to the effect that it was a tragedy to have the fate of the league rest in the hands of the two most obstinate men in American public life. Lodge was plainly harassing Wilson with several of his reservations; many of them scarcely made a difference. But the issues between them were clear enough at a meeting at the White House on 19 August 1919 between Wilson and the Senate Committee on Foreign Relations, including Lodge, Borah, Johnson, Knox, Hitchcock, and Warren G. Harding. The crux of the issue, as usual, was Article X. Wilson clearly did not know how to handle the most significant question of the whole treaty debate. On grounds that could only be called quintessentially progressive, Wilson insisted that Article X "is a moral, not a legal, obligation, and leaves our Congress absolutely free to put its own interpretation upon it in all cases that call for action. It is binding in conscience only, not in law." He insisted that it constituted "the very backbone of the whole covenant. Without it the league would be hardly more than an influential debating society."

The distinction between a moral and a legal obligation, so clear to an evangelical like Wilson, proved invisible to Lodge and other conservatives like Harding. It has also proved meaningless to most legal scholars and historians ever since. But Wilson tried over and over again to make himself clear. Congress did not have to take any specific action in a crisis as far as Article X was concerned. While moral absolutists like Borah and Johnson might easily convince themselves that under Wilson, at least, America would involve itself in crisis after crisis, those outside the progressive frame of reference were quite mystified. Under prodding from Harding, Wilson elaborated: "When I speak of a legal obligation I mean one that specifically binds you to do a particular thing under certain sanctions. That is a legal obligation." A moral obligation, on the other hand, "is of course superior to a legal obligation, and, if I may say so, has a greater binding force." With a moral obligation, there always remains "the right to exercise one's judgment as to whether it is indeed incumbent upon one in those circumstances to do that thing. In every moral obligation there is an element of judgment. In a legal obligation there is no element of judgment." Harding was never an expert at any time in his life on moral obligations, and the point was clearly lost on him. Indeed, Wilson came nowhere near convincing the progressives in the room, either. He was talking moral theology to himself.[55]

Lodge had already thrown the gauntlet down in the Senate the week before. "There is to me no distinction whatever in a treaty between what some persons are pleased to call legal and moral obligations." If America promised to do something in a treaty, she should do it; Lodge had no intention of allowing Article X to promise anything except the due attention of Congress to any crisis. Wilson's remarks were thus a direct response to Lodge's question of 10 August: "Is it too much to ask that provision should be made that American troops and American ships should never be sent anywhere or ordered to take part in any conflict except after the deliberate action of the American people, expressed according to the Constitution through their chosen representatives in Congress?"[56] It was a clear, reasonable, and significant question, and Wilson's response was a progressive, moralistic fog.

In the meantime, the American government simply fell apart. Wilson's insistence on going to Paris, and his obsession with the league covenant, led to the all but total neglect of normal governmental functions. Domestic issues were pressing, and included the demobilization of the armed forces, women's suffrage, prohibition, and inflation. A vicious "red scare" was in progress. The country was in labor turmoil, and the farmers had convinced themselves that the institution of daylight savings time was a national calamity. While Wilson was out of the country, communications were a constant problem. Poor Joe Tumulty, his embattled private secretary, was all but a *de facto* president, which was to say that no one was president. To make matters worse, the Democrats in Congress continued to be a medio-cre lot; the Democratic leader was dying; Acting Leader Hitchcock had the reputation of being a pacifist and pro-German, quite aside from having few qualities of the sort necessary for running the Senate. Wilson had little confidence in Hitchcock, but had no one better to work with. Vice-Presi-dent Thomas R. Marshall, known chiefly for his enthusiasm for the 5-cent cigar, was not even of cabinet, let alone of presidential, caliber. The cabinet became dispirited, and it, too, all but ceased to function. Some members left, and their replacements proved to be worse. Wilson's penchant for second-raters had finally caught up with him, but he did not seem to notice.[57]

Wilson thus faced determined opposition in the Senate and domestic problems that he never solved. He faced them with a body that was visibly deteriorating. Refreshed by his sea voyage home, he had somehow found the energy to deliver important speeches and to meet with the Senate Foreign Relations Committee. But he suffered in the humid summer heat of Washington. He had severe headaches that rarely went away, he was visibly fatigued. He suffered memory lapses, some of them while making public statements. Cary Grayson warned him repeatedly to relax and not to take his struggle to the people, but by this time, he was courting martyrdom. "In the presence of the great tragedy which now faces the world no decent man can count his own personal fortunes in the reckon-ing," Tumulty recalled Wilson saying. "Even though, in my condition, it might mean the giving up of my life, I will gladly make the sacrifice to save the Treaty." As the train traveled west, his wife later recalled, "he grew thinner and the headaches increased in duration and in intensity until he was almost blind during the attacks." The heat, fetid air, and exhausting routine of speechmaking sapped him, and soon the pain became unbeara-ble. He displayed "definite signs of cardiac decompensation and brain involvement," coughed so violently that he had to try to sleep in a seated position, and suffered from double vision. He began to stumble, his voice weakened, and he lapsed into long pauses while speaking. On 26 September, Tumulty found Wilson paralyzed on his left side and barely able to talk. He recovered slightly but on 2 October had a relapse: he could no longer use his left arm; it was numb. He soon lapsed into unconsciousness. He never recovered fully from this attack, and for the rest of his life, had no power in his left arm, problems with the left side of his face, and only a primitive ability to shuffle around on his own. He was probably permanently blind in his left eye and severely impaired in his right eye; he may never again have seen clearly.[58]

The symptoms were those of a stroke. His mind, under such circumstances, did odd things. It forgot, imagined, and distorted. It denied firmly, often with great lucidity, that he was incapacitated. Stroke patients typically suffer from anosognosia, a denial that the blindness or paralysis is really theirs, and are capable of believing that the problem really belongs to someone else, or that their diseased organ is really an inanimate object separate from their bodies. They speak as if the disease has happened to a third person, a "him" rather than a "me." If they get better, as they often do temporarily, such patients find it harder to dissociate themselves from their problems; they cover their dismay with outbursts of irritability. What was perhaps most true, in Wilson's case, is that such patients are capable of seeming rational, logical, and even witty, especially on subjects not directly relating to their disease. Those who are accustomed to performing certain routines and dealing with certain ideas continue to do so, but they no longer can handle new knowledge effectively and have no flexibility. At some point between his illness of early April and his final stroke of October, Wilson simply froze into whatever opinions, policies, and hatreds he had at that time. He sometimes appeared lucid to his loved ones and to occasional visitors—even hostile ones—but in any functional sense, there was nobody home at the White House.[59]

Ike Hoover left the best description, all the better for being medically uninformed and simply a narrative of impressions. The President was "sicker than the world ever knew, and never afterwards was he more than a shadow of his former self. Even when conscious, he was unreasonable, unnatural, simply impossible. His suspicions were intensified, his perspective distorted." Dr. Grayson and Mrs. Wilson refused to be realistic with him and protected him from outside pressures and the truth. For close to a month, Wilson "just lay helpless" in the White House, "just as helpless as one could possibly be and live." He survived, but "what a wreck of his former self! He did grow better, but that is not saying much." "He has changed from a giant to a pygmy in every wise," barely able to think or articulate clearly. All that fall and winter, government stood still. Mrs. Wilson took it upon herself to filter out the bad news and the threatening visitors, and Grayson backed her up. She read occasional state papers to the blinded man and guided his pen when a signature could not be put off. Any word that came out of the White House filtered through her, Grayson, or Tumulty; it bore the president's name, but no one knew how genuine a given statement or document was. The bare facts of the situation were not in dispute. The only issue was whether or not Wilson's mind remained clear, as Mrs. Wilson insisted to her grave. Under the circumstances, the assertion was ludicrous. He was in a fantasy land where the people still supported him and the league was the key to world peace. He was capable, at best, of repeating his old opinions and loathing his enemies.[60]

The last great progressive idea thus died in an atmosphere of disease, misunderstanding, and vindictiveness. The irony is that virtually all informed commentators agreed that it hardly mattered whether Article X was in the treaty. Léon Bourgeois, Elihu Root, and David H. Miller, one of Wilson's own experts, all shared that view. They were correct. Article X did not matter because a nation had to be willing to join the fight, and no outsider could force that act. During the twenty year history of the league, most countries operated as if Article X carried no obligations at all. They proved unwilling even to employ sanctions, let alone soldiers. If the league was something more than the "quilting society" that Borah liked to sniff at, it was less than an organized system for collective security.

America should have participated in the league. Like many progressive ideas, it was an experiment ready for trial—even if, like most, it failed. Except in a symbolic sense, American participation was unimportant. Domestic opinion would no more have permitted genuine collective security than it did in England or France. Another world war changed those attitudes, thereafter encouraging such interventions; another league arose, this time formed without the weaknesses of Wilson's league. Yet the peace never seemed less secure to citizens in succeeding decades. The United Nations went into Korea and stayed out of Vietnam. It hardly seemed to matter. Lodge proved correct in his cynical estimate of nations; so did Borah and Johnson. Once America entered the world arena, a logic of events led from one crisis to another. The presence of American troops in Russia was only the first of countless interventions against the forces of change. Everyone seemed to think of Wilson as a prophet, but he was no prophet; he was just a preacher. Progressivism died with his failures.[61]

A Methodological Note

THIS BOOK is part of a large-scale project. I have intended it as the first of several volumes detailing the dominant climates of creativity in early modern America. My focus has been on those areas that university custom calls the arts, humanities, and social sciences. Not all of the disciplines normally included today existed in 1900; not all of those that existed had any relevance to progressivism. I have tried to choose those areas and figures most illuminative of progressivism, rather than those that had the most influence in an internal, professional sense. Henry James, for example, had no meaningful connection to progressivism regardless of the influence he had on the novel; thus, he has no important role in this book, and the novel has only a minor one.

I have kept always in mind the future books I intend to write: first, on the coming of modernism to America, signaled as early as the Armory Show in 1913 and reaching a creative climax with the painting, literature, and music of the 1917–1925 period; then on collectivism, clearly visible in the aftermath of the deaths of Sacco and Vanzetti in 1927, and cresting with the first hundred days of the New Deal and the writing of books like *U.S.A.* I have thus not discussed, or have mentioned only in passing, many figures and events that have more meaning for these later concepts than for progressivism. James McNeil Whistler, for example, created in a modernistic way even before the dawn of progressivism, but his appropriate place is in the early pages of a book on modernism.

I have chosen the subtitle of the book with some care. I am not writing political, intellectual, or literary history as those terms normally appear in historical works. I have not attempted anything technical by way of analysis, either of the progressive sonata or of pragmatism. I have not attempted more than the first word on progressive architecture. I have instead postulated the existence of a dominant national mood within which children grew up and made their crucial career decisions. I assume that early experiences common to vast numbers of young people have their impact on their subsequent professional achievements. I have outlined and suggested the framework for any technical analysis of those achievements, but I have not gone beyond this preliminary stage.

I began by surveying all significant areas of creative behavior to see what, if anything, individuals of originality had in common, no matter how different their work seemed on the surface. I settled on one hundred figures who seemed to form a group that was coherent in its values. The group included many minor individuals but also every major figure whom, to my knowledge, scholars had ever called "progressive," whatever they may have meant by the term. This large group seemed to me about the maximum number I could handle in a narrative framework. Of these I found that twenty-one were of major importance in telling the story that presented itself to me over the years that I have been studying the subject.

In alphabetical order those twenty-one are, with birthdates in parentheses: Samuel Hopkins Adams (1871), Jane Addams (1860), Charles A. Beard (1874), William Jennings Bryan (1860), John R. Commons (1862), John Dewey (1859), Richard Ely (1854), Robert Henri (1865), George D. Herron (1862), Charles Ives (1874), Robert M. La Follette (1855), George Herbert Mead (1863), Robert Park (1864), Theodore Roosevelt (1858), Upton Sinclair (1878), John Sloan (1871), William I. Thomas (1863), Frederick Jackson Turner (1861), Harvey Wiley (1844), Woodrow Wilson (1856), and Frank Lloyd Wright (1867).

These major figures proved to be representative of the larger sample. As the birthdates indicate, these twenty men and one woman form a generation of almost precisely twenty years, with birthdates ranging from 1854 to 1874. The only exceptions are Wiley, born a decade early, and Sinclair, born four years late. Thus, in speaking of progressivism in the text of the book, I have followed this rule: an advocate of reform ideas born before 1854 is a liberal precursor of progressivism; a progressive of the first generation was born between 1854 and 1874; a progressive of the second generation was born between 1875 and 1894—the birthdate of Stuart Davis, who is the youngest person I can conceive of hearing labeled "progressive" in any sense relevant to my purposes.

Of my larger sample, the members who were born within the first progressive generation were: Newton D. Baker (1871), Ray Stannard Baker (1870), Emily G. Balch (1867), Carl Becker (1873), Edward W. Bemis (1860), Albert J. Beveridge (1862), Anita McC. Blaine (1866), William E. Borah (1865), Jonathan Bourne (1855), Madeline McD. Breckenridge (1872), Sophonisba Breckenridge (1866), Winston Churchill (1871), Charles H. Cooley (1864), Edward P. Costigan (1874), Coe I. Crawford (1858), Joseph M. Dixon (1867), W. E. B. DuBois (1868), Joseph Folk (1869), Homer Folks (1867), Franklin H. Giddings (1855), Charlotte P. Gilman (1860), William Glackens (1870), Asle J. Gronna (1858), William Bayard Hale (1869), Alice Hamilton (1869), Robert Herrick (1868), Herbert Hoover (1874), Frederic C. Howe (1867), Charles E. Hughes (1862),

A Methodological Note

Edmund J. James (1855), Hiram Johnson (1866), Florence Kelley (1859), William Kent (1864), Robert Lansing (1864), Julia Lathrop (1858), John Lind (1854), Ben B. Lindsey (1869), Robert M. Lovett (1870), George Luks (1867), Charles McCarthy (1873), Mary McDowell (1854), Edgar Lee Masters (1869), Charles E. Merriam (1874), Jerome Myers (1867), Albert Jay Nock (1870), Walter Hines Page (1855), George W. Perkins (1862), David Graham Phillips (1867), Amos Pinchot (1873), Gifford Pinchot (1865), Miles Poindexter (1868), Roscoe Pound (1870), Walter Rauschenbusch (1861), George Record (1859), Margaret Dreier Robins (1868), Raymond Robins (1873), James H. Robinson (1863), Edward A. Ross (1866), Charles E. Russell (1860), Vida Scudder (1861), Albert Shaw (1857), Mary K. Simkhovitch (1867), Albion Small (1854), Ellen Gates Starr (1859), Lincoln Steffens (1866), Henry Stimson (1867), Ida Tarbell (1857), M. Carey Thomas (1857), James H. Tufts (1862), William Allen White (1865), and Art Young (1866). In the process of revising the book, I have had to cut material relating to several of these figures, but they were in my original sample.

I have also made passing references to several second-generation progressives because they were involved in events or were creative in ways that I needed to note to cover the subject. These include George Bellows (1882), Stuart Davis (1894), Raymond Fosdick (1883), Vachel Lindsay (1879), Boardman Robinson (1876), and Carl Sandburg (1878).

Other figures who appear occasionally, from Jacob Riis to Ignatius Donnelly, surely belong in histories of the Progressive Era, but were born too early to share the paradigm experiences, and are thus precursors. A few figures, like Herbert Croly, grew up in nonevangelical homes and should be regarded as "urban liberals" rather than as progressives. Such figures made their contributions, and some of those contributions had their impact on progressive creativity, but they were contributions from outsiders who never quite managed to get into the spirit of the age.

My intention in the first five chapters of the book has been to create a paradigm for the growth of a progressive outlook on life and to give some sense of its achievements in social science, literature, and the fine arts. The last three chapters demonstrate key ways in which this outlook made a difference in American public life. In essence the paradigm includes birth in a home of devout Protestantism, often vague in doctrine and theology but strict in morality; support for the early Republican Party, free soil, and the Union cause; devotion to the figure and example of Lincoln; restlessness under this heritage and discipline, usually expressed in the combination of devotion to parents and an inability to find vocational satisfaction in the ministry or closely related career; solid education at a small, denominational college; rejection of the repeated attempts at causing a religious conversion; a period of seeking a meaningful, morally satisfying career, often accompanied by doubts, indecision, psychosomatic illness, and false starts; a narrowing of focus, usually to journalism, settlement work, higher education, law, or politics; finally, the making of a choice and the investing of that choice with the religious and moral significance that the family environment had placed on the ministry a generation earlier. My thesis is that progressivism was the climate of creativity in which these twenty-one people and others who closely resembled them lived. The corollary is that creative figures who did not fit the pattern found that they had to cooperate with the progressive ethos in order to find places for themselves within American life; the alternatives were to leave the country or be ignored. My assumptions have included the idea that progressivism dominated the entire country, and that by voting for progressives, reading their books, and listening to their sermons, the people participated vicariously in that mood.

Life being as messy as it is, no model is perfect. Some among my sample turned out to have a Catholic parent, Copperhead relatives, or some similar aberration. Where I have discovered these divergences, I have tried to account for them, stating the issue as clearly and briefly as I could. Woodrow Wilson's father supported slavery, and Theodore Roosevelt's mother was proud of everything Georgian. I am convinced that Upton Sinclair was the product of divine whimsy, purposely hatched to frustrate social scientists building models. A number of Jews and Roman Catholics have honored places in most histories of progressive activity, which they gained by cooperating with the progressives and not asserting incompatible views in public. Regardless of his father, Wilson came to value the Union victory; regardless of his mother, Roosevelt revered Lincoln. One should keep in mind that historical models tend to be descriptive rather than causative. Someone else with the same parents could have become a quite different adult, and every reader of these pages can think of families in his own experience in which similar backgrounds have produced adults with diametrically opposed personalities.

I found only one person who was important to national progressivism who transcended my categories: Louis D. Brandeis, the great lawyer and Supreme Court Justice, and a Jew as well. Brandeis was, to my way of thinking, an urban liberal who cooperated at times with progressives and their organizations. I believe he will play a significant role in a subsequent book for the "Brandeis brief" and the sociological jurisprudence beginning with *Muller vs. Oregon.* For my purposes, he was as far in advance of his time as Whistler. Other Jews played minor roles: Lillian

Wald, Rabbi Stephen Wise, Oscar Straus, and the like, usually in social work or politics.

Roman Catholics played slightly different roles. More numerous, they were also Christian and thus shared more with Protestants. Never central to the formation of the progressive ethos, they were nevertheless acutely conscious of the need for many of the specific reforms that progressives advocated. As progressivism became the dominant national mood, figures such as Al Smith, Robert Wagner, John Ryan, and Thomas Walsh cooperated with Protestant reformers and thus became significant factors in what remained to them an intrinsically alien world. I have no objection to labeling these Jewish and Catholic figures "progressives" in the sense that they cooperated with the civil religion of Protestant reform, but I personally prefer to think of them as precursors of the social democracy of the New Deal. They were ahead of the rest of America in advocating collectivist values, and collectivism had to wait another generation before it dominated the country. My focus is always on the central figures, and the national climate they created.

The dating of the generations is, of course, arbitrary. A few figures like Graham Taylor (1851) or Henry Carter Adams (1851) could easily be included because they otherwise fit the paradigm, but I prefer to stick to my decades: the Civil War should be legend; Lincoln should be a figure of tragic death and myth; and the problems of vocation those of a world in which the ministry was unattractive and the university a place of fabled possibilities, long dreamed of but as yet untried. At the other end of the spectrum, a young person growing up in the 1890s and afterward did not face the same problems; more vocations were open; more role models were available. Raymond Fosdick idolized Woodrow Wilson; George Bellows admired Robert Henri; Vachel Lindsay never forgot William Jennings Bryan. The problems were different, and they deserve a book of their own.

Acknowledgments

THE AUTHOR of a book as broad as this of necessity picks the brains of colleagues and friends. I first conceived the book in 1963 and have been at work on it since 1970; I am sure that I have profited from any number of long-forgotten conversations and suggestions for which my footnotes make inadequate reference. I am especially grateful to my colleagues in the American Civilization Program at the University of Texas: the students for their fresh skepticism about my attempts to define terms across disciplinary boundaries and the faculty for helping to create an intellectual environment in which new ideas seem to happen.

I owe a great debt to several libraries. It is now possible to use xerox and microfilm for a startling amount of research, but some institutions are far more cooperative about scholars and their needs than others. I recall especially the hospitality of the Grinnell College Library and the Hull-House librarians at the University of Illinois at Chicago Circle. The Smith College Library and the Swarthmore College Peace Collection helped considerably. I have to single out the University of Chicago Library for special mention. It is a rare librarian who can point out to a scholar some papers that he ought to see, but Chicago has such librarians and should be proud of them.

Friends outside American history helped with portions of my work within their areas of competence. Thanks to the encouragement of Gilbert Chase, I was a participant in the Charles Ives Centennial in New York and New Haven in 1974, and neither I nor anyone else there will forget the atmosphere and the wealth of ideas in the air. Chance took me abroad frequently during the late 1970s, and I am sure that many of my insights and emphases changed under the polite but skeptical questions of my colleagues in Helsinki, Melbourne, and Würzburg. I delivered the essence of my argument at the Australia–New Zealand American Studies Association Meeting at Christchurch, N.Z., in August 1978.

As the manuscript took final form, I profited from the sharp critiques of Frank Freidel; John M. Cooper, Jr.; Gail Minault; Jeffrey Meikle; Michael Stoff; and Steve Fraser on the entire manuscript; and from Thomas S. Hines, Alexander Vucinich, and Robert Abzug on substantial portions. I am sure they would be the first to dissociate themselves from several of my views and emphases, but I remain grateful even for knowing of those areas about which they are most skeptical.

Notes

Prologue

1. Theodore Roosevelt, *An Autobiography* (New York, 1927), pp. 2, 7–11; William Jennings Bryan with Mary Baird Bryan, *The Memoirs of William Jennings Bryan* (Philadelphia, 1925), pp. 44, 33, 24, 34. Compare these accounts with, Edward A. Ross, *Seventy Years of It* (New York, 1936); Ray Stannard Baker, *Native American* (New York, 1941); Richard T. Ely, *Ground Under Our Feet* (New York, 1938); David J. Danielski and Joseph S. Tulchin, eds., *The Autobiographical Notes of Charles Evans Hughes* (Cambridge, 1973); Raymond B. Fosdick, *Chronicle of A Generation* (New York, 1958); and Robert Morss Lovett, *All Our Years* (New York, 1948).

2. Charlotte Perkins Gilman, *The Living of Charlotte Perkins Gilman* (New York, 1963, c.1935), pp. 16, 3, 6; Brand Whitlock, *Forty Years Of It* (New York, 1924, c.1914), pp. 12–14; William Allen White, *The Autobiography of William Allen White* (New York, 1946), pp. 6–7, 10–11.

3. White, *Autobiography,* 72; Whitlock, *Forty Years,* pp. 18, 61, 138; Marian C. McKenna, *Borah* (Ann Arbor, Mich., 1961), quote from p. 38; Claude G. Bowers, *Beveridge and the Progressive Era* (Boston, 1932), quote from p. 63; Jane Addams, *Twenty Years at Hull-House* (New York, 1961, c.1910), chap. 2.

4. Addams, *Twenty Years,* 33; Ida Tarbell, *All in the Day's Work* (New York, 1939), pp. 11, 161, 178; Hughes, *Autobiographical Notes,* p. 10.

5. Bowers, *Beveridge,* pp. 4, 10; Whitlock, *Forty Years,* p. 27; Roosevelt, *Autobiography,* p. 55.

6. John R. Commons, *Myself* (Madison, Wis., 1964, c.1934), pp. 7, 8, 16, 21, 26, 182, 184.

7. John Barnard, *From Evangelicalism to Progressivism at Oberlin College, 1866–1917* (Columbus, Ohio, 1969), pp. 27, 118, 122.

8. The chief secondary surveys I have found useful for Wilson's early life are John M. Mulder, *Woodrow Wilson: The Years of Preparation* (Princeton, N.J., 1978) and Henry W. Bragdon, *Woodrow Wilson: The Academic Years* (Cambridge, Mass., 1967). The quotation is from Arthur S. Link *et al.,* eds., *The Papers of Woodrow Wilson* (Princeton, 1966 *et seq.*), XVIII, p. 631; hereafter cited as PWW.

9. *PWW,* I, 618; II, 500–4; III, 248; IV; 266.

10. *PWW,* I, 591, 185; II, 10, 336.

11. *PWW,* III, 25–6; II, 479–80.

12. John M. Cooper, *Walter Hines Page* (Chapel Hill, N.C., 1977), p. 32.

13. Richard Ely, *Ground,* pp. 74, 77, 136; Benjamin Rader, *The Academic Mind and Reform: The Influence of Richard T. Ely on American Life* (Lexington, Ky., 1966), pp. 22, 36.

14. Rader, *Academic Mind,* p. 106 ff.; David Thelen, *The New Citizenship* (Columbia, Mo., 1972), *passim;* Ely, *Ground,* p. 216.

15. Frederic C. Howe, *Wisconsin: An Experiment in Democracy* (New York, 1912), pp. x–xi, 30, *passim.*

Chapter 1

1. The portion of this chapter on Jane Addams is a severe revision of my fully documented contribution to John Buenker et al., *Progressivism* (Cambridge, Mass., 1977). I used three key depositories: the University of Illinois, Chicago Circle, which has the papers of Addams and several related figures, including rare pamphlet literature; the Swarthmore College Peace Collection, especially reels 1.1, 1.17 and 1.24; and the Smith College Library, which has the correspondence between Addams and Ellen Starr. In all cases I have tried to leave Addams'

idiosyncratic spelling and punctuation as I have found them in her handwriting, making exceptions only when my preliminary readers were frankly mystified.

This section was written before much of the recent Addams material appeared. At the time I found Christopher Lasch, ed., *The Social Thought of Jane Addams* (Indianapolis, Ind., 1965), and Anne Firor Scott, ed., *Democracy and Social Ethics* (Cambridge, Mass., 1964), most useful. The best recent work is Allen F. Davis, *American Heroine* (New York, 1973). I can strongly recommend Elizabeth H. Carrell, "Reflections in A Mirror: The Progressive Women and the Settlement Experience" (Ph.D. diss., University of Texas, Austin, Tex., 1981) for biographies of Starr and Vida Scudder. The best brief treatment of Starr is that of Allen Davis in *Notable American Women* (Cambridge, Mass., 1971), III, pp. 351–3.

2. Mead to Castle, 21 August 1883, George Herbert Mead Papers, University of Chicago Library. All subsequent quotes will be from these papers. Mead wrote a very difficult hand that makes accurate transcription problematical in places. His capitalization was often whimsical; he used dashes as exuberantly as Emily Dickinson; he often left out commas and periods; he separated words into two or three sections in some places while joining two or three as if they were one word in others; and his paragraphing was inconsistent. To minimize these problems and to further my own purposes, I have paraphrased wherever possible and have regularized those aberrations of style that seem to me matters of convention. Scholars should note that these letters may well contain one of the most detailed available accounts of the evolution of an important philosophical mind, and that I do not discuss most of the strictly professional issues discussed. Students of the influence of Darwin and of Kant in America will find important material here. The letters deserve publication for the history both of civilization and of philosophy.

3. Mead to Castle, 14 September 1883 and 17 October 1883.

4. Mead to Castle, fragment, 15 November 1883.

5. Mead to Castle, n.d., and Mead to Castle, 28 February 1884.

6. Mead to Castle, 5 March, 7 March, and 12 March 1884.

7. Mead to Castle, 16 March 1884.

8. Mead to Castle, 23 April, 3 May, 10 June, 18 July, 7 August, 16 August, and 14 September 1884.

9. Mead to Castle, 31 January, 8 February, 22 February, 30 March, continued 31 March 1885.

10. Mead to Castle, 29 October, 30 November 1885; 28 February, 6 June, 28 October 1886; 3 April, 14 April, 5 May, and 17 June 1887.

11. Mead to Castle, 24 July, 7 August, 21 August, 28 August, and 31 August 1887.

12. Mead to Castle, 19 June and 1 July 1888.

13. Mead to Castle, 1 July and 18 July 1888.

14. Castle to parents, 3 February 1889, Henry Castle Papers, University of Chicago. All citations to Castle's letters will be to this collection. Many were published privately in Mary Castle, ed., *Henry Northrup Castle: Letters* (London, 1902); the work on this book seems to have been done by George and Helen Castle Mead.

15. Mead to Castle, fragment, August 1890(?).

16. Mead to Castle, 19 October and 21 October 1890.

17. Mead to Castle, fragment of letter, probably from Berlin, 1890; another letter, the 24th of an illegible month, 1890; and 22 July, 20 October and 19 December 1891, and 29 June 1892.

18. Robert M. Barry, "A Man and A City: George Herbert Mead in Chicago," in Michael Novak, ed., *American Philosophy and the Future,* (New York, 1968), 173–92, contains a few circumstantial details.

Chapter 2

1. The majority of the citations in this chapter on George Herron come from materials in the Herron Papers at Grinnell College. This is not a massive archive but it does include a helpful amount of newspaper material and unpublished materials contributed by faculty and students who knew Herron. I wrote a fully annotated essay when my purposes and principles of organization were somewhat different from what they now are; see "George D. Herron

in the 1890s: A New Frame of Reference for the Study of the Progressive Era," *Annals of Iowa* XLII No. 2 (Fall 1973): 81–113.

Four dissertations and three published articles are basic to study of Herron. Mitchell P. Briggs, *George D. Herron and the European Settlement*, "Stanford University Publications in History, Economics and Political Science," vol. 3, no. 2 (Stanford, Calif., 1932), deals almost entirely with World War I and its aftermath; Robert T. Handy, "George D. Herron and the Social Gospel in American Protestantism, 1890–1901" (Ph.D. diss., University of Chicago Divinity School, 1949), deals chiefly with theology; Phyllis Ann Nelson, "George D. Herron and the Socialist Clergy, 1890–1914" (Ph.D. diss., University of Iowa, 1953, University Microfilms no. 6547) is a long, detailed study of Herron's place in clerical socialism; Herbert R. Dieterich, "Patterns of Dissent: The Reform Ideas and Activities of George D. Herron" (Ph.D. diss., University of New Mexico, 1957) is an admirably concise study of all of Herron's career and the only lengthy study using all the available unpublished papers.

Three published articles developed from these dissertations are: Robert T. Handy, "George D. Herron and the Kingdom Movement," *Church History* XIX no. 2 (June 1950): 97–115; H.R. Dieterich, "Radical on the Campus: Professor Herron at Iowa College, 1893–1899," *Annals of Iowa* XXXVII no. 6 (Fall 1964): 401–15; and "Revivalist as Reformer—Implications of George D. Herron's Speaking," *Quarterly Journal of Speech* (December 1960): 391–99.

Of Herron's published works, "The Message of Jesus to Men of Wealth" is available in *The Christian Society* (New York, 1969, c.1894). I also quote from *The New Redemption* (New York, 1893), *The Larger Christ* (New York, 1891), *Social Meanings of Religious Experiences* (New York, 1969, c.1896), *Between Caesar and Jesus* (New York, 1899), and *A Plea for the Gospel* (New York, 1892).

Since the appearance of my article in 1973, the only secondary source to offer much by way of additional material is Dorothy Ross, "Socialism and American Liberalism: Academic Social Thought in the 1890s," *Perspectives in American History* XI (1977–8): 5–79.

2. The portion of this chapter on John Dewey reworks and condenses my fully annotated contribution to John Buenker et al., *Progressivism* (Cambridge, 1977). The majority of my quotations come from materials now easily accessible in *The Early Works of John Dewey, 1882–1898* (Carbondale, Ill. 1967–72). Of biographical sources George Dykhuizen, *The Life and Mind of John Dewey* (Carbondale, Ill., 1973) remains an essential starting place. Since the appearance of *Progressivism*, J.O.C. Phillips has completed an important dissertation at Harvard on the young Dewey that I consulted in manuscript, and Neil Coughlan has published *Young John Dewey* (Chicago, 1975).

Dewey's key account of his own life is "From Absolutism to Experimentalism," in George P. Adams and William P. Montague, eds., *Contemporary American Philosophy* (New York, 1962, c.1930) II: 13–27.

Many books are available on the history of institutions such as Johns Hopkins and the Universities of Michigan and Chicago. Easy to miss are the following dissertations: Joseph L. Brent III, "The Life of Charles Sanders Peirce," (Ph.D. diss., University of California at Los Angeles, 1960); Willinda Savage, "The Evolution of John Dewey's Philosophy of Experimentalism as Developed at the University of Michigan," (Ed.D. diss., University of Michigan, 1950, University Microfilms no. 1999); and Robert E. Tostberg, "Educational Ferment in Chicago, 1883–1904" (Ph.D. diss., University of Wisconsin, 1960, University Microfilms no. 60–5798).

Darnell Rucker, *The Chicago Pragmatists* (Minneapolis, Minn., 1969) is a pioneering study of the atmosphere that surrounded Dewey in Chicago.

I also searched the University of Chicago Archives collections for material on William Rainey Harper, George Mead, and Dewey, but they yielded little that was relevant to the topics discussed here.

Chapter 3

1. Jane Addams, *Twenty Years at Hull-House* (New York, 1960, c.1910), pp. 123, 179–80, 210.

2. Mary O. Furner, *Advocacy & Objectivity: A Crisis in the Professionalization of American Social Science, 1865–1905* (Lexington, Ky., 1975), pp. 1–22; Thomas L. Haskell, *The Emergence of Professional Social Science* (Urbana, Ill., 1977); Dorothy Ross, "The Development of the Social Sciences,"

in Alexandra Oleson and John Voss, eds., *The Organization of Knowledge in Modern America* (Baltimore, Md., 1979), pp. 107–38.

3. Furner, *Advocacy & Objectivity,* chap. 3; Sidney Fine, *Laissez Faire and the General-Welfare State* (Ann Arbor, Mich., 1956), pp. 79–91; Robert C. Bannister, *Social Darwinism: Science and Myth in Anglo-American Social Thought* (Philadelphia, 1979); Sidney Fine, ed., "The Ely-Labadie Letters," *Michigan History* XXXVI no. 1 (March 1952): 1–32, quote from p. 8.

4. Joseph Dorfman, "The Seligman Correspondence, I–IV," *Political Science Quarterly* LVI (1941): 107–24, 270–86, 392–419, 573–99; Furner, *Advocacy & Objectivity,* chap. 3; Alfred W. Coats, "Methodological Controversy as an Approach to the History of American Economics, 1885–1930," (Ph.D. diss., Johns Hopkins, 1953); "The First Two Decades of the American Economic Association," *The American Economic Review,* L no. 4 (September 1960): 555–74; quote from 558; "Henry Carter Adams: A Case Study in the Emergence of the Social Sciences in the United States, 1850–1900," *Journal of American Studies* II no. 2 (October 1968): 177–97; Daniel M. Fox, *The Discovery of Abundance: Simon N. Patten and the Transformation of Social Theory* (Ithaca, N.Y., 1967), chap. 2; Richard Allen Swanson, "Edmund J. James, 1855–1925: A 'Conservative Progressive' in American Higher Education," (Ph.D. diss., University of Illinois, 1966, University Microfilms, no. 67–06749).

5. Richard T. Ely, *Ground Under Our Feet* (New York, 1936), p. 136.

6. Furner, *Advocacy & Objectivity,* chap. 5 and pp. 150–2; Benjamin G. Rader, *The Academic Mind and Reform: The Influence of Richard T. Ely in American Life* (Lexington, Ky., 1966), pp. 118–121.

7. Rader, *Ely,* pp. 123–7.

8. The standard surveys are John Higham, with Leonard Krieger and Felix Gilbert, *History* (Englewood Cliffs, N.J., 1965); Bernard Crick, *The American Science of Politics* (London, 1959); and Albert Somit and Joseph Tanenhaus, *The Development of American Political Science* (Boston, 1967).

9. R. Gordon Hoxie et al., *A History of the Faculty of Political Science Columbia University* (New York, 1955); Fine, *Laissez Faire,* pp. 92–5 for Burgess' ideas.

10. The best source for Turner and the enormous bibliography that surrounds his name is Ray Allen Billington, *Frederick Jackson Turner: Historian, Scholar, Teacher* (New York, 1973), esp. chap. 3 for this paragraph.

11. Frederick Jackson Turner, "The Significance of History," *Wisconsin Journal of Education,* XXI (October 1891): 230–4, and (November 1891): 253–6. Turner's important early essays have been collected in Ray Billington, ed., *Frontier and Section* (Englewood Cliffs, N.J., 1961).

12. Turner, "The Significance of the Frontier in American History," *The Frontier in American History* (New York, 1920), pp. 1–38. The best studies of Turner's ideas and influence are Howard R. Lamar, "Frederick Jackson Turner," in Marcus Cunliffe and Robin Winks, eds., *Pastmasters* (New York, 1969), pp. 74–109, and Richard Hofstadter, *The Progressive Historians* (New York, 1968), part II.

13. Becker has left the best memoir of Turner in "Frederick Jackson Turner," in Howard Odum, ed., *American Masters of Social Science* (Port Washington, N.Y., 1965, c.1927), pp. 273–320. The best introduction to Becker is Burleigh Taylor Wilkins, *Carl Becker* (Cambridge, Mass., 1961).

14. Wilkins, *Becker,* pp. 47–63; Harry Elmer Barnes, "James Harvey Robinson," in Howard W. Odum, ed., *American Masters of Social Science,* pp. 321–408.

15. Wilkins, *Becker,* pp. 59–63; I have treated the later Becker in more detail in *From Self to Society, 1919–1941* (Englewood Cliffs, N.J., 1972), pp. 187–91. On the issue of academic freedom, see Carol S. Gruber, *Mars and Minerva: World War I and the Uses of the Higher Learning in America* (Baton Rouge, La., 1975).

16. James Harvey Robinson, *The New History* (New York, 1912); the key essay is the opening one, followed by chaps. 4, 5, 8. The quotes are from pp. 9, 130–1, 260, 265.

17. Eric F. Goldman, "Charles A. Beard: An Impression," in Howard K. Beale, ed., *Charles A. Beard: An Appraisal* (Lexington, Ky. 1954), pp. 1–8; Ross E. Paulson, *Radicalism and Reform: The Vrooman Family and American Social Thought, 1837–1937* (Lexington, Ky., 1968), pp. 147–58; Burleigh Taylor Wilkins, "Charles A. Beard on the Founding of Ruskin Hall," *Indiana Magazine of History,* LII (September 1956): 282 ff.; Harlin B. Phillips, "Charles Beard, Walter Vrooman, and the Founding of Ruskin Hall," *South Atlantic Quarterly* L, no. 2 (April 1951): 186–91; and "Charles Beard: The English Lectures, 1899–1901," *Journal of the History of Ideas* XIV, no. 3 (June 1953): 451–6, quote from 451.

18. The most useful secondary materials for this discussion have been Max Lerner,

Notes

"Charles Beard's Political Theory," in Beale, ed., *Beard*, pp. 24–45; and Bernard C. Borning, *The Political and Social Thought of Charles A. Beard* (Seattle, Wash., 1962). For the review of Bentley, see *Political Science Quarterly* XXIII (December 1908): 739–41.

19. Charles A. Beard, *Politics* (New York, 1908).

20. Charles Beard, *An Economic Interpretation of the Constitution of the United States* (New York, 1913). The most useful secondary discussions are in Borning, *Beard;* Richard Hofstadter, *The Progressive Historians;* and Forrest McDonald, "Charles A. Beard," in Winks and Cunliffe, eds., *Pastmasters,* pp. 110–41. On Beard as a progressive reformer, see Luther Gulick, "Beard and Municipal Reform," in Beale, ed., *Beard,* pp. 47–60.

21. Robert L. Church, "The Economists Study Society: Sociology at Harvard, 1891–1902," and David B. Potts, "Social Ethics at Harvard, 1881–1931," in Paul H. Buck, ed., *Social Sciences at Harvard, 1860–1920* (Cambridge, 1965), pp. 18–90, 91–128.

22. Hoxie, *Faculty of Political Science,* has a chapter on sociology by Seymour Martin Lipset; the best recent overview of the field is Anthony Oberschall, "The Institutionalization of American Sociology," in Anthony Oberschall, ed., *The Establishment of Empirical Sociology* (New York, 1972), pp. 187–251. See, in addition, John L. Gillin, "Franklin Henry Giddings," in Odum, ed., *Masters of Social Science,* pp. 191–228; and Leo Davids, "Franklin Henry Giddings: Overview of a Forgotten Pioneer," *Journal of the History of the Behavioral Sciences,* IV, no. 1 (January 1968): 62–73.

23. See especially Lewis A. Coser, "American Trends," in Tom Bottomore and Robert Nisbet, eds., *A History of Sociological Analysis* (New York, 1978), pp. 287–320, quote from 287.

24. For general departmental background, see Steven J. Diner, "Development and Discipline: The Department of Sociology at the University of Chicago, 1892–1920," *Minerva* XIII no. 4, (Winter 1975): 514–53; and Robert E.L. Faris, *Chicago Sociology 1920–1932* (Chicago, 1970, c.1967), chaps. 1–2. For Diner's larger study, see his recent *A City and Its Universities: Public Policy in Chicago, 1892–1919* (Chapel Hill, N.C., 1980).

25. The best background source is Vernon K. Dibble, *The Legacy of Albion Small* (Chicago, 1975).

26. For professional background, see especially Don Martindale, *The Nature and Types of Sociological Theory* (Boston, 1960). For quotations see Albion Small, *General Sociology* (Chicago, 1925), p. 552; *Adam Smith and Modern Sociology* (Chicago, 1907), p. 22; *Introduction to a Science of Society* (Waterville, Maine, 1890), p. 81; "The Significance of a Sociology for Ethics," *Decennial Publications of the University of Chicago: Investigations Representing the Departments. . . .* (Chicago, 1903), First Series, IV, p. 133; *The Meaning of Social Science* (Chicago, 1910), pp. 10–11, 277; Small to Edward C. Hayes, printed in Hayes, "Albion Woodbury Small," in Odum, ed., *Masters of Social Science,* p. 184.

27. Most secondary accounts repeat the data in Paul J. Baker, "The Life Histories of W.I. Thomas and Robert E. Park with an Introduction by Paul J. Baker," *American Journal of Sociology,* LXXIX no. 2 (September 1973): 243–50.

28. In addition to Baker, "Life Histories," see the autobiographical note in Herbert Blumer, "An Appraisal of Thomas and Znaniecki's *The Polish Peasant in Europe and America,"* *Critiques of Research in the Social Sciences I* (New York, 1939): 103–6; the chief early work is included in *Sex and Society* (Chicago, 1907) and *Source Book for Social Origins* (Chicago, 1909).

29. William I. Thomas and Florian Znaniecki, *The Polish Peasant in Europe and America* (New York, 1927, c.1918–20), II, p. 1832; Blumer, "An Appraisal," pp. 103–6. I have also profited here and throughout this section from two anthologies of Thomas' work: Edmund H. Volkart, ed., *Social Behavior and Personality: Contributions of W.I. Thomas to Theory and Social Research* (New York, 1951) and Morris Janowitz, ed., *W.I. Thomas On Social Organization and Social Personality* (Chicago, 1966); and also from the general discussion in John Madge, *The Origins of Scientific Sociology* (New York, 1962), chap. 3.

30. Thomas, "The Persistence of Primary-group Norms in Present-day Society and Their Influence in Our Educational System," in Herbert S. Jennings et al., *Suggestions of Modern Science Concerning Education* (New York, 1917), p. 168; *The Polish Peasant,* I, p. 53; II, p. 1832.

31. The details of the scandal are chiefly from Janowitz, *W.I. Thomas,* pp. xiv–xv.

32. For bibliographical data Fred H. Matthews, *Quest for an American Sociology: Robert E. Park and the Chicago School* (Montreal, 1977) is definitive. Winifred Rauschenbusch, *Robert E. Park* (Durham, N.C., 1979) adds useful primary sources and some new detail. There are autobiographical sketches in Baker, "Life Histories of Thomas and Park," and in Robert E. Park, *Race and Culture* (Glencoe, Ill., 1950), pp. v–ix.

33. Park, *Race and Culture*, p. vi; Matthews, *Park*, pp. 31–5; Alexander Vucinich, *Social Thought in Tsarist Russia* (Chicago, 1976), pp. 125–8 and ff.

34. Park, *Race and Culture*, pp. ix–x; Matthews, *Park*, pp. 58–62; Louis R. Harlan et al., eds., *The Booker T. Washington Papers* (Urbana, Ill., 1972–), I, pp. xxxvii–ix; VIII, pp. 85–90, 203–4, 557.

35. Matthews, *Park*, p. 83 and passim; the quote is from Park, *Race and Culture*, pp. viii–ix; Ralph Turner has supplied one of the best discussions of Park's contribution to sociology in Ralph H. Turner, ed., *Robert E. Park On Social Control and Collective Behavior* (Chicago, 1967), pp. ix–xlvi.

36. Robert E. Park, "The City: Suggestions for the Investigation of Human Behavior in the Urban Environment," *Human Communities* (Glencoe, Ill., 1952), pp. 13–51, which is the most convenient place to find this 1916 essay. For the Chicago School, see Madge, *Origins of Scientific Sociology*, chap. 4, and Robert E.L. Faris, *Chicago Sociology, 1920–1932* (Chicago, 1970, c.1967). I have sketched Park's later career and the impact of William Ogburn, his successor, in *From Self to Society, 1919–1941* (Englewood Cliffs, N.J., 1972), pp. 91–8.

Chapter 4

1. Thomas S. Hines has discussed a number of these issues better than anyone else in *Burnham of Chicago* (Chicago, 1979, c.1974).

2. These paragraphs are based on the reading of perhaps one hundred novels, autobiographies, short-story collections, and related materials. Secondary surveys have not been especially helpful, since scholars are accustomed to treating progressive fiction either as muckraking or as trivial. To understand the larger context, the key works on literary realism are Harold H. Kolb, *The Illusion of Life: American Realism As a Literary Form* (Charlottesville, Va., 1969); René Wellek, "Realism in Literary Scholarship," *Concepts of Criticism* (New Haven, 1963); Everett Carter, *Howells and the Age of Realism* (New York, 1954); and George J. Becker, ed., *Documents of Modern Literary Realism* (Princeton, 1963). Grant C. Knight, *The Strenuous Age in American Literature* (Chapel Hill, N.C., 1954), manages somehow to mention most of the novels at least once in nearly chronological order; Larzer Ziff, *The American 1890s* (New York, 1966), Jay Martin, *Harvests of Change* (Englewood Cliffs, N.J., 1967), and Warner Berthoff, *The Ferment of Realism* (New York, 1965) are more synthetic. Bernard Duffey, *The Chicago Renaissance in American Letters* (East Lansing, Mich., 1954) covers the key city involved.

In many ways, biographies of individual novelists are more useful in this context, since they take the work seriously for historical reasons if not aesthetic ones. See especially Robert M. Crunden, *A Hero in Spite of Himself: Brand Whitlock in Art, Politics and War* (New York, 1969); Blake Nevius, *Robert Herrick* (Berkeley, Calif., 1962); and Robert W. Schneider, *Novelist to a Generation: The Life and Thought of Winston Churchill* (Bowling Green, Ohio, 1976).

3. For historical background, the best book is Duffey's *Chicago Renaissance*. I have also profited from points made in two fine studies: Paul Fussell, *The Great War and Modern Memory* (New York, 1975), and David Perkins, *A History of Modern Poetry: From the 1890s to the High Modernist Mode* (Cambridge, Mass., 1976).

4. Masters has never had a scholarly biography, but his autobiography, *Across Spoon River* (New York, 1969, c.1936), is full and illuminating; quotes from pp. 80, 336.

5. The best sources are Carl Sandburg, *Always the Young Strangers* (New York, 1953), quote from p. 32; and Herbert Mitgang, ed., *The Letters of Carl Sandburg* (New York, 1968).

6. Edgar Lee Masters, *Vachel Lindsay* (New York, 1935) was a pioneer study full of primary sources that remains useful despite the lack of sympathy Masters often felt for aspects of his friend's life; Eleanor Ruggles, *The West-Going Heart* (New York, 1959) is helpful but hardly definitive in some of its emphases; Ann Massa, *Vachel Lindsay* (Bloomington, Ind., 1970), is the best critical study of Lindsay's work, heavily interpretive and corrective of earlier accounts. Vachel Lindsay, *Collected Poems* (New York, 1923) contains a lengthy autobiographical account, pp. 1–24, and prints all the poems I have mentioned: "Abraham Lincoln Walks at Midnight," pp. 53–4; "The Eagle That Is Forgotten," pp. 95–6; "Bryan, Bryan, Bryan, Bryan," pp. 96–105; "General William Booth Enters into Heaven," pp. 123–5; "Why I Voted the Socialist Ticket," pp. 301–2; and "King Arthur's Men Have Come Again," pp. 336–7. Other

poems deal with Andrew Jackson, Jane Addams, and Theodore Roosevelt, pp. 90–2, 380–1, 385–6, 387. See also Marc Chénetier, ed., *Letters of Vachel Lindsay* (New York, 1979).

7. Three studies of the Philadelphia years are essential: William I. Homer, *Robert Henri and His Circle* (Ithaca, N.Y., 1969), Sloan's description of Henri quoted from p. 76; Vincent J. De Gregorio, "The Life and Art of William J. Glackens," (Ph.D. diss., Ohio State, 1955, University Microfilms no. 14, 459); and Van Wyck Brooks, *John Sloan* (New York, 1955). The most useful memoirs are Everett Shinn's "Life on the Press," *The Philadelphia Museum Bulletin* XLI (November 1945): 9; and "William Glackens as an Illustrator," *The American Artist* IX (November 1945): 22–3. Most quotations are from Robert Henri, *The Art Spirit* (Philadelphia, 1960, c.1923), pp. 111, 117, 118, 217, 47, 166, and 188–9.

8. The basic data are in the books by Homer, Brooks, and de Gregorio, cited in note 7. See Brooks, *Sloan,* for the quotations from du Bois and Sloan, pp. 72–3.

9. In addition to works previously cited, see Jerome Myers, *Artist in Manhattan* (New York, 1940). For the 1908 exhibition, see Bruce St. John, ed., *John Sloan's New York Scene* (New York, 1965), pp. 112, 118, 128–9, 179, 194–8.

10. Biographical data on Bellows is from Charles H. Morgan, *George Bellows* (New York, 1965), quoted from pp. 17, 206. See also Emma Goldman, *Living My Life* (New York, 1931), pp. 528–9.

11. Brooks, *Sloan,* p. 22; Ira Glackens, *William Glackens,* p. 107; St. John, *Sloan's New York,* pp. 259, 273, 310, 328, 313, 335–6, 340, 364–6.

12. St. John, *Sloan's New York,* pp. 43, 344, 356–7, 627; Art Young, *Art Young* (New York, 1939), pp. 275–7; Louis Untermeyer, *From Another World* (New York, 1939), p. 41; Max Eastman, *Enjoyment of Living* (New York, 1948), pp. 411–2; Art Young, *On My Way* (New York, 1928), pp. 31–6, 61, 80, 274 ff.; William L. O'Neill, ed., *Echoes of Revolt: The Masses 1911–1917* (Chicago, 1966).

13. Homer, *Henri,* p. 174; Hutchins Hapgood, *A Victorian in the Modern World* (Seattle, 1972, c.1939), p. 297; Morgan, *Bellows,* pp. 154, 196–7; Brooks, *Sloan,* pp. 87, 98, 100; St. John, *Sloan's New York,* p. 554.

14. Eastman, *Enjoyment of Living,* 549–59; Floyd Dell, *Homecoming* (New York, 1933), 25; Richard Fitzgerald, *Art and Politics: Cartoonists of the Masses and Liberator* (Westport, Conn., 1973), quotes Helen Farr Sloan to Fitzgerald, 9 July 1967, p. 29. For Art Young's account, see the *New York Sun,* 8 April 1916, p. 6, as quoted in Charlene S. Engel, "George W. Bellows' Illustrations for the *Masses* and Other Magazines and the Sources of His Lithographs of 1916–17," (Ph.D. diss., University of Wisconsin, 1976, University Microfilms no. 76–20891). For Robinson's position see Robinson to Mike Gold, July 1922, printed in Albert Christ-Janer, *Boardman Robinson* (Chicago, 1946), p. 38.

15. John Sloan, *Gist of Art* (New York, 1939), pp. 40–2. The best general history of the show is Milton W. Brown, *The Story of the Armory Show* (New York, 1963), which has an excellent bibliography. Two key shorter accounts are Meyer Schapiro, "Rebellion in Art," in Daniel Aaron, ed., *America in Crisis* (New York, 1952), pp. 203–242, and Charles Hirschfield, " 'Ash Can' vs. 'Modern' Art in America," *Western Humanities Review,* X, no. 4 (Autumn 1956): 353–73. On Henri see especially Homer, *Henri,* pp. 174–6 and Walter Pach, *Queer Thing, Painting* (New York, 1938), p. 47.

16. Myers, *Artist in Manhattan,* p. 36; Brown, *Armory Show,* pp. 206–8; Diane Kelder, ed., *Stuart Davis* (New York, 1971), pp. 3–14; Morgan, *Bellows,* pp. 162–5; de Gregorio, *Glackens,* pp. 354–73; Theodore Roosevelt, "A Layman's Views of An Art Exhibition," *The Outlook,* CIII (January–March 1913): 718–20.

Chapter 5

1. John Kirkpatrick, ed., *Charles E. Ives Memos* (New York, 1972), pp. 114–5. Hereafter cited as *Memos.* For earlier versions of my analysis of Ives, see "Charles Ives' Innovative Nostalgia," *The Choral Journal* XV no. 4 (December 1974): 5–12; and "Charles Ives' Place in American Culture," in H. Wiley Hitchcock and Vivian Perlis, eds., *An Ives Celebration* (Urbana, Ill., 1977), pp. 4–13.

2. Howard Boatright, ed., *Charles Ives Essays Before A Sonata* (New York, 1962), pp. 110–1. Hereafter cited as *Essays.* Vivian Perlis, *Charles Ives Remembered: An Oral History* (New Haven,

1974), pp. 4–6, 16. Hereafter cited as *Ives Remembered.* Many of the interviews in this book may also be heard on Columbia Record no. M432504 entitled *Charles Ives: The 100th Anniversary.*
   3. *Ives Remembered,* p. 225.
   4. The best biographical treatments are Frank R. Rossiter, "Charles Ives and American Culture: The Process of Development, 1874–1921," (Ph.D. diss., Princeton, 1970, University Microfilms no. 71–14, 410), see especially p. 300; and Henry and Sidney Cowell, *Charles Ives and His Music* (New York, 1969, c.1955), see especially pp. 6, 43. Rossiter's dissertation was published in expanded form as *Charles Ives and His America* (New York, 1975) and is now the best place to begin studying Ives' life. See also *Ives Remembered,* 14, and Leah A. Strong, *Joseph Hopkins Twichell* (Athens, Ga., 1966), a study of Harmony's father.
   5. *Memos,* p. 263, quote from Edwards Park, and p. 66, quote from Ives.
   6. *Essays,* p. 19; *Memos,* p. 129.
   7. *Essays,* pp. 11, 22, 14. For useful background see Daniel Edgar Rider, "The Musical Thought and Activities of the New England Transcendentalists," (Ph.D. diss., University of Minnesota, 1964, University Microfilms no. 65–15, 319); William Anson Call, "A Study of the Transcendental Aesthetic Theories of John S. Dwight and Charles E. Ives and the Relationship of These Theories to Their Respective Work as Music Critic and Composer," (D.MUS.A., University of Illinois, 1971, University Microfilms no. 72-6879); Laurence David Wallach, "The New England Education of Charles Ives," (Ph.D. diss., Columbia, 1973, University Microfilms no. 74-17913); and Charles Wilson Ward, "Charles Ives: The Relationship Between Aesthetic Theories and Compositional Processes," (Ph.D. diss., University of Texas, 1974).
   8. Henry D. Thoreau, *A Week on the Concord and Merrimack Rivers* (Boston, 1867), p. 185; Ives, *Essays,* pp. 84, 130; *Memos,* p. 195.
   9. Henry Seidel Canby, *American Memoir* (Boston, 1947), especially pp. 142, 150, 163, 187. See also: William Lyon Phelps, *Autobiography With Letters* (New York, 1939), p. 279 ff.; George Wilson Pierson, *Yale College An Educational History, 1871–1921* (New Haven, 1952), especially chap. 1; George Santayana, "A Glimpse of Yale," *Harvard Monthly* XV (December 1892): 89–97; Arthur Twining Hadley, *Four American Universities* (New York, 1895); and Henry E. Howland, "Undergraduate Life At Yale," *Scribner's Magazine* XXII (July 1897): 1–22.
   10. William Kay Kearns, "Horatio Parker, 1863–1919: A Study of His Life and Music," (Ph.D. diss., University of Illinois, 1965, University Microfilms no. 65-11, 805), is a first-rate and extraordinarily thorough study; Isabel Parker Semler, *Horatio Parker* (New York, 1942), is a still useful family memoir.
   11. *Memos,* pp. 49, 51, 116, 132. See also David Eiseman, "Charles Ives and the European Symphonic Tradition: A Historical Reappraisal," (Ph.D. diss., University of Illinois, 1972, University Microfilms no. 72-19, p. 823).
   12. *Memos,* pp. 130–1; Cowell, *Ives,* pp. 37–9; *Ives Remembered,* p. 53, reprints an Emersonian ad for the agency.
   13. *Memos,* pp. 268–73; Rossiter, *Ives and His America,* pp. 110–25; John P. Brion, "Mr. Life Insurance: Julian S. Myrick," a pamphlet published in 1967, pp. 3–16, kindly supplied to me by the research library of the Mutual of New York Company; *Ives Remembered,* pp. 34–8, 45–51, and *passim.*
   14. Shepard B. Clough, *Century of American Life Insurance,* (New York, 1946), p. 282; Morton Keller, *The Life Insurance Enterprise, 1885–1910* (Cambridge, Mass., 1963), pp. 28, 302.
   15. *Essays,* pp. 235–9.
   16. *Essays,* pp. 29, 59, 128–9; see also p. 144.
   17. *Essays,* pp. 33, 62, 145, 162–3, 166, and 136–209 *passim.*
   18. *Essays,* p. 4; *Ives Remembered,* p. 161.
   19. Cowell, *Ives,* pp. 6, 18, 106, 169; *Essays,* pp. 71–3. The whole subject receives extended treatment in Rosalie Sandra Perry, *Charles Ives and the American Mind* (Kent, Ohio, 1974), chap. 6.
   20. *Essays,* pp. 22–3.
   21. *Memos,* p. 61.
   22. *Memos,* p. 83; Cowell, *Ives,* pp. 144–5.
   23. *Memos,* p. 84.
   24. See the record notes by John Kirkpatrick on RCA record no. LSC-2959. The problem of Ives' quotations, what they mean and how innovative they are, receives its best treatment

in Clayton W. Henderson, "Quotation As a Style Element in the Music of Charles Ives" (Ph.D. diss., Washington University, 1969, University Microfilms no. 69-22,533).

25. *Ives Remembered,* pp. 147–8.

26. *Memos,* pp. 106–8, 163.

27. *Ibid.,* p. 70.

28. *Ibid.,* p. 126.

29. Cowell, *Ives,* pp. 123, 166 and chap. 7, *passim; Ives Remembered,* pp. 36, 59, 62–3.

30. *Memos,* Appendix 8, pp. 198 ff., 237–8; Cowell, *Ives,* pp. 112–115.

31. Frank Lloyd Wright, *An Autobiography* (New York, 1943), pp. 49, 12–13; Maginel Wright Barney, *The Valley of the God-Almighty Joneses* (New York, 1965), p. 12; Wright, "Architecture and Music," *The Saturday Review,* XL (28 September 1957): 72–3; William Wesley Peters to Thomas S. Hines in 1967, conveyed to me by Hines, early 1981.

32. Wright, *Autobiography,* p. 51; Robert C. Twombly, *Frank Lloyd Wright* (New York, 1973), pp. 3–16; Thomas S. Hines, Jr., "Frank Lloyd Wright—The Madison Years: Records versus Recollections," *Journal of the Society of Architectural Historians,* XXVI no. 4 (December 1967): 227–33.

33. Wright, *Autobiography,* pp. 5–7, 28–9, 32; Twombly, *Wright,* pp. 4–5.

34. Wright, *Autobiography,* pp. 9–14; Grant Manson, "Wright in the Nursery: The Influence of Froebel Education on the Work of Frank Lloyd Wright," *The Architectural Review,* CXIII (June 1953).

35. Richard C. MacCormac, "The Anatomy of Wright's Aesthetic," *Architectural Review,* CXLIII (February 1968): 143–6; Maria Kraus Boelte and John Kraus, *The Kindergarten Guide, an illustrated handbook designed for the self-instruction of kindergartners, mothers and nurses* (New York, 1877).

36. MacCormac, "Wright's Aesthetic"; Frank Lloyd Wright, *A Testament* (New York, 1972, c.1957), pp. 22–3.

37. Grant C. Manson, *Frank Lloyd Wright to 1910: The First Golden Age* (New York, 1958), pp. 15–18; Twombly, *Wright,* pp. 17–19; Wright, *Autobiography,* p. 59. I have consistently regularized the spelling of the Lloyd Jones clan without a hyphen, although it frequently appears in the hyphenated "Lloyd-Jones" form as well.

38. Twombly, *Wright,* pp. 77–9; Mary Ellen Chase, *A Goodly Fellowship* (New York, 1939), chap. 4, pp. 87–121.

39. Wright, *Autobiography,* pp. 68–78, 105–6; Twombly, *Wright,* p. 23.

40. Hines, "The Madison Years," pp. 230–1; Wright, *Autobiography,* pp. 89–93.

41. William H. Jordy, *American Buildings and Their Architects: Progressive and Academic Ideals at the Turn of the Twentieth Century* (New York, 1972), chap. 1; Carl Condit, *The Chicago School of Architecture* (Chicago, 1964), chaps. 1, 4; Henry-Russell Hitchcock, *Architecture: Nineteenth and Twentieth Century* (London, 1958), chap. 13; Winston Weisman, "A New View of Skyscraper History," in Edgar Kaufmann, Jr., ed., *The Rise of an American Architecture* (New York, 1970), pp. 113–60; J. Carson Webster, "The Skyscraper: Logical and Historical Considerations," *Journal of the Society of Architectural Historians* XX (March 1961): 3–19; Sigfried Giedion, *Space, Time and Architecture* (Cambridge, Mass., 1967), pp. 368–96.

42. Hitchcock, *Architecture,* p. 232; Jordy, *American Buildings,* pp. 28–36; Louis Sullivan, *Kindergarten Chats* (New York, 1947), p. 28 ff.

43. Claude Bragdon, *The Secret Springs. An Autobiography* (London, 1938), p. 150; Manson, *Wright,* p. 21; Wright, *Genius and the Mobocracy,* pp. 22, 76, 163; Jordy, *American Buildings,* p. 84. Wright's work on the Wainwright Building may be an exception here. The clarity and articulation of the parts to the whole have suggested to Thomas Hines the influence of Sullivan on Wright's subsequent buildings.

44. The Sullivan quote is from an excerpt reprinted in Lewis Mumford, ed., *Roots of Contemporary American Architecture* (New York, 1972, c.1952), p. 74. For a more extended analysis of the connections to pragmatism, see Donald Drew Egbert, "The Idea of Organic Expression and American Architecture," in Stow Persons, ed., *Evolutionary Thought in America* (New York, 1956, c.1950), pp. 336–96.

45. Edward R. De Zurko, *Origins of Functionalist Theory* (New York, 1957), pp. 4, 10–11.

46. Samuel Taylor Coleridge, "Shakespeare, a Poet Generally," in W.G.T. Shedd, ed., *The Complete Works of Samuel Taylor Coleridge* (New York, 1884), IV, p. 55; Richard P. Adams, "Architecture and the Romantic Tradition: Coleridge to Wright," *American Quarterly,* IX (Spring 1957): 47–8; Peter Collins, *Changing Ideals in Modern Architecture, 1750–1950* (London, 1965), chap. 14.

47. Charles R. Metzger, *Emerson and Greenough* (Berkeley, Calif. 1954), p. 35; the two Emerson quotes are from Edward Waldo Emerson, ed., *The Complete Works of Ralph Waldo Emerson* (12 vols., Boston, 1903–21), VI, p. 289 and IV, pp. 252–3. See also Vivian C. Hopkins, *Spires of Form* (Cambridge, Mass., 1951) and Robert B. Shaffer, "Emerson and His Circle: Advocates of Functionalism," *Journal of the Society of Architectural Historians,* VII (July–December 1948), pp. 17–20.

48. For Thoreau, see *The Writings of Henry David Thoreau* (Boston, 1906), III, pp. 138–40 and II, pp. 50–3, 268–70; for Greenough, see Horatio Greenough to Ralph Waldo Emerson, 28 December 1851, Harvard University Archives, as printed in Theodore M. Brown, "Greenough, Paine, Emerson and the Organic Aesthetic," *The Journal of Aesthetics and Art Criticism,* XIV no. 3 (March 1956): 304–17. Metzger, *Emerson and Greenough,* remains the best discussion of both Greenough alone and the subject in general. For Greenough's ideas as they appeared in published form, see Harold A. Small, ed., *Form and Function: Remarks on Art by Horatio Greenough* (Berkeley, Calif. 1947). Adams, "Architecture and the Romantic Tradition," was helpful for several of these insights as well.

49. Wright, *Autobiography,* pp. 33, 53, 75, 162; *A Testament,* p. 19; Walter Crane, *An Artist's Reminiscences* (New York, 1907), chaps. 8–9; David H. Dickason, *The Daring Young Men* (Bloomington, Ind., 1953), p. 182; H. Allen Brooks, "Chicago Architecture: Its Debt to the Arts and Crafts," *Journal of the Society of Architectural Historians,* XXX no. 4 (December 1971): 312–17; Alan Crawford, "From Frank Lloyd Wright to Charles Robert Ashbee," *Architectural History,* XIII (1970): 64–76. For the seminal treatment of the larger theme from William Morris to Walter Gropius, see Nikolaus Pevsner, *Pioneers of Modern Design* (London, 1974, c.1936), and Alf Bøe, *From Gothic Revival to Functional Form* (Oslo, 1957). On the frequently neglected subject of book design, see Frederic D. Weinstein, "Walter Crane and the American Books Arts, 1880–1915," (D.L.S. diss., Columbia University, 1970, University Microfilms no. 71-17,558).

50. For good general background, two pioneering studies remain useful: Holbrook Jackson, *The Eighteen Nineties* (New York, 1966, c.1913) and Graham Hough, *The Last Romantics* (London, 1947). Three more works of restricted value are Raymond Lister, *Victorian Narrative Paintings* (London, 1966), Graham Reynolds, *Victorian Painting* (London, 1966), and John Rosenberg, *The Darkening Glass: A Portrait of Ruskin's Genius* (New York, 1961).

51. Hough, *Last Romantics,* chap. 1; Collins, *Changing Ideals,* chap. 10; Roger B. Stein, *John Ruskin and Aesthetic Thought in America, 1840–1900* (Cambridge, 1967), chap. 2.

52. E.T. Cook and A. Wedderburn, eds., *The Works of John Ruskin* (39 vols., London, 1903–12), IV, p. 210; III, p. 48; VIII, p. 61. The best general discussion is George P. Landow, *The Aesthetic and Critical Theories of John Ruskin* (Princeton, N.J., 1971).

53. See particularly Kristine Ottesen Garrigan, *Ruskin on Architecture* (Madison, Wis., 1973), chaps. 2–3.

54. I have relied chiefly on Philip Henderson, *William Morris* (London, 1967) and Ray Watkinson, *William Morris As Designer* (New York, 1967), especially pp. 42–68.

55. Hough, *Last Romantics,* chap. 3; May Morris, ed., *The Collected Works of William Morris* (London, 1910–1915), XXII, p. 77.

56. See particularly Robert Judson Clark, ed., *The Arts and Crafts Movement in America 1876–1916* (Princeton, N.J., 1972), especially the chronology, p. 10, the discussion of Chicago, pp. 58–9, and of Wright, pp. 68–74, which includes a number of photographs of furniture and designs that Wright made specifically for his various houses. For the larger background, see Elizabeth Aslin, *Nineteenth Century English Furniture* (London, 1962); Brooks, "Chicago Architecture," and James D. Kornwulf, *M.H. Baillie Scott and the Arts and Crafts Movement* (Baltimore, Md., 1972).

57. Wright, *Autobiography,* p. 79; Leicester Hemingway, *My Brother Ernest Hemingway* (Cleveland, 1962), p. 20; Barney, *Valley of Joneses,* pp. 127–9; Mark L. Peisch, *The Chicago School of Architecture* (New York, 1964), pp. 39–40; Leonard K. Eaton, *Two Chicago Architects and Their Clients* (Cambridge, Mass., 1969), pp. 12–13; Carlos Baker, *Ernest Hemingway* (New York, 1969), chap. 1.

58. *Autobiography,* pp. 112, 118; Barney, *Valley of Joneses,* pp. 131–6; John Lloyd Wright, *My Father Who Is On Earth* (New York, 1946), pp. 54–5, 75–8.

59. Vincent J. Scully, Jr., *The Shingle Style and the Stick Style* (New Haven, Conn., 1971, c.1955), especially pp. 88–9, 99, 158–61; Henry-Russell Hitchcock, *Architecture: Nineteenth and Twentieth Century,* chap. 15; idem, "Frank Lloyd Wright and the 'Academic Tradition' of the Early Eighteen-Nineties," *Journal of the Warburg and Courtauld Institutes* VII (1944): 46–63; idem, "Ruskin

and American Architecture, or Regeneration Long Delayed," in John Summerson, ed., *Concerning Architecture* (London, 1968), pp. 166–208, especially p. 200. For a similar stylistic development in one of Wright's closest friends, George Elmslie, see David Gebhard, "William Gray Purcell and George Grant Elmslie and the Early Progressive Movement in American Architecture from 1900 to 1920" (2 vols., Ph.D. diss., University of Minnesota, 1957), especially pp. 72–4.

60. Wright, *A Testament*, p. 150; for the bulk of this paragraph, I am much indebted to Clay Lancaster, "The American Bungalow," *The Art Bulletin* XL no. 3 (September 1958): 239–53. The articles are "A Home in a Prairie Town," and "A Small House with 'Lots of Room in it,'" *The Ladies' Home* Journal, XVIII (February 1902) and XVIII (July 1901).

61. Sullivan, *Autobiography of an Idea*, pp. 321–4. For the best recent published version of the history of the fair, see Thomas S. Hines, *Burnham of Chicago* (New York, 1974). Three useful unpublished works are David H. Crook, "Louis Sullivan, the World's Columbian Exhibition and American Life" (Ph.D. diss., Harvard, 1964); Titus M. Karlowicz, "The Architecture of the World's Columbian Exposition" (Ph.D. diss., Northwestern, 1965, University Microfilms no. 66-02721); and Robert Knutson, "The White City: The World's Columbian Exposition of 1893" (Ph.D. diss., Columbia, 1956, University Microfilms no. 00-17062).

62. Gebhard, "Purcell and Elmslie," p. 29 and figure 1.

63. Clay Lancaster, *The Japanese Influence in America* (Tokyo, 1963), especially chaps. 8–9; Scully, *Shingle Style*, pp. 21–2.

64. Lancaster, *Japanese Influence*, pp. 86–8; Dimitri Tselos, "Exotic Influences in the Architecture of Frank Lloyd Wright," *Magazine of Art* XLVI no. 4 (April 1953): 160–9, 184; Wright, *Autobiography*, p. 196 and *The Japanese Print* (Chicago, 1912). The development of mission furniture in America, which is often mentioned in scholarly analysis of the Prairie Style, seems to be more of a parallel development to Wright than an influence upon him. He knew of it and occasionally criticized it; it certainly ranks as an unjustly neglected aspect of progressive creativity. See John C. Freeman, *The Forgotten Rebel: Gustav Stickley and His Craftsman Mission Furniture* (Watkins Glen, N.Y., 1966) and David H. Dickason, *The Daring Young Men* (Bloomington, Ind., 1953), chaps. 13–14.

65. Eaton, *Two Chicago Architects*, pp. 67–74; Twombly, *Wright*, pp. 38–9; John Lloyd Wright, *My Father*, pp. 153–72, reprints the text of the book.

66. Ruari McLean, *Modern Book Design from William Morris to the Present Day* (London, 1958); John T. Winterich, "Introduction," *The Works of Geoffrey Chaucer, A Facsimile of the William Morris Kelmscott Chaucer* (Cleveland, 1958); Susan Otis Thompson, "Kelmscott Influence on American Book Design" (D.L.S. diss., Columbia University, 1972, University Microfilms no. 75-9365), subsequently published as *American Book Design and William Morris* (New York, 1977).

67. "The Art and Craft of the Machine" has been reprinted in many places. I have used the version in Edgar Kaufmann and Ben Raeburn, eds., *Frank Lloyd Wright: Writings and Buildings* (Cleveland, 1960), pp. 55–73.

68. Wright, "Preface," to *Ausgeführte Bauten und Entwürfe* (Berlin 1910), as reprinted in Kaufmann and Raeburn, *Frank Lloyd Wright*, pp. 84–106.

69. Wright, *Autobiography*, pp. 139–41.

70. Ibid., pp. 141–5.

71. For the best general summations of the Prairie Style, see H. Allen Brooks, *The Prairie School: Frank Lloyd Wright and his Midwestern Contemporaries* (Toronto, 1972), pp. 4–7, and Marcus Whiffen, *American Architecture Since 1780: A Guide to Styles* (Cambridge, Mass., 1969), pp. 201–7. For the best discussion of a single building, see Jordy, *American Buildings*, chap. 3, on Wright's Robie House.

72. I have accepted the basic terminology of H. Allen Brooks, in *The Prairie School* and in his key article, " 'Chicago School': Metamorphosis of a Term," *Journal of the Society of Architectural Historians*, XXV (1966): 115–118. I have thus intentionally avoided any use of the term "Chicago School," but anyone interested in pursuing the controversy should see, among others, "The Chicago School of Architecture—A Symposium," *Prairie School Review*, IX nos. 1 and 2 (1972); Hugh M.G. Garden, "The Chicago School," Ibid., III no. 1 (1966): 19–22; Titus M. Karlowicz, "The Term Chicago School: Hallmark of a Growing Tradition," Ibid., IV, no. 3 (1967): 26–30, and "The Chicago School and the Tyranny of Usage," Ibid., VI no. 3 (1969): 11–19; and William H. Jordy, "The Commercial Style and the 'Chicago School,' " *Perspectives in American History* I (1967): 390–400.

73. Brooks, *Prairie School*, lists the key figures, p. 8. See also Peisch, *Chicago School*, especially

for Griffin; Gebhard, "Purcell and Elmslie," for that firm; and Sally A. Chappell, "Barry Byrne: Architecture and Writings" (Ph.D. diss., Northwestern, 1968, University Microfilms no. 69-6902).

74. See especially Condit, *Chicago School*, pp. 211–217.

Chapter 6

1. John Chamberlain, *Farewell to Reform* (Chicago, 1965, c.1932), was a caustic, left-wing critique that was well-researched for its day and noteworthy especially for its detailed attention to progressive writing. Louis Filler, *Crusaders for American Liberalism* (New York, 1939), later republished several times under the title *The Muckrakers: Crusaders for American Liberalism*, provided a brisk and sympathetic defense that remains a fine introduction for the general reader. Two subsequent dissertations have added greatly to a more detailed general perspective: David M. Chalmers, "The Social and Political Philosophy of the Muckrakers," (Ph.D. diss., University of Rochester, 1955), examined the thought of the progressive writers and found a broad spectrum of opinion that made facile generalizations about the entire group rather risky; he published a much abbreviated version as *The Social and Political Ideas of the Muckrakers* (New York, 1964); the most recent general study and the one to which I am most indebted for generalizations about what is or is not typical about the group is Judson A. Grenier, "The Origins and Nature of Progressive Muckraking" (Ph.D. diss., University of California, Los Angeles, 1965, University Microfilms no. 65-13,855). My remarks in this paragraph also refer to many more general studies of progressivism and individual progressives.

2. S.S. McClure, *My Autobiography* (New York, 1963, c.1914), p. 18. See also the early pages of Peter Lyon, *Success Story: The Life and Times of S.S. McClure* (New York, 1963); and Grenier, "Muckraking," chap. 1.

3. Grenier, "Muckraking," has a long synthesis, summarized on pp. 82–3.

4. The best guide to the early Sinclair is William A. Bloodworth, Jr., "The Early Years of Upton Sinclair: A Study of the Development of A Progressive Christian Socialist," (Ph.D. diss., Texas, 1972), which covers the period up to about 1912 in detail. The next most important source is *The Autobiography of Upton Sinclair* (New York, 1962), which incorporates material written much earlier. Sinclair's *Love's Pilgrimage* (New York, 1911), contains material on early childhood, but concentrates in extraordinary detail on Sinclair's courtship, marriage, and early writing. Virtually all of Sinclair's early papers burned in a fire at Helicon Hall in 1907. For a guide to what little is left, see Indiana University, Lilly Library, *The Upton Sinclair Archives*, (Bloomington, Ind., 1963?). Sinclair's published letters deal only with letters to him, and only with the later years in his life.

5. Bloodworth, *Sinclair*, chaps. 1–2; Sinclair, *Autobiography, passim*, quote from p. 45; for the "conversion," see p. 54.

6. Sinclair, *Autobiography*, pp. 6–8, 43–5.

7. Sinclair, *Autobiography*, pp. 28, 45 ff.; *Love's Pilgrimage, passim*, especially pp. 74, 471.

8. *Love's Pilgrimage*, p. 23 ff.; *Autobiography*, pp. 28–9.

9. Bloodworth, *Sinclair*, chaps. 2–3; Sinclair, *Autobiography*, quotes from pp. 63, 100.

10. Bloodworth, *Sinclair*, chap. 3; Sinclair, *Autobiography*, pp. 101–2. In *Love's Pilgrimage* Herron appears as Henry Darrell, pp. 416ff., 430 ff., 526. Sinclair also examines Herron in *The Brass Check* (Pasadena, Calif., 1920), chap. 20.

11. Sinclair, *Autobiography*, pp. 93, 102; *Manassas* (Pasadena, Calif., 1923, c.1904), p. 109.

12. Sinclair, *The Industrial Republic* (New York, 1907), especially p. 199.

13. Sinclair, *Autobiography*, p. 108; *The Profits of Religion* (Pasadena, Calif., 1918), pp. 176, 290.

14. Sinclair, *Mammonart* (Pasadena, Calif., 1925), pp. 9–10.

15. Bloodworth, *Sinclair*, chap. 5; see also Sinclair, *Brass Check*, chaps. 4–9.

16. Sinclair, *Appeal to Reason*, 11 February 1905, as quoted in Bloodworth, *Sinclair*, p. 165.

17. Bloodworth, *Sinclair*, chap. 5; Sinclair, *Autobiography*, pp. 114–9.

18. Bloodworth, *Sinclair*, pp. 199; Sinclair, *My Lifetime in Letters* (Columbia, Mo., 1960), p. 20; Mark Sullivan, *America Finding Herself* (New York, 1927), p. 541. Numbers in parentheses indicate pages in *The Jungle* (New York, n.d., c.1906).

19. I am most indebted here and for the next several paragraphs to James Harvey Young,

Notes

The *Toadstool Millionaires* (Princeton, N.J., 1961). For the quotation see Richard H. Shryock, *The Development of Modern Medicine* (Philadelphia, Penna., 1936), p. 255.

20. David L. Dykstra, "Patent and Proprietary Medicines: Regulation and Control Prior to 1906" (Ph.D. diss., Wisconsin, 1951), chaps. 1, 3; Young, *Toadstool Millionaires*, pp. 68, 56.

21. Young, *Toadstool Millionaires*, pp. 113–17.

22. Dykstra, "Patent Medicines," chap. 2.

23. Dykstra, "Patent Medicines," chap. 4; Harvey Wiley, *An Autobiography* (Indianapolis, Ind., 1930), chap. 16, especially pp. 198–9; William E. Mason, *Digest of the Pure Food and Drug Laws of the United States and Foreign Countries* (57 Congress, 1 Session, Senate Report no. 3, Washington, 1901), p. 2:4257.

24. James G. Burrow, *AMA: Voice of American Medicine* (Baltimore, Md., 1963), chaps. 1–3.

25. Samuel Hopkins Adams, *The Great American Fraud* (Chicago, 1912), pp. 5–9, contains the best discussion of the "red clauses."

26. Young, *Toadstool Millionaires*, pp. 212–13; Edward Bok, *The Americanization of Edward Bok* (New York, 1923), chap. 30; Mark Sullivan, *The Education of an American* (New York, 1938), chap. 21, p. 183 ff.; Edward LaRue Weldon, "Mark Sullivan's Progressive Journalism, 1874–1925: An Ironic Persuasion" (Ph.D. diss., Emory, 1970, University Microfilms no. 70-22,-892), 39–58.

27. Young, *Toadstool Millionaires*, pp. 213–14; Sullivan, *Education*, pp. 183–9; Walter Johnson, *William Allen White's America* (New York, 1947), pp. 148–9.

28. Michael D. Marcaccio, "The Earnest Brothers: A Biography of Norman, Hutchins and William P. Hapgood" (Ph.D. diss., University of Virginia, 1972, University Microfilms no. 72-33,244), chap. 6, especially p. 188 ff.; subsequently published as *The Hapgoods* (Charlottesville, Vir., 1977); Sullivan, *Education*, pp. 188–92; Young, *Toadstool Millionaires*, pp. 214–15; Dykstra, "Patent Medicines," chap. 6, especially p. 162.

29. Harold S. Wilson, *McClure's Magazine and the Muckrakers* (Princeton, 1970), pp. 148–9; Young, *Toadstool Millionaires*, pp. 214–15; Will Irwin, *The Making of a Reporter* (New York, 1942), p. 155.

30. Young, *Toadstool Millionaires*, pp. 216–18; Samuel Hopkins Adams, *The Great American Fraud* (5th ed. revised, Chicago, 1912), pp. 3–11.

31. Adams, *Fraud*, pp. 3–5 (7 October 1905).

32. Ibid., pp. 12–22 (28 October 1905).

33. Ibid., p. 23 ff. (18 November 1905).

34. The best survey of material in the *Congressional Record* is still Thomas A. Bailey, "Congressional Opposition to Pure Food Legislation, 1879–1906," *The American Journal of Sociology*, no. 1 (July 1930): 52–64. Stephen Wilson, *Food and Drug Regulation* (Washington, 1942), chap. 2, adds data; Dykstra, "Patent Medicines," lists all the relevant bills between 1890 and 1906, p. 229. See also James R. Mann, *Pure Food* (59 Congress, 1 Session, House Report no. 2118, vol. 1, part 1, Washington, 1906), and Gustavus A. Weber, *The Food, Drug and Insecticide Administration: Its History, Activities and Organization* (Baltimore, Md., 1928).

35. Oscar E. Anderson, *The Health of a Nation: Harvey W. Wiley and the Fight for Pure Food* (Chicago, 1958); see also Wiley, *Autobiography*, p. 27.

36. Anderson, *Wiley*, p. 26.

37. Wiley, *Autobiography*, pp. 189–91; Anderson, *Wiley*, pp. 71–7.

38. Anderson, *Wiley*, p. 120; Dykstra, "Patent Medicines," chap. 7.

39. Anderson, *Wiley*, pp. 120–9.

40. Harvey W. Wiley, *The History of a Crime Against the Food Law* (Washington, 1929), conveniently reprints much key congressional testimony; see pp. 1–56, especially pp. 24, 31, 47. Wilson, *Food and Drug Regulation*, pp. 20–21; Wiley, *Annual Report of the Chief Chemist*, Division of Chemistry, House Documents (1899), vol. 32, p. 3992.

41. Wilson, *Food and Drug Regulation*, pp. 21–2; Anderson, *Wiley*, p. 149 ff.; Sullivan, *Our Times*, II, p. 520; Wiley, *History of Crime*, chap. 2; Wiley, *Autobiography*, pp. 215–20.

42. H.W. Wiley, M.D., *Influence of Food Preservatives and Artificial Colors on Digestion and Health*, U.S. Department of Agriculture, Bureau of Chemistry, Bulletin No. 84, Part 1 (Washington, 1904), pp. 225, 758, 1039, 1293, 1499; Wilson, *Food and Drug Regulation*, pp. 21–2; Wiley, *History of Crime*, pp. 61–3.

43. The legislative history of the bill has been reviewed many times. The best recent versions are by John Braeman, in *Albert J. Beveridge: American Nationalist* (Chicago, 1971), pp. 101–9, and "The Square Deal in Action: A Case Study in the Growth of the 'National Police

Power,' " in John Braeman et al., eds., *Change and Continuity in Twentieth-Century America* (Columbus, O., 1964), pp. 35–80.

44. Wiley, *Autobiography*, especially pp. 225–6; Anderson, *Wiley*, chap. 8.

45. Based on a study of Morison et al., eds., *Letters of Roosevelt*, V; the appropriate letters to and from Roosevelt in the Theodore Roosevelt Papers, Library of Congress, hereafter cited as TRP-LC; and what I assume to be four leading, representative studies of Roosevelt in this period: William Henry Harbaugh, *Power and Responsibility* (New York, 1961), pp. 255–60; Henry F. Pringle, *Theodore Roosevelt* (rev. ed., New York, 1956), pp. 300–2; John M. Blum, *The Republican Roosevelt* (Cambridge, 1954), p. 74; and George Mowry, *The Era of Theodore Roosevelt* (New York, 1958), pp. 207–8.

46. Hermann Hagedorn, ed., *The Works of Theodore Roosevelt* (20 vols., New York, 1926), XV, p. 326; Young, *Toadstool Millionaires*, pp. 235–6; Roosevelt to R.S. Baker, 13 September and 13, 20, 22, and 28 November 1905, *Letters of Roosevelt*, V, pp. 25, 76–7, 83–5, 88–9, 100–1; Roosevelt to Baker, 8 September and 16 October 1905, and Roosevelt to Steffens, 26 September, 7 October, and 9 November 1905, all TRP-LC.

47. William Allen White to Roosevelt, 15 January 1906; Attorney General William H. Moody to Roosevelt, 25 January; Roosevelt to Moody, 26 January, all TRP-LC.

48. Doubleday to Roosevelt, 23 March 1906; Roosevelt to Sinclair, 9 March 1906, TRP-LC.

49. Roosevelt to Sinclair, 9 March 1906; Roosevelt to Secretary Wilson, 12 March 1906, TRP-LC.

50. Roosevelt to Sinclair, 9 April 1906, confirmed in Roosevelt to Charles P. Neill, 9 April 1906, TRP-LC.

51. Sinclair to Roosevelt, 10 April 1906, followed by telegram, 10 April, followed by letter, 10 April, the last establishing the chronology; Roosevelt to Sinclair, 11 April 1906, printed in *Letters*, V, pp. 208–9; memo, phone call, Sinclair to Roosevelt, 11 April 1906, and Roosevelt to Baker, 9 April 1906, all TRP-LC.

52. Roosevelt to Sinclair, 21 March 1906; Roosevelt to James Wilson, 11 April; James Wilson to Roosevelt, also 11 April; Roosevelt to Wilson, 16 April; and Roosevelt to Neill, 16 and 27 April, 4 and 23 May 1906, all TRP-LC.

53. Roosevelt to Taft, 15 March 1906; and to Owen Wister, 27 April, in *Letters*, V, pp. 183–4, 229; see also the long description of Phillips, pp. 262–9. Roosevelt to S.S. McClure, 11 April 1906, TRP-LC; R.S. Baker to Roosevelt, 7 April 1906; Roosevelt to Baker, 9 April, both printed in Baker, *American Chronicle*, pp. 202–3.

54. *Presidential Addresses and State Papers* (8 vols., New York, 1910), V., pp. 712–24. For detailed discussion of the speech, see John Semonche, "Theodore Roosevelt's 'Muck-Rake Speech': A Reassessment," *Mid-America*, XLVI (April 1964): 114–25.

55. Roosevelt to Wilson, 11 and 16 April 1906; Roosevelt to Neill, 16 and 27 April, 4 and 23 May 1906; Roosevelt to Beveridge, 23 May 1906, all TRP-LC. See also *Letters*, V. p. 282, and the rather plaintive L.F. Swift (president of Swift & Co.) to William Loeb, Jr. (Roosevelt's secretary), 23 May 1906, TRP-LC.

56. Roosevelt to Wadsworth, 26 May 1906, in *Letters*, V pp. 282–3.

57. Roosevelt to Upton Sinclair, 2 June 1906, TRP-LC; Roosevelt to Wadsworth, 31 May 1906, printed in *Letters*, V, pp. 291–2; U.S. Congress, House, *Conditions in Chicago Stock Yards: Message from the President . . . Transmitting the Report of Mr. James Bronson Reynolds and Commissioner Charles P. Neill . . .* , 59th Congress (1st Session, 1906), House Document 875.

58. Roosevelt to Wadsworth, 29 May 1906, TRP-LC; Roosevelt to Beveridge, 29 May, and to Wadsworth, 31 May 1906, in *Letters*, V, pp. 289, 291–2; Roosevelt to Wadsworth, 8 June 1906, in *Letters*, V, pp. 294–6; Roosevelt to Wadsworth, 14 June, and Wadsworth to Roosevelt, 15 June 1906, both TRP-LC.

59. Roosevelt to Wadsworth, 15 June 1906, in *Letters*, V, pp. 298–9.

60. Roosevelt to Wadsworth, 15 June 1906, in *Letters*, V., pp. 298–9; Roosevelt to Senator Redfield Proctor, 18 June 1906, TRP-LC.

61. The most detailed account is Dykstra, "Patent Medicine," pp. 246–66.

62. Young also makes a few of these points, *Toadstool Millionaires*, p. 244.

63. The report is included in an extended quotation within a letter, Roosevelt to Wadsworth, 8 June 1906, *Letters*, V, pp. 294–5.

64. Wiley, *History of A Crime* and *Autobiography*, the quote from the latter, p. 325; Anderson, *Wiley*, provides an excellent overview. For Roosevelt's instructions see his letter to Neill, 28 June 1906, TRP-LC.

65. The most prominent example is Gabriel Kolko, *The Triumph of Conservatism* (New York, 1963), especially pp. 98–112.

66. See especially James Harvey Young, *The Medical Messiahs* (Princeton, N.J., 1967), and Charles O. Jackson, *Food and Drug Legislation in the New Deal* (Princeton, N.J., 1970).

Chapter 7

1. Unless otherwise noted, circumstantial detail is from a group of national newspapers for early August 1912. I have most frequently used the *New York Times*, the *Chicago Tribune*, the *San Francisco Examiner*, and the *New York Tribune*.

2. Theodore Roosevelt to Booker T. Washington, 8 June 1904, TRP-LC, Letterbook vol. 47, p. 439. For background see Dewey W. Grantham, Jr., "The Progressive Movement and the Negro," *South Atlantic Quarterly*, LIV no. 4 (October 1955): 461–77, especially 468–9; and Pearl Kluger, "Progressive Presidents and Black Americans," (Ph.D. diss., Columbia, 1974, University Microfilms no. 76-29,298).

3. Arthur S. Link, "Theodore Roosevelt and the South in 1912," *North Carolina Historical Review*, XXIII no. 3 (July 1946): 313–24, especially 314. Much of the relevant correspondence has been printed in Arthur S. Link, ed., "Correspondence Relating to the Progressive Party's 'Lily White' Policy in 1912," *Journal of Southern History*, X (November 1944): 480–90.

4. William H. Maxwell to Roosevelt, 28 July 1912; Rev. Mr. J. Gordon McPherson to Roosevelt (night letter), 6 August 1912; John M. Parker to Roosevelt, 24 July 1912, reel no. 150, TRP-LC.

5. Link, "Roosevelt and the South," pp. 315, 321; B.F. Fridge to J.M. Dixon, 24 July 1912, reel no. 150, TRP-LC. South Carolina was the one state without an organization of some kind by the opening of the convention: see "Official Report of the Proceedings of the Provisional National Progressive Committee," reel no. 447.

6. "Official Proceedings . . . ," reel no. 447. See also George Mowry, "The South and the Progressive Lily White Party of 1912," *Journal of Southern History*, VI (May 1940): 237–47; and *Theodore Roosevelt and the Progressive Movement* (New York, 1960, c.1946), pp. 267–8.

7. Elting Morison et al., eds., *The Letters of Theodore Roosevelt*, VII (Cambridge, Mass., 1954), pp. 584–90.

8. Alice Roosevelt Longworth, *Crowded Hours* (New York, 1933), p. 223; Oscar K. Davis, *Released for Publication* (Boston, 1925), pp. 324–5; John Garraty, *Right-Hand Man, The Life of George W. Perkins* (New York, 1960), passim.

9. Frank Munsey to Roosevelt, 13 July 1912, enclosing Munsey to C.H. Stoddart, 13 July, both TRP-LC, reel no. 149; Garraty, *Right-Hand Man*, p. 252 ff.; Frank A. Munsey, "What Mr. Roosevelt's Election Will Mean to the Business World," *Munsey's Magazine*, XLVII no. 4 (July 1912), pp. 506–12; Longworth, *Crowded Hours*, p. 224; Henry L. Stoddard, *As I Knew Them* (New York, 1927), pp. 422–3; Davis, *Released for Publication*, pp. 261–6.

10. Garraty, *Right-Hand Man*, p. 256 ff.; Stoddard, *As I Knew Them*, pp. 407–8.

11. Helene Maxwell Hooker, ed., *History of the Progressive Party 1912–1916* by Amos R.E. Pinchot (New York, 1958), pp. 1–36.

12. Hooker, *Progressive Party*, pp. 93–117; Garraty, *Right-Hand Man*, p. 282.

13. David W. Levy, "The Life and Thought of Herbert Croly, 1869–1914," (Ph.D. diss., University of Wisconsin, 1967, University Microfilms no. 67-12,446), passim., especially pp. 378–82. Levy has effectively corrected numerous errors in the better-known published accounts of Croly.

14. Edward A. Fitzpatrick, *McCarthy of Wisconsin* (New York, 1944), chap. 12, especially pp. 157–60. For Roosevelt's announced position on trusts just before these events, see "What a Progressive Is," *Works*, XIX: 240–54. For other pressures see George Record to Roosevelt, 12 July, Chester Rowell to Roosevelt, 12 July; Sen. Joseph Bristow to Roosevelt, 15 July, reel no. 149, TRP-LC.

15. Arthur Ruhl, "The Bull Moose Call: A New Sound in American Politics and Those Who Answered it," *Collier's* (24 August 1912): 21; Edward S. Lowry, "With the Bull Moose in Convention," *Harper's Weekly* (17 August 1912): 9.

16. Ruhl, "Bull Moose Call."

17. William Menkel, "The Progressives at Chicago," *The American Review of Reviews* XLVI (September 1912), pp. 310–17.

18. Lowry, "With the Bull Moose."

19. Claude G. Bowers, *Beveridge and the Progressive Era* (Boston, 1932), pp. 416–24; John Braeman, *Albert J. Beveridge* (Chicago, 1971), pp. 213–22.

20. "Official Proceedings," pp. 13–41; Walter Johnson, *William Allen White's America* (New York, 1947), p. 206.

21. Richard Harding Davis, "The Men At Armageddon," *Collier's* (24 August 1912): 10–11; Lowry, "With the Bull Moose in Convention," 9; Abbott, "The Progressive Convention," p. 859; Bowers, *Beveridge,* p. 425; Stoddard, *As I Knew Them,* p. 410.

22. Ruhl, "Bull Moose Call"; "Official Proceedings," reel no. 447, TRP-LC.

23. "Official Proceedings," passim, with quotes from pp. 58, 136, and 138; there are many published versions of the "Confession of Faith," as in *Works,* XIX, pp. 358–411.

24. The literature on the platform fight and its odd results on Wednesday is large and contradictory. The unpublished minutes and platform proposals are in the Theodore Roosevelt Association Collection, Harvard University. The best narrative account is Gable, "Bull Moose Years," pp. 242–50. I have supplemented Gable with Nelson McGeary, *Gifford Pinchot* (Princeton, N.J. 1960), pp. 228–9; Hooker, ed., *History of the Progressive Party,* p. 177; Edward A. Fitzpatrick, *McCarthy of Wisconsin* (New York, 1944), pp. 160–3; Fausold, *Gifford Pinchot,* pp. 98–100; Garraty, *Right-Hand Man,* p. 268; White, *Autobiography,* p. 484; Johnson, *William Allen White's America,* p. 205; White, article in *Chicago Tribune,* 7 August, p. 5; Abbott, "The Progressive Convention," p. 861.

25. "Official Proceedings," pp. 194–6; the speech was widely reprinted. One copy is in the Addams Papers, Swarthmore College Peace Collection.

26. Jane Addams, *The Second Twenty Years at Hull-House* (New York, 1930), p. 39; Davis, *Spearheads for Reform* (New York, 1967), p. 205; Menkel, "The Progressives at Chicago," p. 314.

27. "Official Proceedings," pp. 243–55; Kirk H. Porter and Donald B. Johnson, eds., *National Party Platforms, 1840–1964* (Urbana, Ill., 1966), pp. 175–82; the offending paragraph is the last on p. 178. See also Gable, "Bull Moose Years," pp. 248–50; Davis, *Released for Publication,* chap. 61; Hooker, ed., *History of Progressive Party,* p. 177; Fitzpatrick, *McCarthy of Wisconsin,* pp. 162–3; Fausold, *Gifford Pinchot,* pp. 98–100; Garraty, *Right-Hand Man,* pp. 268–70.

28. The key primary sources are in A. Lincoln, ed., "My Dear Governor: Letters Exchanged by Theodore Roosevelt and Hiram Johnson," *California Historical Society Quarterly,* XXXVIII (September 1959): 229–47. For useful background see A. Lincoln, "Theodore Roosevelt, Hiram Johnson, and the Vice-Presidential Nomination of 1912," *Pacific Historical Review* XXVIII No. 3 (August 1959); 267–83; Richard C. Lower, "Hiram Johnson and the Progressive Denouement, 1910–1920" (Ph.D. diss., University of California, Berkeley, 1969, University Microfilms no. 70–6158); John F. Fitzpatrick, "Senator Hiram W. Johnson: A Life History, 1866–1945" (Ph.D. diss., University of California at Berkeley, 1975, University Microfilms no. 75-26691); Spencer C. Olin, Jr., *California's Prodigal Sons: Hiram Johnson and the Progressives, 1911–1917* (Berkeley, Calif., 1968); Miles C. Everett, "Chester Harvey Rowell: Pragmatic Humanist and California Progressive" (Ph.D. diss., University of California at Berkeley, 1966, University Microfilms no. 66-8310); Mowry, *The California Progressives.*

29. "Official Proceedings," p. 295 ff.; A. Lincoln, "Theodore Roosevelt . . . Nomination of 1912"; and "My Dear Governor."

Chapter 8

1. The literature is vast and some historians have changed their positions on key issues. For savage treatment of Bryan in post-World War II historiography, see Richard Hofstadter, "The Democrat As Revivalist," in *The American Political Tradition and the Men Who Made It* (New York, 1948), and Ray Ginger, ed., *William Jennings Bryan: Selections* (Indianapolis, Ind., 1967). Bryan has never lacked for defenders, however, even in early assessments. J.V. Fuller's anonymous "William Jennings Bryan" in Samuel Flagg Bemis, ed., *The American Secretaries of State and Their Diplomacy,* X (New York, 1929) was remarkably fine for its day; Ray Stannard Baker, himself a distinguished progressive, understood Bryan quite well in *Woodrow Wilson, Life and Letters,* IV (New York, 1931) and V (New York, 1938). The best recent short treatment

Notes

is Richard Challener, "William Jennings Bryan," Norman Graebner, ed., *An Uncertain Tradition: American Secretaries of State and Their Diplomacy* (New York, 1961); the best lengthy treatment is Paolo E. Coletta, *William Jennings Bryan,* II (Lincoln, Neb., 1969). Ernest R. May, "Bryan and the World War, 1914–1915" (Ph.D. diss., University of California, Los Angeles, 1951) is unjustly neglected. Arthur S. Link, *Wilson,* II (Princeton, N.J., 1956) and III (Princeton, 1960) provides a useful synthesis.

2. The best study of Bryan's positive political contributions is Louis W. Koenig, *Bryan* (New York, 1971). On his negative image, see especially Coletta, *Bryan,* II, p. 120. For House's remarks, see Charles Seymour, ed., *The Intimate Papers of Colonel House* (Boston, 1926), I, pp. 282–3.

3. Anne W. Lane and Louise H. Wall, eds., *The Letters of Franklin K. Lane* (Boston, 1922), p. 167; Josephus Daniels, *The Wilson Era: Years of Peace, 1910–1917* (Chapel Hill, N.C., 1944), pp. 114, 139, 157, 399–400, 413; Koenig, *Bryan,* pp. 529–30, and Coletta, *Bryan,* II, p. 91, contain key Wilson letters, which should be supplemented by the material in Arthur S. Link et al., eds., *The Papers of Woodrow Wilson,* XXVII (Princeton, N.J., 1978), pp. 128, 253, hereafter cited as *PWW* with volume number; on Moore, see May, "Bryan and the World War," pp. 31–2, and Richard Megargee, "The Diplomacy of John Bassett Moore: Realism in American Foreign Policy" (Ph.D. diss., Northwestern, 1936, University Microfilms no. 64-2504).

4. Stephen Gwynn, ed., *The Letters and Friendships of Sir Cecil Spring Rice* (London, 1929), II, pp. 198, 220, 274, 431–3; Constantin Dumba, *Memoirs of A Diplomat* (London, 1933), pp. 226–7; *PWW,* XXVII, p. 171; Count Bernstorff, *My Three Years in America* (New York, 1920), p. 26.

5. Coletta, *Bryan,* II, p. 94; Mary Baird Bryan, *The Memoirs of William Jennings Bryan* (Philadelphia, 1925), p. 327; Arthur S. Link, *Wilson The Diplomatist* (Chicago, 1963, c.1957), pp. 10–11, and *Wilson,* II, 69–70.

6. Gwynn, ed., *Spring Rice,* pp. 202–223; Johann Bernstorff, *The Memoirs of Count Bernstorff* (London, 1936), pp. 109–110 and *Three Years in America,* pp. 61–3; Dumba, *Memoirs,* pp. 191–208.

7. In addition to the works cited in succeeding footnotes, see David F. Houston, *Eight Years with Wilson's Cabinet* (New York, 1926); John Wesley Payne, "David F. Houston: A Biography" (Ph.D. diss., University of Texas, Austin, 1953); and Adrian N. Anderson, "Albert Sidney Burleson: A Southern Politician in the Progressive Era" (Ph.D. diss., Texas Technological College, 1967, University Microfilms no. 68-2606).

8. Rupert N. Richardson, *Colonel House: The Texas Years* (Abilene, Tex., 1964) has useful detail; Arthur D. Howden Smith, *Mr. House of Texas* (New York, 1940) covers the whole life but is unreliable. For other details see *PWW,* XXVII, p. 163; and Seymour, ed., *Intimate Papers,* passim, especially pp. 39, 49–60, 88, 176–7.

9. Baker, *Wilson,* V, p. 308; *Washington Wife: Journal of Ellen Maury Slayden from 1897–1919* (New York, 1963), p. 195; Bernstorff, *Memoirs,* pp. 107–9; *Three Years,* p. 70; Dumba, *Memoirs,* pp. 210 ff.; Daniels, *Wilson Era,* p. 181; Charles Callan Tansill, *America Goes to War* (Gloucester, Mass., 1963, c.1938), pp. 144–5, 209, 313; Ernest R. May, *The World War and American Isolation* (Chicago, 1966, c.1959), pp. 80, 84, 40; Albert Pingaud, *Histoire diplomatique de la France pendant la grande guerre* (Paris, 1938), II, p. 244.

10. The key source for matters of fact is Daniel M. Smith, *Robert Lansing and American Neutrality, 1914–1917* (Berkeley, Calif. 1958). In terms of judgment, I have found May, "Bryan and the World War," most persuasive. See also Link, *Wilson The Diplomatist,* p. 26, and Tansill, *America Goes to War,* pp. 66, 179, 182 ff.

11. Henry F. Pringle, *The Life and Times of William Howard Taft* (New York, 1939), II, pp. 678–99; Walter Scholes, "Philander C. Knox," in Graebner, ed., *Uncertain Tradition,* pp. 59–78.

12. Bryan to Wilson, 15 January 1914, Wilson-Bryan Correspondence, National Archives.

13. Ray Stannard Baker and William E. Dodd, eds., *The Public Papers of Woodrow Wilson: The New Democracy* (New York, 1929), I, pp. 64–9.

14. The best account is Robert E. Quirk, *An Affair of Honor: Woodrow Wilson and the Occupation of Veracruz* (New York, 1967, c.1962); a useful brief synthesis is Arthur S. Link, *Woodrow Wilson and the Progressive Era* (New York, 1954), chap. 5.

15. Bryan to Wilson (Memorandum), 20 July 1913, and Wilson to John Lind, 4 August 1913, *PWW,* XXVIII, pp. 49–52, 110–1; Link, *Wilson,* II, chap. 11; A.D.H. Smith, *House,* p. 91.

16. *Papers Relating to the Foreign Relations of the United States, 1914* (Washington, 1922), pp. 443–4; *PWW,* XXVIII, pp. 431–2.

17. *PWW,* XXVII, pp. 335, 483, 536–52; for other Hale communications, see especially

XXVIII, pp. 7–8, 27–34. For Hale's background the best source is Larry D. Hill, *Emissaries to a Revolution: Woodrow Wilson's Executive Agents in Mexico* (Baton Rouge, La., 1973), chap. 2, pp. 21–39.

18. Wilson to Bryan, 3 July 1913, Wilson-Bryan Correspondence, National Archives; *Foreign Relations, 1913,* pp. 821–2.

19. *Foreign Relations, 1913,* pp. 821–2; Delbert J. Huff memo, 12 May 1913, *PWW,* XXVII, p. 419; Bryan to Wilson, 1 September 1913, Wilson-Bryan Correspondence, National Archives.

20. Slayden, *Washington Wife,* p. 236.

21. George M. Stephenson, *John Lind of Minnesota* (Minneapolis, Minn., 1935), especially pp. 226, 228, 239–40; Lind to Bryan, 19 September 1913, PWW, XXVIII, pp. 293–300. There is much valuable detail in Peter Calvert, *The Mexican Revolution, 1910–1914: The Diplomacy of Anglo-American Conflict* (Cambridge, Mass., 1968).

22. Lind to Bryan, 28 August 1913, in Stephenson, *Lind,* pp. 234–5.

23. Lind to Bryan, 19 September 1913, and Wilson to Bryan, 6 October 1913, Wilson-Bryan Correspondence, National Archives.

24. Stephenson, *Lind,* p. 241 ff.

25. Moore to Bryan, 14 May 1913, *PWW,* XXVII, pp. 437–40.

26. Moore to Wilson, 28 October 1913, *PWW,* XXVIII, pp. 458–63.

27. Henry Cabot Lodge, *The Senate and the League of Nations* (New York, 1925), chiefly pp. 15–18.

28. Arthur Walworth, *America's Moment: 1918* (New York, 1977), pp. 134, 138; Ray Stannard Baker, *American Chronicle* (New York, 1945), p. 372; Lord Cecil, *A Great Experiment* (London, 1941), p. 63; *Lord Riddell's Intimate Diary of the Peace Conference and After, 1918–1923* (London, 1933), p. 34; Charles Seymour, *Letters from the Paris Peace Conference,* ed., Harold B. Whiteman (New Haven, Conn., 1965), pp. 221–2.

29. For a sampling of French press and diplomatic views, see the contribution by Jean-Baptiste Duroselle to J. Joseph Huthmacher and Warren I. Susman, eds., *Wilson's Diplomacy: An International Symposium* (Cambridge, Mass., 1973), especially pp. 21, 33, 109–111; and Palmer, *Bliss,* p. 400.

30. John Maynard Keynes, *The Economic Consequences of the Peace* (New York, 1971, c.1920), pp. 37–53. Keynes was even harsher in his private letters; see R.F. Harrod, *Life of John Maynard Keynes* (London, 1966, c.1951), pp. 249–52; for Smuts, p. 260.

31. Harold Nicholson, *Peacemaking: 1919* (New York, 1939), pp. 7, 15, 28, 36–7, 42, 69 ff., 146, 164, 170, 184, 52–3, 72, 196–9.

32. David Lloyd George, *Memoirs of the Peace Conference* (New Haven, Conn., 1939), I, pp. 38, 44–5, 50, 112–115, 120–1, 139–60.

33. The standard treatment is Alfred W. Crosby, Jr., *Epidemic and Peace, 1918* (Westport, Conn., 1976), chap. 10. See also *Lord Riddell's Diary,* p. 51; Harrod, *Life of Keynes,* p. 234; and William K. Hancock, *Smuts: The Sanguine Years 1870–1919* (Cambridge, Mass., 1962), pp. 508–9.

34. The key secondary treatments are Crosby, *Epidemic and Peace*; and three works by Edwin A. Weinstein that change interpretations over the years: "Woodrow Wilson's Neurological Illness," *The Journal of American History,* LVII no. 2 (September 1970), pp. 324–51; *Woodrow Wilson: A Medical and Psychological Biography* (Princeton, N.J., 1981); and with James W. Anderson and Arthur S. Link, "Woodrow Wilson's Political Personality: A Reappraisal," *Political Science Quarterly,* XCIII no. 4 (Winter 1978–9), pp. 585–98. The chief primary account is Grayson to Tumulty, 10 April 1919, printed in Joseph P. Tumulty, *Woodrow Wilson As I Know Him* (New York, 1921), pp. 350. In addition see Ray Stannard Baker, *Woodrow Wilson and World Settlement,* II (New York, 1923), pp. 42–3; *PWW,* XXIX, p. 377; Irwin H. Hoover, *Forty-Two Years in the White House* (Boston, 1934), pp. 80–1, 95–9; Cary T. Grayson, *Woodrow Wilson* (New York, 1960), pp. 81–5; Herbert Hoover, *The Ordeal of Woodrow Wilson* (New York, 1958), p. 198, *America's First Crusade* (New York, 1942), pp. 40, 64, and *The Memoirs of Herbert Hoover,* I (New York, 1951), p. 468; Edith Bolling Wilson, *My Memoir* (Indianapolis, Ind., 1939), pp. 116, 130–1, 134, 158, 248–50; Colonel Edmund W. Starling with Thomas Sugrue, *Starling of the White House* (New York, 1946), pp. 139–40, 145–6; and Nicholas Murray Butler, *Across the Busy Years* (New York, 1939), II, p. 201.

35. The standard source for much of this is Warren F. Kuehl, *Seeking World Order* (Nashville, Tenn., 1969). See also Ruhl J. Bartlett, *The League to Enforce Peace* (Chapel Hill, N.C., 1944);

Notes

Baker, *Woodrow Wilson, Life and Letters*, VIII, p. 38; Roland Stromberg, "Uncertainties and Obscurities about the League of Nations," *Journal of the History of Ideas*, XXXVIII no. 1 (January–March 1972): 140.

36. See especially Kuehl, *Seeking World Order*, p. 255 and passim; and Lansing, *Peace Negotiations*, pp. 11, 30, 122–4, 164 ff., 190–1, 199–201, 204–5, 212.

37. Roland N. Stromberg, *Collective Security and American Foreign Policy* (New York, 1963), p. 22 ff.—one of the few genuinely sensible accounts of these matters. See also Henry R. Winkler, *The League of Nations Movement in Great Britain* (New Brunswick, N.J., 1952); Nevins, *Henry White*, p. 412; Lansing, *Peace Negotiations*, pp. 82, 92; Baker, *Wilson and World Settlement*, I, p. 215; Seymour, *Letters*, p. 180; Cecil, *Great Experiment*, pp. 47–100.

38. Lloyd George, *Memoirs*, I, p. 50; Cecil, *Great Experiment*, p. 68; Hancock, *Smuts*, I, pp. 464, 500–3; David Hunter Miller, *The Drafting of the Covenant* (London, 1928), II, document 5; William K. Hancock and Jan van der Poel, eds., *Selections from the Smuts Papers*, IV (Cambridge, 1966), pp. 9, 41–2, 48–50; the quotation conflates two separate letters to M.C. Gillett, 19 and 20 January 1919. Bonsal, *Unfinished Business*, has related material, pp. 34–6.

39. For these and a number of related issues, I am much indebted to John C. Vinson, *Referendum for Isolation: Defeat of Article Ten of the League of Nations Covenant* (Athens, Ga., 1961), especially pp. 35 ff., 57.

40. Lansing to Wilson, 20, 21, 23, and 29 December 1918, all printed in Lansing, *Peace Negotiations*, pp. 48–61; other material passim.

41. Lodge to White, 30 April and 20 May 1919, in Nevins, *White*, pp. 450–1; Hitchcock to Wilson, printed under the 23 March 1919 entry in Bonsal, *Unfinished Business*, pp. 145–6; Cecil, *Great Experiment*, pp. 82–3; Denna F. Fleming, *The United States and the League of Nations* (New York, 1932), p. 194.

42. Lansing prints the key sections in the appendices to *The Peace Negotiations*; for full texts of all versions, see Baker, *Wilson and World Settlement*, III, documents no. 10, 12, 14, 18, 20.

43. Hoover, *Ordeal of Wilson*, p. 267; Fleming, *U.S. and League*, pp. 227–30; Vinson, *Referendum*, pp. 70–81; Stromberg, *Collective Security*, pp. 36–7; Philip C. Jessup, *Elihu Root* (New York, 1938), II, pp. 383–401; and Martin D. Dubin, "Elihu Root and the Advocacy of a League of Nations," *The Western Political Quarterly*, XIX no. 3 (September 1966): 439–55.

44. The key documents are Smuts to Wilson (and Lloyd George), 14 May 1919; Wilson to Smuts, 16 May 1919; Smuts to Wilson, 30 May 1919; Wilson to Smuts, 31 May 1919; and Smuts to A. Clark, 10 July 1919, most readily available in Hancock and van der Poel, *Smuts Papers*, IV, pp. 157–8, 160–1, 208–9, 256–9, 263. See also the 5 May 1919 memo to Lloyd George, pp. 148–50. For biographical details see Hancock, *Smuts*, I, pp. 521–3, 539. For Clemenceau, see *Grandeur and Misery of Victory*, pp. 173–5, 179–80.

45. Alice Roosevelt Longworth, *Crowded Hours* (New York, 1933), p. 285.

46. John A. Garraty, *Henry Cabot Lodge* (New York, 1953) began modern assessments of Lodge; several footnotes, e.g., p. 379 ff., contain relevant remarks added by Lodge's grandson, Henry Cabot Lodge, Jr., who ironically enough later represented America in the United Nations. The most cogent brief summary of recent scholarship is James E. Hewes, Jr., "Henry Cabot Lodge and the League of Nations," *Proceedings of the American Philosophical Society*, CXIV no. 4 (August 1970): 245–55. The most extensive reevaluation is William C. Widenor, "Henry Cabot Lodge and the Search for an 'American' Foreign Policy," (Ph.D. diss., University of California at Berkeley, 1975, University Microfilms no. 76-15,445), subsequently published as *Henry Cabot Lodge and the Search for an American Foreign Policy* (Berkeley, Calif., 1980). Other relevant studies are Sheldon M. Stern, "Henry Cabot Lodge and Louis A. Coolidge in Defense of American Sovereignty, 1898–1920," *Proceedings of the Massachusetts Historical Society* LXXXVII (1975): 118–34; and John A.S. Grenville and George B. Young, *Politics, Strategy, and American Diplomacy* (New Haven, Conn., 1966), chap. 8, pp. 201–38.

47. Much of what Lodge says about his relationship with Wilson in the Senate and the League of Nations (New York, 1925) has withstood later historical examination; compare his account to that of Garraty, *Lodge*, pp. 294–312; and Widenor, "Lodge," pp. 121–2, 212, 265–6, 272–81.

48. *Congressional Record*, 65th Congress, 3rd Session (1918), p. 727; Widenor, "Lodge," passim, especially pp. 451, 486, 454, 515–18; Garraty, *Lodge*, passim, especially pp. 353, 356, 363–77, 385, 388.

49. In addition to William E. Borah of Idaho and Hiram Johnson of California, the list included Asle J. Gronna of North Dakota, Robert M. La Follette of Wisconsin, George Norris

of Nebraska, Miles Poindexter of Washington, Frank B. Brandegee of Connecticut, Albert B. Fall of New Mexico, Bert M. Fernald of Maine, Joseph France of Maryland, Philander C. Knox of Pennsylvania, Joseph M. McCormick of Illinois, George H. Moses of New Hampshire, James A. Reed of Missouri, Lawrence Y. Sherman of Illinois, and Charles S. Thomas of Colorado. Reed and Thomas were Democrats.

50. Hiram Johnson to his sons, 23 August 1919, quoted extensively in Richard C. Lower, "Hiram Johnson and the Progressive Denouement, 1910–1920" (Ph.D. diss., University of California at Berkeley, 1969, University Microfilms no. 70-6158), p. 386. Other useful studies of Johnson are Peter G. Boyle, "The Story of an Isolationist: Hiram Johnson" (Ph.D. diss., University of California at Los Angeles, 1970, University Microfilms no. 71-9265); John J. Fitzpatrick III, "Senator Hiram W. Johnson: A Life History, 1866–1945" (Ph.D. diss., University of California, Berkeley, 1975, University Microfilms no. 75-26,691); and Howard A. DeWitt, "Hiram Johnson and American Foreign Policy, 1917–1941," (Ph.D. diss., University of Arizona, 1972, University Microfilms no. 72-15,602).

51. See especially Johnson to son Jack, 7 December 1918, as quoted in Fitzpatrick, "Johnson," p. 76; Johnson to Hiram, Jr., 20 May 1919, quoted in DeWitt, "Johnson," p. 120; Johnson to C.K. McClatchey, 7 and 12 April 1919, in Boyle, "Johnson," pp. 146–7. For key speeches on these themes, see *Congressional Record,* 66th Congress, 1st Session, pp. 63, 501–9, and the sharp grilling of Lansing in *U.S. Congress, Senate Treaty of Peace with Germany. Hearing Before the Committee on Foreign Relations,* 66th Congress, 1st Session, Senate Document No. 106.

52. Key speeches include those in the *Congressional Record,* 65th Congress, 3rd Session, pp. 124, 195–6, 1386–7, 2425; and in William E. Borah, *American Problems,* ed. by Horace Green (New York, 1924), pp. 105–30, 131–42. See also key secondary treatments: John C. Vinson, *William E. Borah and the Outlawry of War* (Athens, Ga., 1957), p. 23 and passim; Robert J. Maddox, *William E. Borah and American Foreign Policy* (Baton Rouge, La., 1969), p. 8 and passim; Ralph Stone, *The Irreconcilables* (New York, 1973, c.1970), pp. 42–3 and passim. On Borah's place as a progressive, see John M. Cooper, Jr., "William E. Borah, Political Thespian," *Pacific Northwest Quarterly,* LVI No. 4 (October 1965), pp. 145–53, and *The Vanity of Power: American Isolationism and the First World War, 1914–1917* (Westport, Conn., 1969), pp. 143–5.

53. Belle Case and Fola La Follette, *Robert M. La Follette* (New York, 1953), II, passim, especially pp. 955, 968–9, 976–81, 991.

54. The best overview in many ways is still Thomas A. Bailey, *Woodrow Wilson and the Great Betrayal* (Chicago, 1963, c. 1945). For the quote see Walworth, *Wilson,* II, p. 276. On Lodge's willingness to compromise, see especially Bonsal, *Unfinished Business,* pp. 274–5, 279.

55. For key transcripts see Baker and Dodd, eds., *War and Peace,* I, pp. 574–80; and Lodge, *Senate and League,* Appendix IV, pp. 297–379. For the quote see Walworth, *Wilson,* II, p. 276. On Lodge's willingness to compromise, see especially Bonsal, *Unfinished Business,* pp. 274–5, 279.

56. Lodge reprints his speech in its entirety in *Senate and League,* Appendix V, pp. 380–410, see especially pp. 388–91. For later opinion on Article X, see Vinson, *Referendum,* pp. 93, 108; and Kuehl, *Seeking World Order,* p. 338.

57. See particularly John M. Pyne, "Woodrow Wilson's Abdication of Domestic and Party Leadership: Autumn 1918 to Autumn 1919" (Ph.D. diss., Notre Dame, 1979, University Microfilms no. 79-19,956).

58. Weinstein, "Woodrow Wilson's Neurological Illness," especially pp. 345–6; Edith Wilson, *My Memoir,* p. 280 ff.; Tumulty, *Wilson,* p. 435 ff.; Starling, *Starling,* p. 152.

59. The medical terminology is from Weinstein, *op.cit.,* supplemented by personal observation of stroke victims; see also Walter C. Alvarez, *Little Strokes* (Philadelphia, 1966).

60. Ike Hoover, *Forty-Two Years,* pp. 92–107, especially 95, 100–5; Edith Wilson, *My Memoir,* pp. 289–90.

61. See especially Vinson, *Referendum,* p. 108 ff., Stromberg, *Collective Security,* and David H. Miller, *My Diary at the Conference of Paris* (21 vols., New York, 1924-6), XX, pp. 569–93.

Index

Abolitionism, 4, 6–8, 40, 44, 68, 166–67, 170–71, 201, 203, 212, 214
"Abraham Lincoln Walks at Midnight" (Lindsay), 101, 285n6
Academia, *see* Education
Academic freedom, 71, 73
Adams, Henry Carter, 69, 195, 278, 283n4
Adams, Herbert Baxter, 10–11, 72–73, 81
Adams, Samuel Hopkins, 179–82, 188, 195, 197–98, 275, 292n4
Addams, Jane, x, 38–39, 42, 47, 58–59, 71, 83–85, 89, 100, 134–35, 138, 150, 172, 275, 280, 282, 286, 295; education of, 20, 23; family, relations with, 5, 16–21; and Hull-House, 5, 17, 24–25, 66–68, 144, 149, 156; ill health of, 21; and Negro rights, 217, 220; and the Progressive Party, 211, 216–17, 219, 221; religious views of, 18, 20, 22–23; and Ellen Gates Starr, 22–24; *Twenty Years at Hull-House*, 66–67; and women's rights, x, 17, 19, 211, 220
Addams, John, 17–21, 24
Adler, Dankmar, 139–40
Aesthetic ideas, 91–94, 103–7, 112–17, 120–21, 126–29, 136–37, 140, 143–49, 152–53, 156–71
Altgeld, John P., 39–40, 96, 99–101, 110
American Economic Association, 13, 70–71, 283n
American Historical Association, 74
American Medical Association, 177, 180, 188
American Social Science Association, 68
Ames, Edward Scribner, 60
Angell, James Rowland, 60
Anthropology, 81, 83–84
Architecture, 90–92, 100, 116–17, 133–61, 286, 288–91
Armory Show, 112–14, 286n15
Art, functionalism in, 141–44, 147, 161; *see also* Architecture, Music, Painting
"Art and Craft of the Machine" (Wright), 156–57
Arts and Crafts Movement, 144–57
Ashbee, Charles Robert, 145, 149
Ashcan School, 107
Association of American Painters and Sculptors, 112

Baker, Newton D., 12, 275
Baker, Ray Stannard, 164, 180, 188, 190
Barton, Bruce, 150
Beard, Charles A., 8, 77–79, 275, 283n17, 284n19, n20
Becker, Carl, 74–77, 275, 283
Bellows, George W., 101, 107–14, 276, 278, 286n10, n14, n16
Beveridge, Albert J., 5–6, 95, 187, 189, 192–94, 197, 206, 212–14
Blaine, Anita McCormick, 60, 275
Boas, Franz, 80, 83, 89
Bok, Edward, 178–79, 186, 292n26
Booth, General William, 96, 101
Borah, William E., 4, 265, 267–70, 273, 275, 280n3 298n49, 299n52
Brandeis, Louis D., 209
Bryan, William Jennings, xi, 3–4, 39, 96, 99–102, 109, 170–71, 225–47, 265, 275, 278, 280n1, 295n1, 296n, 297
Bungalow, 151–52, 290n60
Bureau of Chemistry, 183, 185–87, 196
Burnham, Daniel, 91–93, 152, 285n1

Call, The, 109–10
Capitalism, 43, 50, 167 68, 170, 198, 233–34, 265
Castle, Henry Northrup, 25–27, 30–36, 281
Cather, Willa, 93
Catholicism, ix, 23, 146, 235, 241, 277–78
Cecil, Lord Robert, 249, 258
Charity, 3, 81, 219
Chase, William Merritt, 100, 106, 108
Chicago Exposition of 1893, 151–53
Chicago Institute: Academic and Pedagogic, 60
Chicago Normal School, 60
Chicago, University of, 26, 38, 55, 58–62, 68, 73–74, 81–83, 86–89
"Christianity and Democracy" (Dewey), 56–57
Christian sociology, 8, 13, 44, 70; American Institute of, 8, 70
Churchill, Winston, 95, 275, 285n2
Clark, John Bates, 69
Cleveland, Grover, 39
Collectivism, xi, 110, 278

Index

Columbia University, 59, 63, 72, 75–80, 87, 169, 283n9
Committee on Medical Legislation, 177
Common Lot, The (Herrick), 94–95
Commons, John R., 6–8, 12, 14, 20, 47, 70, 209, 275, 280n6
Communism, 50–51
Competition, 12, 43
Comrade, The, 170
Conflict, The (Phillips), 95
Congressional Government (Wilson), 11, 72
Coniston (Churchill), 95
Cooley, Charles H., 83, 85, 89, 275
Crane, Stephen, 93, 161
Crane, Walter, 103, 105, 145, 149, 289n49
Criminality, 35, 67, 95, 101
Croly, Herbert, 209, 276, 294n13

Daniels, Josephus, 227, 229–30, 241
Darwinism, 52, 54, 60, 69, 73, 81–82, 125, 141, 162, 281n2
Debs, Eugene, 47, 51, 109–10, 205
Dell, Floyd, 110–11, 286n14
Democratic Party, 6–9, 98–99, 101, 109, 184, 204–14, 226, 230–31, 271
Dewey, John, x, xi, 37, 64, 76, 83, 128, 150, 157, 275, 282; educational views of, 58–63, 67, 75, 138, 149; family background of, 52; philosophical views of, 26, 28–29, 37–38, 40, 52–55, 59–60, 74, 78, 85, 134, 141, 149, 156; religious views of, 52, 54–58, 61; social and political concerns of, 56–62; "Christianity and Democracy," 56–68; "Ethical Principles Underlying Education," 61–62; "From Absolutism to Experimentalism," 54; *Psychology*, 55
Divorce, 48, 79, 86, 95, 135, 217
Dixon, Joseph M., 202–3, 214, 275
Donnelly, Ignatius, 94, 276
Doolittle, Hilda, 98
Dreiser, Theodore, 94, 172
Dunning, William A., 72

"Eagle That Is Forgotten, The" (Lindsay), 101, 285n6
Eakins, Thomas, 103–4, 161
Eastman, Max, 110–12, 286n12
Economic Interpretation of the Constitution of the United States, An (Beard), 79
Economics, 12–13, 42, 65, 69–71, 74, 77–80, 284
Education, xi, 4, 6–9, 12–17, 20, 26–30, 32–33, 38–39, 52–89, 136–38, 149, 156, 277; *see also* Anthropology; Economics; History; Law; Literature; Philosophy; Political Science; Progressives, educational background of; Psychology; Sociology
Eliot, T. S., 97–98
Elmslie, George Grant, 160, 291n73
Ely, Richard, 10, 12–14, 45, 47, 69–73, 81, 275, 280n13, 283n5
Emerson, Ralph Waldo, 91, 102, 104–5, 108, 120, 136–40, 143–44, 146, 149, 161–62, 205, 289
"Ethical Principles Underlying Education" (Dewey), 61–62
Europe, relations with, ix, xi, 10, 21–27, 34–35, 42, 47, 52, 69, 72–73, 82, 88, 91, 96–97, 105, 112–14, 133–37, 145–49, 154–55, 157, 225, 228–32, 241–73

"First Symphony" (Ives), 123
Foreign policy, xi, xii, 100–1, 188, 225–73, 296, 297, 298, 299; see also Europe, relations with; Latin America; League of Nations (Wright), 145
Frederic, Harold, 93
Froebel, Friedrich, 136–37, 152–53, 159, 288n34
"From Absolutism to Experimentalism" (Dewey), 54
Frost, Robert, 98

Garden, Hugh M., 160
Garland, Hamlin, 94
Gates, George Augustus, 43, 45–48
"General William Booth Enters Into Heaven" (Lindsay), 101, 285n6
George, Henry, 42, 50, 59
George Helm (Phillips), 95
Giddings, Franklin H., 80–81, 275
Gilman, Charlotte Perkins, 4, 275
Gilman, Daniel Coit, 11–13, 52
Glackens, William, 103, 105–7, 113, 275, 286n7
Gladden, Washington, 8, 45
Glasgow, Ellen, 93–94
Goldman, Emma, 108–9, 111, 286n10
Greenough, Horatio, 143–44, 147, 153, 161, 289n47, n48
Griffin, Walter Burley, 160

Hale, William Bayard, 238–39, 241, 275
Hall, G. Stanley, 35, 53
Harding, Warren G., 269–70
Harper, William Rainey, 58–59, 73, 81
Hartley, Marsden, 114

Harvard University, 29, 32–34, 60, 74, 76, 79–80, 87–88, 94, 184
Hassam, Childe, 106
Health, 166, 173–90, 195–99, 293, 296; *see also* Medicine; Pure Food and Drug Act; Progressives, health problems of
Hegelian idealism, 53–54, 56, 58, 60–61, 78
Hemingway, Ernest, 150
Henderson, Charles R., 81
Henri, Robert, 100–1, 104–15, 275, 278, 286*n*7
Herrick, Robert, 24–25, 275, 285*n*2
Herron, George D., 40–52, 58, 70, 134, 170–71, 275, 282*n*1
Hillside Home School (Wright), 138
History, 7, 10, 61, 65, 71–79, 82; institutional, 10, 74; new, 74–76, 78; "scientific," 72, 75–76
Homer, Winslow, 103, 161
Hoover, Herbert, 125, 260, 275, 297*n*34
Hopper, Edward, 108
House, Edward M., 225–31, 237, 248, 253–55, 257, 296*n*2
Howe, Frederic C., 12–14, 109, 276
Howells, William Dean, 93, 161
Huerta, Victoriano, 236–46
Hughes, Charles Evans, 6–7, 220, 275, 280*n*1
Hull-House, 5, 17–18, 20–25, 38, 58–62, 64–68, 71, 77, 84, 88–89, 101, 144, 149, 156, 172

Immigrants, 68, 84–85, 96, 99, 135
Innovative nostalgia, 90, 115–17, 121, 127–28, 133, 162
Inside of the Cup, The (Churchill), 95
Instrumentalism, r. 53–55, 59, 67
Iowa College, 43, 48
Ives, Charles, *x*, 92, 141, 144, 146, 150, 155, 275, 286, 287, 288; business career of, 117, 124–26; educational and musical background of, 121–23, 139; family life of, 116–18, 123; "The Fourth Symphony," 119; health problems of, *xi*, 161; political views of, 126–27; religious views of, 119–20, 126, 130–31; technical innovations of, 127–33, 162; "Three Pieces," 130; and Transcendentalism, 120–21, 131

James, Edmund J., 69, 276, 283*n*4
James, Henry, 93
James, William, 30, 33–34, 37, 55, 59–60, 86–87, 127; *Principles of Psychology*, 37, 39
Japanese influence on art, 133, 152–54, 159, 290
J. Hardin and Son (Phillips), 96
Johns Hopkins, 8, 10–13, 29, 32–33, 35, 52, 68–73, 82

Johnson, Hiram, 205–6, 223–24, 265–67, 270, 273, 276, 295*n*28, 299*n*
Jones, Jenkin Lloyd, 137–39, 145, 149–50, 155
Jones, Samuel "Golden Rule," 4, 47, 100
Journalism, 6–8, 11, 15, 73, 87, 103, 105–6, 109–11, 161, 163, 165, 169, 177–82, 186, 191–92, 277, 292; *see also* Muckrakers; Socialism
Judaism, *ix*, 69, 84, 215, 235, 241, 277–78
Jungle, The (Sinclair), 94–95, 165, 171–74, 185, 187, 189, 193

Kantian philosophy, 37, 52–53, 56, 281*n*2
Kent, Rockwell, 101, 108
Keynes, John Maynard, 250, 254, 262, 297*n*30
"King Arthur's Men Have Come Again" (Lindsay), 101, 285*n*6
Kingdom, School of the, 45–46
Kirchwey, George W., 217–18
Knox, Philander C., 232–35, 269, 299*n*49
Kuhn, Walt, 113

Labor protection, 14, 38, 77, 172, 197, 213–14, 216, 219, 221–22
Labor unions, 172–73, 222, 271
La Follette, Robert M., 8, 14, 208–9, 212, 223, 268, 275, 299*n*53
Lane, Franklin K., 100, 227, 296*n*3
Lansing, Robert, 225, 227, 229, 231, 237, 278, 296*n*10, 299*n*51
Latin America, relations with, *ix*, 230, 232–39, 243, 267; *see also* Mexican War
Law, *ix*, 6, 10–11, 15, 36, 65, 72, 78–79, 277
Lawson, Ernest, 106
League of Nations, 232, 248, 251, 252–63, 266, 271, 273, 297, 298
Lewis, William Draper, 209, 217, 220, 222–24
Light-fingered Gentry (Phillips), 96
Lincoln, Abraham, *ix*, 4–6, 68, 96–100, 151, 162, 221, 224, 233, 252–53, 276–77
Lind, John, 240–43, 276, 296*n*15, 297
Lindsay, Vachel, 96–102, 276, 278, 285*n*6, 286
Lindsey, Ben B., 205, 276
Lippmann, Walter, 253
Literature, 90–102, 106, 112, 115, 117, 164–66, 169–74, 185, 187, 189, 276
Lloyd George, David, 248, 250–52, 254, 258, 297*n*32, 298*n*38
Lodge, Henry Cabot, 245–46, 256, 259–65, 267, 269–70, 273, 297*n*27, 298*n*46, *n*47
London, Jack, 8, 94, 171, 174
Lord Phillimore's Report, 258

Index

Lowell, Amy, 99
Luks, George, 101-3, 105-6, 276

McCarthy, Charles, 74, 209-10, 217-19, 222, 276
McClure, S. S., 164-65, 178-80, 191, 291n2, 293n53
McDowell, Mary, 276
Machine, attitudes toward the, 148-49, 156-57
McKinley, William, 39, 102, 169, 185, 192
Madero, Francesco I., 236, 238-39, 244
Mahony, Marion, 160
Manassas (Sinclair), 170-71
Marin, John, 114
Marriage, 10, 17, 22, 26, 48, 55, 95-96, 139, 169
Masses, The, 109-11
Masters, Edgar Lee, 96-98, 276, 285n4, n6
Maurer, Alfred, 114
Mead, George Herbert, x, 55, 60, 76, 83, 85, 276, 281; Henry Northrup Castle, relationship with, 25-27, 30-34, 36; educational background of, 26; family life of, 26-27, 37; health problems of, 26, 32; philosophical views of, xi, 25, 27-31, 33-38; religious views of, 27-31, 33, 35-37; and socialism, 35, 37, 51; urban reform, concern with, 25
Medicine, 173-88, 195-96, 221, 292
Memoirs of an American Citizen, The (Herrick), 95
Merriam, Charles E., 217, 276
Mexican War, 230, 232, 236-46, 263, 267
Michigan, University of, 26, 37, 53, 55, 59, 63, 87
Minor, Robert, 110-11
Mission, American sense of, ix, 226, 252
Missionary work, 8, 15, 20, 26, 28-29
"Mobile Speech" (Wilson), 233-34
Modernism, 97-98, 112-17, 133, 145-46, 157, 162, 169, 274
Moore, Addison, 60
Moore, John Bassett, 72, 227, 230-31, 243-44, 247, 296n3
Morality, ix, 9, 15, 18-19, 21, 23, 25, 28-31, 33, 35-37, 42-44, 50, 56, 58, 61-62, 67, 69, 70, 78, 80, 82, 86, 91, 94-95, 97-98, 102-5, 117, 122, 126, 133, 141, 145-46, 149, 165, 172, 174, 179, 182-83, 185, 189-91, 196-97, 199, 204-9, 216, 223-24, 235, 237-40, 244, 247-49, 252, 268, 270
Morgan, J. P., 207-8
Morris, George Sylvester, 53-55
Morris, William, 92, 103, 144, 147-57, 161, 289, 290n66
Mr. Crewe's Career (Churchill), 95

Muckrakers, 124, 164-66, 177, 179, 182-83, 188, 191-92, 291
Munsey, Frank A., 206-8, 212, 294n9
Music, 90, 92, 116-34, 139, 141, 160-61, 287-88, 293
Myers, Jerome, 107, 113-14, 276, 286n9, n16

Nast, Thomas, 103
National Association for the Advancement of Colored People, 68
National Conference of Charities, 68
National-Health Service, 221
National Social Reform Union, 46-47
Naturalism, 93-94, 161
Nature, 98, 113-14, 117, 120-21, 128, 130, 136, 143, 145-46, 148, 161
Negroes, 4, 68, 77, 88, 202-6, 210, 216-17, 294— Newcomb, 65
New Deal, x, xi, 51, 63, 95, 278
New Freedom, 97, 226, 228
New History, The (Robinson), 76-77
Nicholson, Harold, 251, 297n31
Nock, Albert Jay, 276
Norris, Frank, 93

Oberlin College, 7, 8, 20, 26-28, 30-31, 37, 83, 280n7
Old Wives for New (Phillips), 96
O'Neill, Eugene, 108
Organicism, 60, 62, 121, 137, 142-44, 146-48, 157-61, 289
Osgood, Herbert L., 72

Pach, Walter, 113
Pacifism, 80, 86, 101, 109
Page, Walter Hines, 11, 12, 173, 224, 227, 230, 238, 276, 280n12
Painting, 100-15, 286; representational, 102-3
Park, Robert, x, 82, 86-89, 275, 284n32, 285
Parker, Francis W., 60-61
Parker, Horatio, 122-23, 287n10
Patten, Simon, 69
Peffer, William A., 14
Peirce, Charles S., 52
People's Palace, 25
Perkins, George W., 125, 206-9, 212, 218-19, 222, 276, 294n8, n9, n10
Phillips, David Graham, 95-96, 191, 276
Philosophy, xi, 25-33, 52-60, 69, 74, 78-79, 87-88, 149, 156
Pinchot, Amos, 109, 208-9, 276
Pinchot, Gifford, 208-9, 217-19, 276, 295n24, n27

Plum Tree, The (Phillips), 96
Poetry, 96–102
Poindexter, Miles, 276
Polish Peasant in Europe and America, The (Thomas), 84–86
Political economy, 12–13, 69–70
Political science, 10, 71–72, 77–80, 283
Politics, ix, 8, 10, 36, 39–40, 46–47, 51, 56, 70, 72, 74–75, 77–78, 90–93, 95–96, 99–105, 109–12, 117, 126, 134, 148–49, 164–65, 169–73, 269–73, 294, 295, 296
"Politics" (Beard), 77–79
Polyharmony, 129
Polyrhythms, 129, 131
Polytonality, 118, 129
Populism, 13, 40, 46, 77, 94, 110, 170, 207
Pound, Ezra, 98, 100
Pound, Roscoe, 276
Poverty, 22, 35, 39, 40–41, 58, 167
Pragmatism, 26, 28, 37–38, 55, 77–78, 141, 274
Prairie School, 137–38, 145, 149–52, 154, 158–60, 290n71, n73
Prendergast, Maurice, 106–7, 114
Presbyterianism, xii, 3, 8–9, 13, 20, 165, 184, 206, 225–26, 228, 231–32, 235, 238, 240, 250
Prison reform, 68, 95
Program music, 121, 127–28, 131
Progressive Party, 6, 89, 200–14, 217, 220–24, 266, 294, 295
Progressives: educational backgrounds of, 7–9, 13–14, 17, 20–21, 26–30, 33, 42, 75, 100, 104, 108, 135–39, 184, 276; family relationships of, 3–9, 12, 15, 17–28, 32, 41, 47–48, 52, 60, 98–100, 104–5, 117, 123–24, 127, 133–39, 150, 161, 166–70, …84, 276; health and psychological …ns of, 7, 11, 15–17, 20–21, 24, 26– …42, 48, 132, 168–69, 253–56, …7, 297, 299
… 101, 168, 218, 223,
…erica, 176, 179

(Lind-
197-98,
e (Wright)
…241, 244
8-9, 241, 244
69, 71, 73, 80,
207, 280, 295n1,
nal background of,
of, 8-9; and foreign
73; health problems
269, 271-73; and the

Railroads, 31, 95, 223; regulation of, 165, 188–90
Rauschenbusch, Walter, 28, 276
Realism, 52, 93–95, 97, 102–4, 106, 112–13, 130–31, 161–62, 285, 296
Record, George, 276
Reform: educational, 38, 58–63, 72, 77, 206, 220; moral, 37, 77; political, 37, 103, 216, 219–21; social, 36, 39, 44, 49, 51, 65, 68, 70–71, 80–83, 89, 92, 94–97, 100–1, 155, 162, 165–68, 171–79, 182–99, 208, 213–17, 221–22, 228, 234, 291; urban, 36, 65, 219
Reinsch, Paul J., 14, 74
Relativism, 74–75, 127
Religion, *see individual religions*
Republican Party, ix, 5–7, 68, 72, 77, 92, 99, 101, 107, 109–10, 184, 191, 195, 201–5, 208, 212, 223–26, 238, 256–57, 259–61, 276
Richardson, Henry Hobson, 140–41, 151, 153
Riis, Jacob, 71, 276
Robins, Margaret Dreier, 276
Robins, Raymond, 276
Robinson, Boardman, 110–11, 276
Robinson, Edwin Arlington, 98
Robinson, James H., 75–78, 85, 276, 283n14
Romanticism, 94, 143, 150, 288
Roosevelt, Franklin D., 6, 226
Roosevelt, Theodore, 8, 40, 71–72, 89, 92–93, 95, 99, 102, 109, 114–15, 128, 161, 171, 262–63, 266, 275, 277, 286, 293, 294, 295; family background of, 3–4; and food and drug legislation, 188–98; and foreign affairs, 188, 264; and the muckrakers, 96, 164–66, 183, 188–93; and Negro rights, 202–5, 210, 215, 217; and the Progressive Party, 6, 201–12, 215–24; religious views of, 3, 216; and the Republican Party, 5–6, 223; and trusts, 206–10, 213, 218–22
Root, Elihu, 247, 260–64, 273
Ross, Edward A., 12, 14, 45, 71, 276, 280n1
Rowell, Chester H., 217–18
Royce, Josiah, 33–34, 55
Ruskin, John, 25, 77–79, 91–92, 104, 140, 144–51, 156–57, 161, 289
…College, 77
…278

Army, 96
Franklin S., 68–69
…rg, Carl, 96, 98–99, 276, 285n
…er, Vida, 276, 281n1
…man, Robert A., 69–70, 72,
283n4

Settlement movement, 8, 15, 17, 24–25, 39, 68, 79–81, 101, 172, 220, 276; *see also* Hull-House; People's Palace; Toynbee Hall
Shaw, Albert, 12, 69, 71, 276
Sherman Anti-Trust Act, 209, 218
Shingle Style, 151, 289n59
Shinn, Everett, 103–4, 106, 109, 286n7
Silsbee, Joseph Lyman, 138–39, 151, 153
Simkhovitch, Mary K., 276
Sinclair, Upton, 47, 94–96, 100, 165–75, 189–91, 197, 275, 291, 292; *Jungle, The,* 94–95, 165, 171–74, 185, 187, 189, 193
Sloan, John, 105–15, 275, 286
Small, Albion, 12, 73, 81–83, 86, 88, 276, 284n26
Smith, Al, 278
Smuts, Jan C., 254, 258, 261–62, 298
Social Gospel movement, 8, 40–41, 45–46, 70, 79, 95, 168
Socialism, 8, 13, 35, 37, 46–52, 71, 77, 84, 94, 99, 101, 108–12, 115, 145, 148, 156, 166, 169, 170–74, 189–93, 282; *The Call,* 109–10; *The Comrade,* 170; *The Masses,* 109–11
Society of Independent Artists, 107
Sociology, 7, 13, 48, 65, 73, 77–89, 284n; *see also* Christian sociology
South, the, 8–9, 11, 166–67, 170, 173, 202–5, 210, 215, 217, 226
Spargo, John, 8, 170
Spencer, Herbert, 42, 80, 83, 161
Spring Rice, Sir Cecil, 227–30, 296
Starr, Ellen Gates, 22–24, 149, 276, 281
Starr, Frederick, 81
Steffens, Lincoln, 8, 164, 188, 223, 276
Stieglitz, Alfred, 114
Stimson, Henry, 276
Straus, Oscar, 278
Strong, Josiah, 8, 45
Stuckenberg, J. H. W., 8
Student Christian Association, 55–56
Sullivan, Louis, 139–41, 144, 150–53, 160, 288n42, n44, 290n61
Summer, William Graham, 69–70, 81, 89
Susan Lenox: Her Fall and Rise (Phillips), 96

Taft, William Howard, 109, 191–206, 208, 212–13, 223, 232–36, 243, 247, 260, 264, 296n11
Talbot, Marion, 81–82
Tarbell, Ida, 5, 13, 164, 276
Tariff, 165, 188, 208, 213, 216, 219, 221, ▪28
    14, 218, 221
    ▪. Graham, 8, 47, 278
    ▪ymphony" (Ives), 132–33
    ▪M. Carey, 276

Thomas, William I., x, 8, 76, 82–89, 275, 284
Thoreau, Henry David, 121, 143, 289n48
"Three Pieces" (Ives), 130
Together (Herrick), 95
Toynbee Hall, 24–25, 80, 97, 149
Transcendentalism, 117, 120–22, 124–26, 131, 136, 143, 146, 154, 161, 287
Treason of the Senate, The (Phillips), 96
Trusts, 165, 186, 206–9, 212–13, 218–19, 222, 228; *see also* Sherman Anti-Trust Act
Tufts, James H., 55, 58, 60, 276
Turner, Frederick Jackson, 12, 73–76, 283
Turn of the Balance, The (Whitlock), 95
Twain, Mark, 93
Twenty Years at Hull-House (Addams), 66–67

Union Reform League, 46
Unions, labor, 172–73, 222, 271
"Universe Symphony," 130–31

Van Hise, Charles R., 14, 74
Versailles, Treaty of, 249–54, 259–61, 264–70
Vincent, George, 81

Wagner, Robert, 278
Wald, Lillian, 278
Walsh, Thomas, 278
Weber, Max, 114
Weir, J. Alden, 106, 112
West, the, 74, 205, 210, 219, 221–22, 266, 269
Wharton, Edith, 94
Whistler, James A. McNeill, 91, 107, 146, 152, 276
White, William Allen, 4, 6, 100, 179, 213, 217–19, 276, 280, 293n47, 295n20, n24
Whitlock, Brand, 4, 6, 95–96, 100, 106, 280n2, n5, 285n2
Whitman, Walt, 97, 99–100, 103–4, 140
"Why I Voted the Socialist Ticket" ▪say), 101, 285n6
Wiley, Harvey, 183–88, 192, 195, 275, 292, 293
William Herman Winslow Ho▪ 154
Wilson, Henry Lane, 236, 2▪
Wilson, Woodrow, 6, 12–1▪ 109, 117, 132, 161, 205▪ 296, 298, 299; educati▪ 8–11, 72; family life▪ affairs, xi, 100, 22▪ of, 7, 11, 253–56▪

League of Nations, 232, 248, 251–52, 256–65, 270–71, 273; and Mexico, relations with, 230, 232, 236–47, 267; "Mobile Speech" of, 233–34; and New Freedom, 228; religious views of, 8–9, 39, 228, 234–35, 248; and Versailles, The Treaty of, 249–51, 253–54, 259–61, 264–65, 267–73; and women's rights, 220; and World War I, 225, 229, 245–48, 252, 261–62

Wisconsin, University of, 13–14, 73–74, 76

Wise, Rabbi Stephen, 278

Women's Rights, 7, 16–17, 19, 25, 94–97, 201, 211, 213, 216–17, 219, 221, 270

World War I, 63, 100–1, 112, 124, 132, 225–31, 244, 246–48, 251–52, 296n10, 299

World War II, xi, 86, 225, 263

Wright, Frank Lloyd, x, xi, 99–100, 275, 289, 290n; architectural innovations of, 92–93, 116, 133, 151–53, 159–62; architectural theory of, 141–44, 147–48, 154, 156–58; "The Art and Craft of the Machine," 156–58; and the Arts and Crafts Movement, 144–49, 153–57; family and educational background of, 116–17, 133–41, 150, 161; Japanese influence on, 133, 152–53, 160; Prairie Style of, 137, 145, 149–50, 152, 158–60; religious views of, 134–38, 144, 154

"Yale-Princeton Game, A" (Ives), 128

Yale University, 60, 117, 121–23, 128, 139, 208, 287n9

Young, Art, 110–12, 276, 286n12

Zueblin, Charles, 81